Gas Sensors Based on Conducting Metal Oxides

Metal Oxides Series

The Metal Oxides Book Series Edited by Ghenadii Korotcenkov

Forthcoming Titles

- Metal Oxide Powder Technologies, Yarub Al-Douri, 9780128175057
- Metal Oxide Glass Nanocomposites, Sanjib Bhattacharya, 9780128174586
- Tin Oxide Materials, Marcelo Ornaghi Orlandi, 9780128159248
- Cerium Oxide, Salvatore Scire, Leonardo Palmisano, 9780128156612
- Solution Processed Metal Oxide Thin Films for Electronic Applications, Zheng Cui, 9780128149300
- Metal Oxides for Non-volatile Memory, Panagiotis Dimitrakis, Ilia Valov, 9780128146293 Colloidal Metal Oxide Nanoparticles, Sabu Thomas, Anu Tresa Sunny, Prajitha V, 9780128133576
- Gallium Oxide, Stephen Pearton, Fan Ren, Michael Mastro, 9780128145210
- Metal Oxide Nanostructures, Daniela Nunes, Lidia Santos, Ana Pimentel, Pedro Barquinha, Luis Pereira, Elvira Fortunato, Rodrigo Martins, 9780128115121
- Metal Oxides in Energy Technologies, Yuping Wu, 9780128111673
- Gas Sensors Based on Conducting Metal Oxides, Nicolae Barsan, Klaus Schierbaum, 9780128112243
- Metal Oxide Nanostructured Phosphors, H. Nagabhushana, Daruka Prasad, S.C. Sharma, 9780128118528

Recently Published Titles

- Metal Oxide-Based Photocatalysis, Adriana Zaleska-Medynska, 9780128116340
- Metal Oxides in Heterogeneous Catalysis, Jacques C. Vedrine, 9780128116319
- Magnetic, Ferroelectric, and Multiferroic Metal Oxides, Biljana Stojanovic, 9780128111802
- Iron Oxide Nanoparticles for Biomedical Applications, Sophie Laurent, Morteza Mahmoudi, 9780081019252
- The Future of Semiconductor Oxides in Next-Generation Solar Cells, Monica Lira-Cantu, 9780128111659
- Metal Oxide-Based Thin Film Structures, Nini Pryds, Vincenzo Esposito, 9780128111666
- Metal Oxides in Supercapacitors, Deepak Dubal, Pedro Gomez-Romero, 9780128111697
- Transition Metal Oxide Thin Film-Based Chromogenics and Devices, Pandurang Ashrit, 9780081018996

Metal Oxides Series

Gas Sensors Based on Conducting Metal Oxides

Basic Understanding, Technology and Applications

Edited by

Nicolae Barsan

Institute for Theoretical and Physical Chemistry, University of Tübingen, Tübingen, Germany

Klaus Schierbaum

Department of Materials Science, Institute of Experimental Condensed Matter Physics, University of Düsseldorf, Düsseldorf, Germany

Series Editor

Ghenadii Korotcenkov

ELSEVIER

Elsevier
Radarweg 29, PO Box 211, 1000 AE Amsterdam, Netherlands
The Boulevard, Langford Lane, Kidlington, Oxford OX5 1GB, United Kingdom
50 Hampshire Street, 5th Floor, Cambridge, MA 02139, United States

British Library Cataloguing-in-Publication Data
A catalogue record for this book is available from the British Library

Library of Congress Cataloging-in-Publication Data
A catalog record for this book is available from the Library of Congress

ISBN: 978-0-12-811224-3

Publisher: Matthew Deans
Acquisition Editor: Kayla Dos Santos
Editorial Project Manager: Andrae Akeh
Production Project Manager: Debasish Ghosh
Cover Designer: Miles Hitchen

Typeset by MPS Limited, Chennai, India

Contents

List of contributors

Nicolae Barsan Institute for Theoretical and Physical Chemistry, University of Tübingen, Tübingen, Germany

David Degler European Synchrotron Radiation Facility, Grenoble, France

Ulrich Guth Technical University of Dresden, Institute of Physical Chemistry, Dresden, Germany

Anne Hemeryck Equipe M3, LAAS-CNRS, Toulouse, France

Kuniyuki Izawa Figaro Engineering Inc., Mino, Japan

Jong-Heun Lee Department of Materials Science and Engineering, Korea University, Seoul, Korea

Alexandru Oprea Institute for Physical and Theoretical Chemistry, University of Tübingen, Tübingen, Germany

Julia Rebholz Innovative Sensor Technology IST AG, Ebnat-Kappel, Switzerland

Klaus Schierbaum Department of Materials Science, Institute of Experimental Condensed Matter Physics, University of Düsseldorf, Düsseldorf, Germany

Anna Staerz University of Tübingen, Tübingen, Germany

Heiko Ulmer Sensirion AG, Stäfa, Switzerland

Udo Weimar University of Tübingen, Tübingen, Germany

Hans-Dieter Wiemhöfer Institute of Inorganic and Analytical Chemistry, University of Münster, Münster, Germany

Series editor biography

Ghenadii Korotcenkov

Research Professor (2008–present)

School of Materials Science and Engineering, Gwangju Institute of Science and Technology, South Korea

Ghenadii Korotcenkov received his PhD in Physics and Technology of Semiconductor Materials and Devices from Technical University of Moldova in 1976 and his PhD in Physics of Semiconductors and Dielectrics from Academy of Science of Moldova in 1990 (Highest Qualification Committee of the USSR, Moscow). He has more than 40 years' experience as a teacher and scientific researcher. For a long time he was leading a gas sensor group and managed various national and international scientific and engineering projects carried out in the Laboratory of Microelectronics and Optoelectronics at the Technical University of Moldova. His research was financially support by the international foundations and programs, such as CRDF, MRDA, ICTP, INTAS, INCO-COPERNICUS, COST, and NATO. In 2007 and 2008 he was an invited scientist at the Korea Institute of Energy Research (Daejeon). Currently (starting from 2008) Korotcenkov is a research professor in the School of Materials Science and Engineering at Gwangju Institute of Science and Technology (GIST) in Korea.

Specialists from the former Soviet Union know Korotcenkov's research results in the fields of Schottky barriers, MOS structures, native oxides, and photoreceivers on the basis of III-Vs compounds, such as InP, GaP, AlGaAs, and InGaAs. His scientific interests, which started in 1995, include material sciences, focusing on metal oxide film deposition and characterization, surface science, and the design of thin film gas sensors and thermoelectric convertors. Korotcenkov is the author and/or editor of 35 books and special issues, including the 11-volume *Chemical Sensors* series published by Momentum Press (USA), 15-volume *Chemical Sensors* series published by Harbin Institute of Technology Press (China), 3-volume *Porous Silicon: From Formation to Application* issue published by CRC Press (USA), and the 2-volume *Handbook of Gas Sensor Materials* published by Springer (USA). Korotcenkov is the author and co-author of more than 550 scientific publications including 20 review papers, 35 book chapters, more than 250 articles published in peer-reviewed scientific journals, and more than 50 refereed papers published in international conference proceedings (h-factor = 38 [Scopus] and h = 44 [Google scholar citation]). He holds 17 patents and presented more than 200 reports during national and international conferences, including 15 invited talks. Korotcenkov was coorganizer of several international conferences. His research activities have been honored by the Award of the Supreme Council of Science and Advanced Technology of the Republic of Moldova (2004), The Prize of the Presidents of Ukrainian, Belarus and Moldovan Academies of Sciences (2003), Senior Research Excellence Award of Technical University of Moldova (2001, 2003, 2005), and National Youth Prize of the Republic of Moldova in the field of science and technology (1980), among others. Korotcenkov also received a fellowship from the International Research Exchange Board (IREX, USA, 1998), Brain Korea 21 Program (Korea, 2008–12), and Brainpool Program (Korea, 2015–17).

Editors' biographies

Dr. Nicolae Barsan, Senior Research Scientist, Institute of Physical and Theoretical Chemistry, University of Tübingen, Germany. Dr. Barsan is the founding member of the International Society for Olfaction and Chemical Sensing, United Kingdom, and the senior researcher at the Institute of Chemistry at the University of Tübingen, Germany. He has written over 200 papers, 6 book chapters, and 13 patents. His research focuses on the understanding and application of surface interactions of materials with gases. Since 1984 when he started his scientific career at the Institute of Materials Physics and Technology, Romania, he has focused on the understanding of phenomena taking place at the surface of metal oxides operated in realistic conditions for gas sensing applications. He has also explored different technologies for devising better gas-sensing materials and innovative sensors. He has invented the hybrid microsensor production technology based on the combination of sensing layers based on preprocessed powders and micromachined Si transducers. In 2001 he cofounded Advancing Sensing Devices (ASD), which merged with Applied Sensors and established itself as a leading European company in the field of gas sensor technology and application developments.

Prof. Dr. Klaus Schierbaum, Heinrich Heine University Düsseldorf, Germany. Dr. Schierbaum has been professor of Materials Science at the Heinrich Heine University Düsseldorf, Germany, since 1998. In 2000 he was a locum professor at the Applied Physics Department of the University GieBen, Germany. He has been and is actively involved in large national projects dealing with sensor development and EC funded projects. He is a member of the Deutsche Physikalische Gesellschaft (DPG), German Bunsen Society for Physical Chemistry (DBG), German Society for Materials Science, German Association of University Professors and Lecturers, Association of Electrochemistry Research Institutions, and many others.

Preface to the series

The field of synthesis, study, and application of metal oxides is one of the most-rapidly progressing areas of science and technology. Metal oxides are one of the most ubiquitous compound groups on Earth, which have a large variety of chemical compositions, atomic structures, and crystalline shapes. In addition, metal oxides are known to possess unique functionalities that are absent, or inferior, in other solid materials. In particular, metal oxides represent an assorted and appealing class of materials, properties of which exhibit a full spectrum of electronic properties—from insulating to semiconducting, metallic, and superconducting. Moreover, almost all the known effects including superconductivity, thermoelectric effects, photoelectrical effects, luminescence, and magnetism can be observed in metal oxides. Therefore, metal oxides have emerged as an important class of multifunctional materials with a rich collection of properties that have great potential for numerous device applications. Specific properties of the metal oxides, such as the wide variety of materials with different electrophysical, optical, and chemical characteristics, their high thermal and temporal stability, and their ability to function in harsh environments, make metal oxides very suitable materials for designing transparent electrodes, high-mobility transistors, gas sensors, actuators, acoustical transducers, photovoltaic and photonic devices, photo- and heterogeneous catalysts, solid-state coolers, high-frequency and micromechanical devices, energy harvesting and storage devices, nonvolatile memories, and many others in the electronics, energy, and health sectors. In these devices, metal oxides can be successfully used as sensing or active layers, substrates, electrodes, promoters, structure modifiers, membranes, and fibers—i.e. they can be used as active and/or passive components.

Among the other advantages of metal oxides are the low fabrication cost and robustness in practical applications. Furthermore, metal oxides can be prepared in various forms such as ceramics, thick films, and thin films. Thin film deposition can be used for deposition techniques that are compatible with standard microelectronic technology. The latter factor is important for large-scale production, because the microelectronic approach promotes low cost for mass production, offers the possibility of manufacturing devices on a chip, and guarantees good reproducibility. Various metal oxide nanostructures, including nanowires, nanotubes, nanofibers, core-shell structures, and hollow nanostructures also can be synthesized. As it is known, the field of metal-oxide nanostructured morphologies (e.g., nanowires, nanorods, nanotubes, etc.) has become one of the most active research areas within the nanoscience community.

The ability to create a variety of metal oxide-based composites and the ability to synthesize various multi-component compounds significantly expand the range of properties that metal oxide-based materials can have, making metal oxides a truly versatile, multifunctional material with widespread applicability. Small changes in their chemical composition and atomic structure can be accompanied by a spectacular variation in the properties and behavior of metal oxides. Advances in synthesizing and characterizing techniques are revealing numerous new functions of metal oxides.

Taking into account the importance of metal oxides for progress in microelectronics, optoelectronics, photonics, energy conversion, sensor and catalysis, a large number of books devoted to this class of materials have been published. However, one should note that some books from this list are too general, some are collections of various original works without any generalizations, and others were published many years ago. During past decade, great progress has been made on the synthesis as well as on the structural, physical and chemical characterization and application of metal oxides in various devices, and a large number of papers have been published on metal oxides. However, many important topics related to the study of metal oxides and their application have not been discussed. To remedy the situation, we decided to generalize and systematize the results of research in this direction and to publish a series of books devoted to metal oxides.

One should note that for the proposed book series "Metal Oxides", this book is the first one and is devoted to the consideration of metal oxides only. We believe that combining books on metal oxides in a series will help readers in their search for information on the subject. In particular, we plan that the books from our series, which will be specialized by their content, will provide interdisciplinary discussion for various oxide materials with a wide range of topics, from material synthesis and deposition, to characterizations, processing and then to device fabrications and applications. This book series is prepared by a team of highly qualified experts, which guarantees it a high quality.

I hope that our books will be useful and comfortable to use. I also hope that readers will consider the "Metal Oxides" book series like an encyclopedia of metal oxides that enables us to understand the present status of metal oxides, to estimate the role of multifunctional metal oxides in design of advanced devices, and then, based on observed knowledge, formulate new goals for the further research.

The intended audience of the series are scientists and researchers working, or planning to work, in fields of materials related to metal oxides, that is, scientists and researchers whose activities are related to electronics, optoelectronics, energy, catalysis, sensors, electrical engineering, ceramics, and biomedical designs, etc. The "Metal Oxides" book series will also be appealing to practicing engineers or project managers in industries and national laboratories, who would like to design metal oxide-based devices, but don't know how to do it, or how to select optimal metal oxides for specific applications. With many references to the vast resource of literature on the subject, this series will serve as a significant and insightful source of valuable information, providing scientists and engineers with new insights for understanding and improving existing metal oxide-based devices, and for designing new metal oxide-based materials with new and unexpected properties.

The “Metal Oxides” book series would also be very helpful for university students, post-graduates, doctors, and professors. The structure of these books offer a basis for courses in the field of material sciences, chemical engineering, electronics, electrical engineering, optoelectronics, energy technologies, and environmental control, among others. Graduate students could also find the book series useful in their research and understanding of the features of metal oxides synthesis, study, and application of this multifunctional material in various devices. We are sure that all of them will find information that will be useful for their activity.

Finally, I thank all contributing authors and book editors who have been involved in the creation of these books. I am thankful that they agreed to participate in this project and for their efforts in the preparation of these books. Without their participation, this project would have not been possible. I also express my gratitude to Elsevier for giving us the opportunity to publish this series. I especially thank all the team members of the editorial office at Elsevier for their patience during the development of this project and for encouraging us during the various stages of its preparation.

Ghenadii Korotcenkov

Preface to the volume

In 1953, using the example of germanium, Brattain and Bardeen first reported that the electrical properties of semiconductors varies as a result of changes in the surrounding gaseous atmosphere, in this specific case, humidity [1]. This was one of the arguments for the development of packaging for semiconductor-based devices, and indicated a way to gain more information about the environment. This cannot be done with classical semiconductors because they will change irreversibly, for example, from Si to SiO_2. Around the same time, Heiland began examining the electrical characteristics of ZnO and, in 1954, found that oxygen apparently led to electron depletion of zinc oxide with detectable effects on its electrical conductivity [2]. Less than a decade later, the number of explosions due to increased domestic liquefied petroleum gas use in Japan reached a critical level and a leak detection method was desperately needed. In 1962, when Seiyama et al. at Kyushu University published that the resistance change of ZnO thin films as a result of gas exposure could be used to detect gaseous components and, in an elegant solution, coupled this with a gas chromatogram [3]. In the same year, Taguchi, an independent inventor, filed a patent for a gas sensor based on a metal oxide [4].

Interestingly, research on the electrical properties of metal oxides was also relevant for heterogeneous catalysts. Already in the 1930s, Wagner and Hauffe had reported N_2O decay and CO oxidation over NiO and CuO, and such experiments were interpreted theoretically within the framework of the band model as a consequence of electron exchange reactions [5]. Later, Shaver described the sensitivity increase of tungsten oxide-based gas detectors by adding small amounts of catalytic metals [6]. In 1957, Bielanski et al. examined the catalytic activity and resulting variation in conductivity of ZnO and Cr_2O_3 upon exposure to water–alcohol mixtures [7].

In a completely different research field, with yttrium-stabilized zirconium oxide as the key metal oxide, ion conduction was used for the potentiometric determination of oxygen in the 1950s and 1960s by Kiukkola and Wagner and by Peters and Möbius [8,9]. The ion conduction of this oxide had already been studied by Nernst in 1900 [10]. The Nernst glower, formerly used as a continuous source of infrared radiation, was one of the first outcomes in this field of study. The well-known λ–probe, a key component to optimize combustion in car engines, is an example of the current use of this material. λ–probes have a tremendous impact on emission reduction in cities. The in-cabin car air-quality control SnO_2 and WO_3 sensor array stands for the semiconducting gas sensor. Those devices are manufactured in the millions and are expected to solve more and more ambitious applications. Advanced exhaust gas composition monitoring devices are obvious examples, while the chemical sensors that are required for the Internet of Things are more recent examples.

The approach we are undertaking combines chapters on basic aspects of functioning principles with chapters that describe technologies and challenges of present and future sensors based on metal oxides. The understanding of charge carrier transport in metal oxide gas sensors and how charge transfer reactions couple to fluxes of molecules and electrons is essential from a fundamental point of view. Development of materials with respect to producible fabrication technologies, behavior, and failure under specific conditions are of equal relevance from an application point of view.

Accordingly, contributors are from academia and from industry. We would like to thank our colleagues and friends who contributed decisively to this book: Dr. David Degler, Prof. Ulrich Guth, Dr. Anne Hemeryck, Kuniyuki Izawa, Prof. Jong-Heun Lee, Dr. Alexandru Oprea, Dr. Julia Rebholz, Anna Staerz, Dr. Heiko Ulmer, Prof. Dr. Udo Weimar, and Prof. Hans-Dieter Wiemhöfer.

Nicolae Barsan and Klaus Schierbaum
June 19, 2018

References

[1] W.H. Brattain, J. Bardeen, Surface properties of germanium, Bell Syst. Tech. J. (1953) 1–41.
[2] G. Heiland, Zum Einfluss von Wasserstoff auf die elektrische Leitfähigkeit von ZnO-Kristallen, Z. Physik. 138 (1954) 459–464.
[3] T. Seiyama, A. Kato, K. Fujiishi, M. Nagatani, A new detector for gaseous components using semiconductive thin films, Anal. Chem. 34 (1962) 1502.
[4] N. Taguchi, Japan. Patent 45-38200 (1962).
[5] C. Wagner, K. Hauffe, Untersuchungen über den stationären Zusatnd von Katalysatoren bei heterogenen Reaktionen, Z. Elektrochem, Ber. Bunsenges. Phys. Chem. 44 (1938) 172.
[6] P.J. Shaver, Activated tungsten oxide gas detectors, Appl. Phys. Lett. 11 (1967) 255.
[7] A. Bielanski, J. Deren, J. Haber, Electric conductivity and catalytic activity of semiconducting oxide catalysts, Nature 179 (1957) 668–669.
[8] H. Peters, H.-H. Möbius, Procedure for the gas analysis at elevated temperatures using galvanic solid electrolyte elements, DD-Patent 21673, August 1961.
[9] (a) K. Kiukkola, C. Wagner, Galvanic cells for the determination of the standard molar free energy of formation of metal halides, oxides, and sulfides at elevated temperatures, J. Electrochem. Soc. 104 (1957) 308–316. (b) Measurements on galvanic cells involving solid electrolytes, J. Electrochem. Soc. 104, 379–387.
[10] W. Nernst, Z. Electrochem. 6 (1900) 41.

Book outline

Chapter 1: Introduction (Nicolae Barsan and Klaus Schierbaum)

This chapter provides useful definitions for the sensors and their analytical performance. It also discusses the need and impact of sensors and provide information about the volume and structure of existing and future markets. The rationale of the book is explained and a short overview of the chapters are provided.

Chapter 2: Gas sensors based on oxygen ion conducting metal oxides (Ulrich Guth and Hans-Dieter Wiemhöfer)

This chapter deals with solid-state electrolyte gas sensors. The basic principle of electrochemical cells with solid electrolytes and the electrochemical notations and formulas are described for the widely used oxygen cell based on yttria-stabilized zirconium oxide. Concepts used for electrode and auxiliary phases to measure various gases by means of potentiometry, mixed-potential measurements, amperometry, and coulometry are explained. Furthermore, conventional and developing technologies for the practical realization of devices are outlined together with present and novel materials applied in solid-state electrolyte gas sensors.

- *Potentiometric sensors*

  A brief review is given on the physicochemical treatment of oxygen sensors, based on electrochemical potentials and the Nernst equation. This concept can be extended to include the detection of other gases which are in equilibrium with oxygen. Important applications include high combustion control, lambda probes, steel hardening, steel preparation, porcelain, brick, and glass production. Examples of the design of commercially available sensors are presented.
- *Amperometric sensors*

  Amperometric sensors operate at a certain voltage in which the electrode reaction is determined by the gas transport. A well-known example is the wideband lambda probe. The underlying equations for the calibration curves of such sensors are given. This concept can be also applied for simultaneous determination of oxygen and nitric oxides in exhaust gases.
- *Coulometric sensors*

  The coloumetric principle applies Faradays law and can be used in coulometric devices to measure oxygen, hydrogen, hydrocarbons, and other oxidizable and reducible gases.

The main advantage of these sensors is their long-term stability without calibration. Combinations with a pump cell and a potentiometric cell makes it possible to monitor gases under continuous gas flows.

- *Mixed-potential sensors*

  Mixed-potential sensors combine an oxygen reference electrode with a second electrode of low-oxygen sensitivity, but a high sensitivity to combustibles or gases like nitric oxides. A brief overview on the catalytic behavior of electrode materials is presented and common and promising electrode materials for mixed-potential sensors are discussed.

- *Solid electrolyte sensors in dynamical mode*

  The final section of the chapter deals with sensors in which time-dependent voltage is applied and time-dependent current response is measured. A distinction must be drawn between periodical and continuous change in potential or in current with the time (voltammetry). The first mode is well-known as an impedance measurement; the second mode, less known for solid-state electrolytes, refers to voltammetric experiments and reports on the use of this concept for hydrogen determination in air at high temperatures.

Chapter 3: Basics of semiconducting metal oxides–based gas sensors (Alexandru Oprea, David Degler, Nicolae Barsan, Anne Hemeryck, and Julia Rebholz)

This chapter reviews the basic knowledge needed to understand gas sensing with semiconducting metal oxides. It explains the factors controlling semiconductivity in oxides and how interaction with atmospheric gases changes both surface-charge and free-charge carriers concentration. It also explains the different concepts applied to obtain sensor signals and the different technological approaches used in current gas sensors. A presentation of the most interesting gas sensitive materials and their specificities are included.

- *Semiconducting metal oxides* (Alexandru Oprea, University of Tübingen)

  A short review of the significant characteristics of semiconductors is provided that introduce certain parameters, such as concentration and mobility of free-charge carriers, work function, and electronic affinity, etc. The origin of semiconductivity in metal oxides will be examined for both *n*-type and *p*-type oxides. The state-of-the-art knowledge of the bulk properties of the most significant oxides—e.g., SnO_2, In_2O_3, ITO, WO_3, ZnO, Cr_2O_3, CuO, and Co_2O_3, etc.—are presented.

- *Modeling the gas sensing* (Nicolae Barsan, Anne Hemeryck, Julia Rebholz)

  The first step is the examination of the surface reactions between the atmospheric gases and the oxides—the so-called reception function. The latest experimental findings are presented as well as theoretical modeling. It is shown how this knowledge can be formalized with the help of the quasi-chemical reactions. Furthermore, an explanation of how charging of the surface changes the properties of the oxide and how this can be quantified with the help of the Poisson and electro-neutrality equations for both *n*-type and *p*-type materials is provided. Finally, the transduction function is described and the impact of semiconductor bulk properties and sensing layer structure and morphology are explained.

- *Experimental investigations* (Nicolae Barsan, David Degler)
  The understanding of sensing is based on experimental inputs. These need to provide information about the structural properties, including the chemical state of the dopants and sensitizers, and the impact of gas exposure on the sensor's parameters. The latest developments of the operando characterization techniques are presented and the most important of these are described; this includes the kind of information that can be gained, but also the limitations and caveats.

Chapter 4: Technological realization of semiconducting metal oxide–based gas sensors (Jong-Heun Lee)

This chapter presents approaches for the synthesis of sensitive materials, the methods employed for their deposition on the transducers, and the evolution of transducers themselves. The evolution of the different concepts for the synthesis are presented in correlation with the impact of a better understanding of sensing as well as the changing demands of leading applications. Besides the approaches employed to obtain the "classical" noble metal-decorated oxides, more modern routes—such as the engineering of receptor/transducers units based on homo- and hetero-junctions, core/shell structures or integrated filter/sensing layers combinations—are presented. The evolution of the transducers from pellets to MEMS hotplates/thick film sensing layers combinations to flexible sensors realized by additive technologies are reviewed.

Chapter 5: Application of SMOX-based sensors (Kuniyuki Izawa, Heiko Ulmer, Anna Staerz, Udo Weimar, and Nicolae Barsan)

In this chapter the most important current applications of semiconducting metal oxides–based gas sensors are explained. The impact of applications on technology and the lessons learnt from the past are reviewed. Additionally, the most promising future applications are critically examined.

Chapter 6: Conclusion and outlook (Nicolae Barsan and Klaus Schierbaum)

The concluding chapter is devoted to basic principles of metal oxide gas sensors with ion- and electron-conducting solids. Charge carrier transport and charge transfer coupled to fluxes are essential key phenomena that must be understood and tailored for conventional but also for advanced materials. The future prospects of our field of research depend both on the understanding of these fundamental phenomena and on the development and application of new concepts, as described in the preceding chapters.

Introduction

1

Nicolae Barsan[1] *and Klaus Schierbaum*[2]
[1]Institute for Theoretical and Physical Chemistry, University of Tübingen, Tübingen, Germany, [2]Department of Materials Science, Institute of Experimental Condensed Matter Physics, University of Düsseldorf, Düsseldorf, Germany

Air components have become important metrics, not only for professionals, but also for public perception. A prominent example is the increase of nitric oxides in cities from the exhaust gases of diesel vehicles [1]. An increasing number of people understand how crucial "pure" breathing air is for one's health and comfort.

In addition to particles of varying sizes, air contains many different gaseous components in different concentrations, shown in Fig. 1.1. Oxygen and water vapor ("moisture") are life-essential gases present in the atmosphere. In addition, carbon dioxide—normally present in our air without harm—affects humans when having a high concentration, first causing tiring, then anesthetizing, and cab ultimately be lethal if its concentration exceeds the value of 8% over a period of 30–60 minutes [3].

Other gases such as hydrocarbons, CO, NOx, NH_3, H_2, or SO_2 also present hazards: they are detrimental to health, an explosion hazard, and can even be lethal. News of carbon monoxide poisoning, often called the "silent killer," in highly developed industrialized countries is regularly reported [4].

Specifically, there are a number of dangerous gases present in an industrial setting for which qualitative and quantitative detection methods are needed [5]. The field of sensor technology originates from the need for simple determination without the use of complex, analytical measuring methods and spectrometers in an industrial setting. Solid-state gas sensors based on metal oxides have been developed and used for five decades [6]. They have clear advantages: they are small, are highly sensitive allowing the detection of many different gases in the ppm or even the ppb range, they can be operated online, and they are cost-effective compared to methods using classical instrumental analysis.

Ideally, sensors should be cheap, allow mobile application, be easy to use with a fast response time, and be selective [7]. In reality, however, sensors—and not only the solid-state gas sensors with electron- and ion-conducting metal oxides treated in this book—often suffer from limited measuring accuracy, are not selective with the simultaneous occurrence of several gases and have problems with long-term stability [8]. Today, sensors are still far from this ideal, but it is justifiable to say that the considerable efforts in this field over the past few decades have led to substantial progress. A variety of materials and technologies have been developed and the fundamental understanding of the complex, mechanistic processes involved in the interaction of gases with the sensors and their surfaces has been constantly refined.

Gas Sensors Based on Conducting Metal Oxides. DOI: https://doi.org/10.1016/B978-0-12-811224-3.00014-7

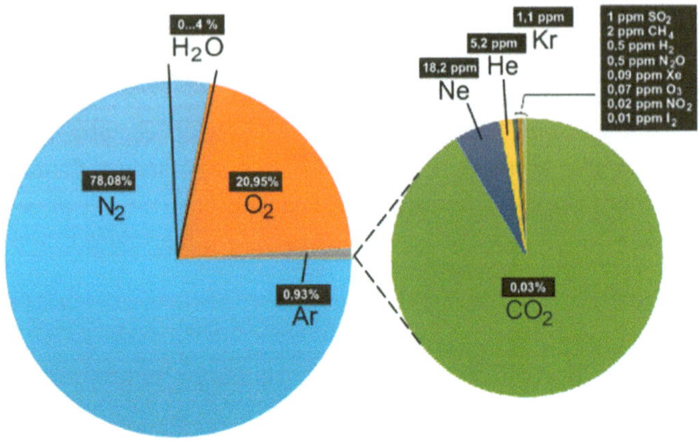

Figure 1.1 The composition of air near the ground.
Source: Adopted from B. Klose, Meteorologie, Springer-Lehrbuch, https://doi.org/10.1007/978-3-662-43622-6_2, Springer-Verlag Berlin Heidelberg. 2016 [2].

The central question has two parts: how the transport of charges (electrons and/or ions) and potential-generating processes are influenced by the molecules from the gas phase, and, in turn, which physical and chemical effects ultimately control the "communication" between the charge carriers and the molecules. As a result, there is a differentiation between the receptor and the transducer part of the sensor. This terminology can be explained using biosensors in which molecular receptors have been deposited as self-contained units on a physical, for example, optical or gravimetrically operating transducer. For solid-state gas sensors, the separation between the receptor and transducer is more difficult because the functions are interrelated. This becomes obvious from Fig. 1.2 which shows the basic types of sensors.

Fig. 1.2A shows a semiconductor gas sensor in a planar arrangement. The heater is deposited on the backside of an insulating support (substrate). The sensitive layer is typically heated to between 300°C and 500°C. The substrate contains a topside interdigital electrode on which the sensitive layer, for example, of tin dioxide, is applied. As can be schematically seen as the current between the grounded electrode and the voltage source is amplified. Changes in the resistance of the metal oxide layer as a result of variations in the surrounding gas composition are used as sensor signals. Reactions between atmospheric gases and the surface of grains in the sensitive layer result in changes of the current flow at the grain-to-grain boundaries ("percolation path") [7].

Fig. 1.2B uses an ion-conducting metal oxide ("solid electrolyte"), frequently yttrium-stabilized zirconium dioxide, in which oxygen ions are mobile at temperatures above 500°C; the electron conduction can be neglected [8]. Different electrode materials allow two separate, partial reactions to take place at the working (or sensitive) and counter electrodes. One can distinguish between anodic oxidation (electron emission of a molecule) and a cathodic reduction (electron absorption of a molecule). It is important that the molecule to be detected is involved in the reaction at the working electrode and that the ions moving in the ionic conductor are

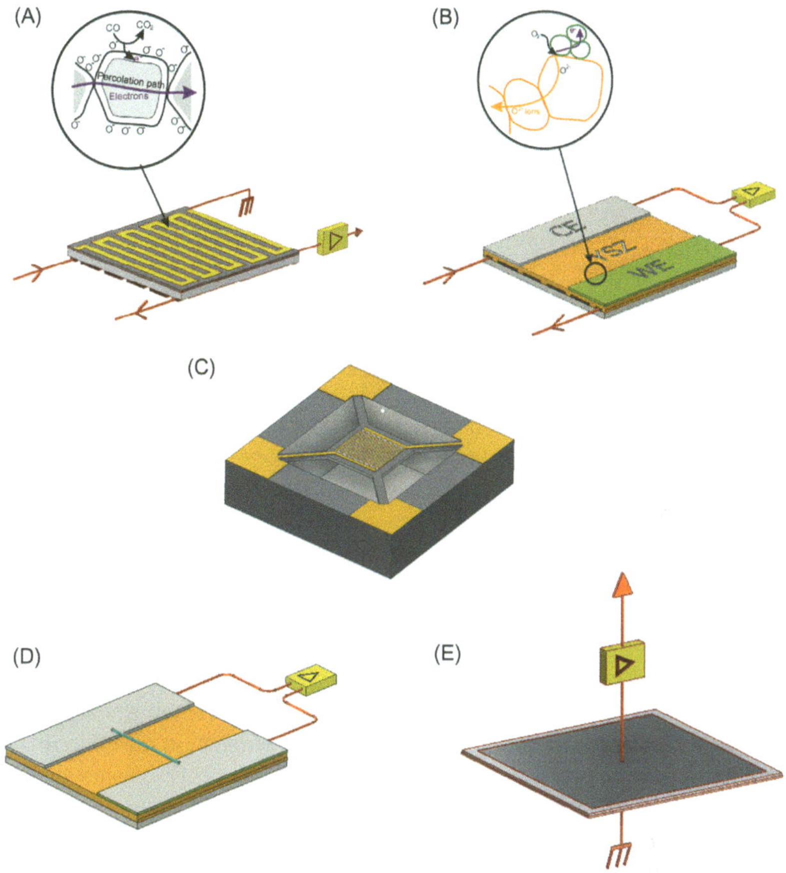

Figure 1.2 Schematic representations of solid-state gas sensors. (A) semiconductor, (B) mixed-potential, (C) microhot plate, (D) nanowire, and (E) TiO_2-based sensor. Further explanations are given in the text.

involved in both reactions. The spatial separation of the two reactions does not have to be perfect; it is sufficient if they proceed with a sufficiently different rate on both electrodes. In this way, a potential is formed between the working electrode and the counter electrode. It is customary that the working electrode is the sensitive electrode and a molecule that is present in large excess (often oxygen) is reacted at the counter electrode. This arrangement is called a mixed-potential sensor; fuel cells of analogous construction are referred to as single-chamber fuel cells [9]. In the potentiometric variant of the solid-electrolyte sensor, the two porous electrodes of platinum are arranged separately from each other in two different "half cells" and the difference in the chemical potential of a molecule (oxygen in the lambda probe) or the partial pressures associated therewith, determines the potential difference [10]. In the amperometric variant, the mass flow of the molecules to be detected is limited by a small opening in a cover or membrane, and the charge carrier flow through the ion conductor is, thus, limited and itself a function of the partial pressure of the molecule to be detected [11]. In all these variants, a high temperature is necessary.

The high temperatures of the metal oxide-based sensor means that portable applications can hardly be realized. The constraint for power reduction and the need to reduce the dimensions of the sensors, coupled with the advantages of mass production at a low cost, drives the development and implementation of MEMS technology for gas sensors (Fig. 1.2C) [12]. The advantages of "micro heating plates" are obvious and will be discussed later in this Chapter 4. Nanotechnology pursues a completely different approach. Individual nanowires from the sensitive metal oxide between two electrodes are explored and developed (Fig. 1.2D). They are heated at the lowest electrical power in the microwave range. Their enormous surface-to-volume ratio is determined by the nanometer dimension of the cross section.

Another solution is to develop heaterless sensors, shown in Fig. 1.2E. Very porous titanium dioxide layers on a titanium support together with a porous graphite layer containing catalyst particles form a stacked structure in which the sensor signal is determined amperometrically, impedanceometrically, or calorimetrically [13,14]. This technology is still in the early stages, but the proof-of-concept for the detection of carbon monoxide and nitrogen dioxide has already been provided. This type of sensor may mark the transition to liquid electrolyte sensors, where the use of catalyst-loaded graphite layers as the electrode material is common, all of which are without external heating.

When working with and describing sensors, a standard terminology is used. Table 1.1 summarizes important variables accompanied by short definitions.

Specifically, sensor performance is determined based on certain analytical parameters; some of the most important of which will be discussed here. Firstly the sensor response X [16] reflects the change of the measured sensor characteristics when a stimulus is applied—for example, the electrical resistance or conductance of a metal oxide gas sensor under exposure to a certain partial pressure p_1 or concentration c_1 of the target gas component 1, which corresponds to the steady-state type of R in Fig. 1.3. For low concentrations, ideal gas equation $p_1 = c_1\ RT$ can be assumed in which air is taken as a pure component mixed ideally with the pure target gas 1 (R and T denote the universal gas constant and the absolute temperature, respectively).

Ideally, sensors show steady-state responses X^{stat} to a constant stimulus, that is, the presence of a constant value p_1 of the target gas component 1, after a certain interaction time with the sensor is passed. In many cases, X follows a simple time law for which $X(t) - X_0 \sim (1 - e^{-t/\tau})$ holds with τ as a characteristic time constant for the sensor/stimulus interaction. Practically, X^{stat} would then be virtually attained after a time 10τ is passed and X would be 99.3% of X^{stat}.

$X^{stat}(c_1)$ represents the transfer function of the sensor with respect to 1 which is commonly denoted as calibration curve. Note that X^{stat} is the time-averaged signal that shows statistical fluctuations as other measuring values. In most cases, the sensor exhibits cross sensitivities to other gas components. The presence of a constant value of p_2 of such an interfering gas leads to a steady-state responses X^{stat} in the

Table 1.1 Terms used in the field of measurement with chemical sensors, biosensors, and sensor arrays [15]

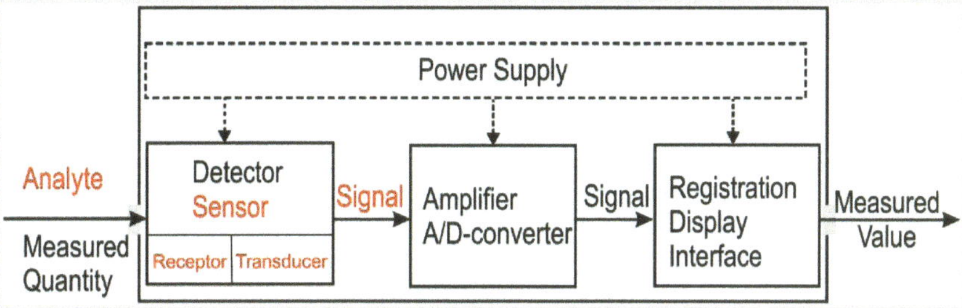

Analyte	(Bio)chemical quantity intended to be measured.
Measured value	The observed value of the measured quantity.
Detector	A device for detection of electromagnetic radiation. A smoke detector is a device that senses smoke by means of the change of optical transmission or ionization. An odor detector senses correspondingly odors.
Sensor	A device whose purpose is to detect events or changes in its environment and then provide a corresponding output (signal). The sensor may internally consist of a receptor and a transducer as separate components.
Receptor	A component of a sensor that interacts with the analyte.
Transducer	A component of a sensor (a device) that converts one form of energy to another.
Amplifier	An electronic device that increases the power of a signal.
A/D-converter	A device that converts the continuous signal to a digital number that represents the signal's amplitude.
Chemical sensor	A device that transforms chemical information, ranging from the concentration of a specific sample component to total composition analysis, into a useful analytical signal (IUPAC).
Chemical information	The information associated with all relevant measured quantities (denoted as analytes). The information is gained from a chemical reaction of the analyte or from the physical property of the system in which the analyte is present.
Analytical signal	The quantity from which the analytical information is deduced.
Chemical sensor array devices ("electronic tongue")	An analytical instrument comprising an array of nonspecific, low selective chemical sensors with high stability and cross sensitivity to different species in solution, and an appropriate method of pattern recognition and/or multivariate calibration for data processing. If properly configurated and trained (calibrated), the electronic tongue is capable of recognizing the qualitative and quantitative composition of multispecies of different natures.
Electronic nose	A device intended to detect odors or flavors.

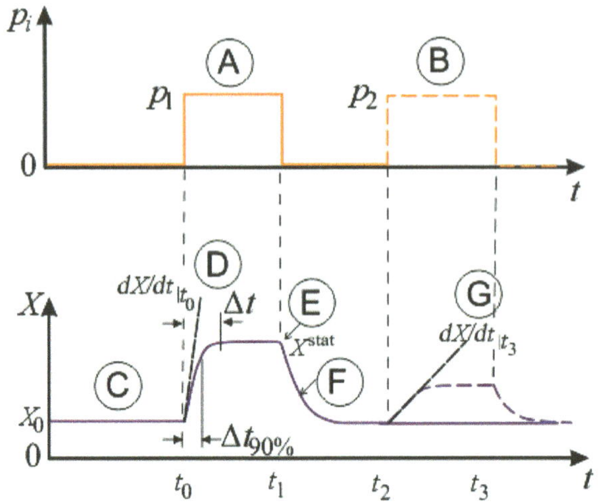

Figure 1.3 Upper curve: A stimulus p_1 of a gas component 1 (A) is applied at time t_0 and removed after a certain period at t_1. A second stimulus p_2 of a gas component 2 (B) is applied at t_2 and removed at t_3. Lower curve: The sensor response X as shown as a function of time t. (C) Mostly, the sensor exhibits a null-point signal X_0 for $p_i = 0$ ($i = 1$ or $i = 2$). (D) shows that the sensor response characterized by a finite slope dX/dt at t_0. After a response time Δt a stationary value of X^{stat} is eventually obtained (E); for practical reasons, often a value $\Delta t_{90\%}$ is given after the signal X has reached 90% of the value X^{stat}. (F) shows the decay of the signal when the stimulus is removed. (G) shows the response of a selective sensor (solid blue line) and a nonselective sensor (dashed blue line). The response time upon the stimulus (B) may differ from (A), as, for example, given by a different slope dX/dt.

absence of p_1. It is common to compare the effect of p_1 and p_2 in terms of a "sensitivity." If c_1 and c_2 are identical, relative values $X^{stat}(c_1)/X_0$ and $X^{stat}(c_2)/X_0$ reflect the different performance of the sensor with respect to gas components 1 and 2 at fixed, constant, and identical values c_1 and c_2. Alternatively $[X^{stat}(c_1)-X_0]/X_0$ and $[X^{stat}(c_2)-X_0]/X_0$ can be used. The specification of such values is, however, limited to sensors whose zero-point signal X_0 is finite and different from zero. In general, these quantities are not consistent with a proper definition of sensitivity S. The true sensitivity to, for example, gas 1, is defined as $dX^{stat}(1)/dc_1$ which describes the incremental change of X^{stat} to small changes of c_1. It is obvious that only in special cases with linear calibration curves $X^{stat} = S\ c_1 + X_0$ the true sensitivity is independent on c_1 within the operational range of the sensor. In most cases, for example, for calibration curves following power-laws, S depends on c_1. In these cases, it is a common, but nevertheless invalid, practice to use the values $X^{stat}(c_1)/X_0$ and $[X^{stat}(c_1)-X_0]/X_0$ to compare different sensors and to describe their apparent "sensitivity" for a specific target gas. A careful distinction must be made, however, between these quantities.

Real sensors show a drift of the baseline X_0 or of the signal X in the presence of a constant partial pressure or concentration of the target gas. In order to be able to determine useful quantities in such situations, the effect of the drift and the signal for the gas concentration of interest must be taken into account and should be validated for a proper sensor application. Usually, specified calibration intervals are required.

Statistically relevant quantities of sensors are shown in Fig. 1.4, which depicts the hypothetical calibration curve (signal X versus concentration dependency) of a sensor.

In addition to the static sensor parameters, the dynamic behavior of sensors is characterized by response and decay times as well as short- and long-term drifts. Neglecting drifts, the dynamics of the signal X is related to the time required to reach the steady-state signal X^{stat} or a given percentage of X^{stat} for a step change of the concentration in the measurement conditions. First-order behavior is commonly found experimentally. This can then be described with a time constant τ. The above presented calculations provide an understanding of τ and the so-called t_{63} and t_{95} times. Other times such as t_{90} (determined analogously) are commonly used. For further details, see [16].

Other variables are often used in colloquial terms because precise definitions from other scientific disciplines are hardly applicable in sensor research. An example is the term selectivity, defined in analytical chemistry as a quantitative value of

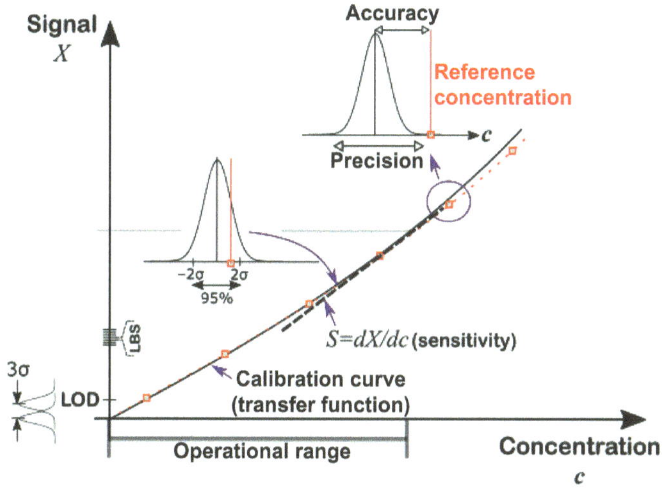

Figure 1.4 Signal-versus-concentration dependence of a sensor. The signal is determined by calibration with reference standards (red squares) from which the steady-state calibration curve (transfer function) is derived (red dotted line). Within the operational range the sensor signal can accurately measure the concentration c of the analyte at a certain resolution. Assuming an ideal sensor without drift and with negligible noise, the sensor's resolution can be expressed as bit resolution of the analog-to-digital converter, used in the amplifier stage. The transfer function represents the average measuring values that show a Gaussian distribution. Beyond the operational range, the sensor is still precise, but does not exhibit sufficient accuracy. The limit of detection determines the lower side of the transfer function.

a method to determine different analytes [17]. Selectively in sensor research is commonly used as a qualitative value only.

Engineering science also provides important terms which are often not rigorously applied in research, but which can be crucial for development in order to validate the technical readiness level of a sensor technology. Technical readiness is an important parameter to achieve market access and addresses the economic importance of gas sensor technology. Market studies are typically not fully published, but their essential conclusions are usually accessible. Their assessments and conclusions will only be summarized here, without taking into account the procedures of the studies. They are organized according to different sensor technologies, target gases, application fields, end users, and economic areas. Here, important relevant assessments for the semiconductor gas sensors and solid electrolyte sensors are to be considered and discussed in the context of general trends and market forces. According to a report, "Gas Sensors Market – Global Industry Size, Share, Trends, Analysis and Forecast, 2012–2018," the global gas sensors market is expected to reach a value of US$2.32 billion by 2018, at a compound annual growth rate of 5.7% from 2012 to 2018 [18,19]. It is segmented on the basis of target gases (oxygen, carbon dioxide, carbon monoxide, nitrogen oxides, and others), technology, end-users (industrial, environmental, medical, building automation, domestic appliances, petrochemical, automotive, and others) and geography (North America, Europe, Asia Pacific, Middle East and Africa, and Latin America). North America, Europe, and the Asia-Pacific region have the largest shares with respect to global demand and growth rate, but in all economic areas a stable growth prognosis for basic sensor technologies is reported. The key parameters for future growth of existing and newly developed sensor technologies are of both technical and economical nature and relate to sensitivity, selectivity, response time, energy consumption, lifetime, wireless/wired connectivity, environment, package size, price, regulations, and market volume. This holds true for all gas sensor technologies that can be broadly classified into electrochemical sensors (liquid as well as solid-state electrolytes), semiconductor sensors (i.e., chemical field-effect transistor, ChemFET), solid state and metal oxide semiconductor sensors, photoionization devices, catalytic, and infrared devices. Both metal oxide semiconductor ("electronic conductivity") and solid-state electrochemical ("ionic conductivity") sensors have a very significant share in the overall gas sensor market. Future growth is also related to the main market drivers such as advancements in embedded electronics, but also due to the implementation of government legislations and regulatory initiatives in various countries.

This book provides a comprehensive guide on metal oxide-based gas sensors. It focuses on the two most important, but distinct, types of gas sensors based on conducting metal oxides. Ion conduction, applied in so-called solid-state electrolytic sensors, and electronic conduction which is used in semi-conductivity gas sensors. The well-known λ –probe, a key component used to optimize combustion in car engines, is an example of the former type; the in-cabin car air-quality control SnO_2 and WO_3 sensor array is an exemplary semiconductivity type. Those sensors are manufactured in the millions; they are nowadays expected to solve more and more ambitious applications. Advanced exhaust gas composition monitoring devices are an example, but also the chemical sensors that are required for the Internet of Things. The approach we are undertaking combines chapters on basic aspects of functioning principles with chapters that describe technologies and challenges of present and future sensors based on metal oxides. Accordingly, the contributors are from academia and from industry.

As the world becomes more connected and digitalized, sensors will begin to play an indispensable role in numerous applications. Sensors based on metal oxides are an attractive option. This book provides relevant information for those working in, or looking to enter, the field.

Section 2 concentrates on the basics of ion conducting metal oxide-based gas sensors and their applications. At the beginning the basic principle of electrochemical cells with solid electrolytes and the electrochemical notations and formulas are described for the widely used oxygen cell based on yttria-stabilized zirconium oxide. Concepts to use electrodes and auxiliary phases to measure various gases by means of potentiometry, mixed-potential measurements, amperometry, and coulometry are explained. Furthermore, conventional and more recently developed technologies for practical realization of devices will be outlined, together with present and novel materials applied in solid-state electrolyte gas sensors.

Section 3 provides a comprehensive overview of all the aspects involved in sensing with semiconducting metal oxide-based sensors. The complex theory behind the semiconductivity of metal oxides is reviewed and the bulk properties of the most significant oxides—for example, SnO_2, In_2O_3, and WO_3, are presented. An chapter into how theoretical modeling can aid in understanding the reception of metal oxide-based sensors is provided. Different analytical methods which can be used to examine sensors during operation are detailed. The information gained from these methods is useful in fine tuning the sensor's characteristics. In particular, the effect of different synthesis and preparation methods are discussed in detail. The technological evolution of the transducers from pellets to MEMS hotplates/thick film sensing layers combinations to flexible sensors realized by additive technologies are reviewed. The section concludes with a chapter describing the historical origin of gas sensors, a review of their current applications, and a prediction of future ones.

The role of sensors will only become more significant with time. There are experts from a wide spectrum of fields, engineering to material science, working in the field of metal oxide gas sensors. As their application becomes more common and the technology becomes more sophisticated, it is important that each member of the field maintain a basic understanding of all aspects of these sensors. Due to the wide breadth of this book, it is an ideal guide for experts of one field to review aspects from other relevant areas.

References

[1] Compare, e.g., air protection plans and exhaust gas standards as stipulated by national and international legislation.

[2] B. Klose, Available from: https://doi.org/10.1007/978-3-662-43622-6_2Meteorologie, Springer-Lehrbuch, Springer-Verlag Berlin Heidelberg, 2016.

[3] Compare data published by the European Industrial Gases Association under https://www.eiga.eu/

[4] Environmental Health Criteria 213 CARBON MONOXIDE (SECOND EDITION), World Health Organization Geneva, 1999, http://apps.who.int/iris/bitstream/10665/42180/1/WHO_EHC_213.pdf

[5] Compare oxygen gas detection as an example S. Akbar, P. Dutta, C. Lee, High-temperature ceramic gas sensors: a review, Int. J. Appl. Ceram. Technol. 3 (2006) 302–311.

[6] History e.g. W. Göpel, T.A. Jones, T. Seiyama, J.N. Zemel, Historical remarks, in: W. Göpel, J. Hesse, J.N. Zemel (Eds.), Sensors - A Comprehensive Survey *Vol. 2* Chemical and BiochemicalSensors (Part I), VCH-Verlag Weinheim, New York, Basel, Cambridge, 1991.

[7] C. Wang, L. Yin, L. Zhang, D. Xiang, R. Gao, Metal oxide gas sensors: sensitivity and influencing factors, Sensors 10 (2010) 2088–2106. Available from: https://doi.org/10.3390/s100302088.

[8] F. Schubert, S. Wollenhaupt, J. Kita, G. Hagen, R. Moos, Platform to develop exhaust gas sensors manufactured by glass-solder-supported joining of sintered yttria-stabilized zirconia, J. Sens. Sens. Syst. 5 (2016) 25–32. Available from: https://doi.org/10.5194/jsss-5-25-2016. Available from: www.j-sens-sens-syst.net/5/25/2016/.

[9] M. Kuh, T.W. Nappora, Single-chamber solid oxide fuel cell technology – from its origin to today's state of the art, Energies 3 (2010) 57–134. Available from: https://doi.org/10.3390/en3010057.

[10] J.R. Stetter, W.R. Penrose, S. Yang, Sensors, chemical sensors, electrochemical sensors, and ECS, J. Electrochem. Soc. 150 (2003) S11–S16.

[11] K.S. Alber, J.A. Cox, P.J. Kulesza, Solid-state amperometric sensors for gas phase analytes: a review of recent advances, Electroanalysers 9 (1997) 97–101.

[12] I. Simon, N. Barsan, M. Bauer, U. Weimar, Micromachined metal oxide gas sensors: opportunities to improve sensor performance, Sens. Actuat. B 73 (2001) 1–26.

[13] M. El Achhab, K. Schierbaum, Gas sensors based on plasma-electrochemically oxidized titanium foils, J. Sens. Sens. Syst. 5 (2016) 1–9. Available from: https://doi.org/10.5194/jsss-5-1-2016.

[14] K. Schierbaum, M. El Achhab, Generation of an electromotive force by hydrogen-to-water oxidation with Pt-coated oxidized titanium foils, Phys. Status Solidi A 208 (12) (2011) 2796–2802. Available from: https://doi.org/10.1002/pssa.201127400.

[15] K. Schierbaum, in: N. Barsan, G. Gauglitz, A. Oprea, E. Ostertag, G. Proll, K. Rebner, K. Schierbaum, F. Schleifenbaum, U. Weimar (Eds.), Chemical and Biochemical Sensors, 1. Fundamentals, Ullmann's Encylopedia of Industrial Chemistry, Wiley-VCH Verlag, Weinheim, 2016published online: 29 SEP 2016 . Available from: https://doi.org/10.1002/14356007.b06_121.pub2.

[16] A. D'Amico, C. Di Natale, A contribution on some basic definitions of sensors properties, Sens. J. 1 (3) (2001) 183–190. IEEE.

[17] H. Kaiser, Zur Definition von Selektivität, Spezifität und Empfindlichkeit von Analysenmethoden, Fresenius Z. Anal. Chem. 260 (1972) 252.

[18] http://www.transparencymarketresearch.com.

[19] UK-MNT-Gas-Sensor-Roadmap', Robert Bogue et al, 2006 Technologie-Roadmap 'Prozess-Sensoren 2005–2015', Abschlußbericht, NAMUR und VDI/VDE-GMA 'Der westeuropäische Markt für Chemosensoren bis 2016', SENSOR MAGAZIN, Ausgabe 3, 2011, pp. 6-7, und 'Sensor-Märkte bis 2016', N. Schröder, SENSOR MAGAZIN, Ausgabe 1, 2012, pp. 6–8.

Further reading

Gas Sensors Market Analysis By Product (Oxygen, Carbon Dioxide, Carbon Monoxide, NOx), By Technology (Electrochemical, Semiconductor, Solid State, PID, Catalytic, Infrared), By End-Use (Medical, Building Automation & Domestic Appliances, Environmental, Petrochemical, Automotive, Industrial) And Segment Forecasts To 2020, Published: March 2014 I ISBN Code: 978-1-68038-083-5.

Gas sensors based on oxygen ion conducting metal oxides

2

Ulrich Guth[1] *and Hans-Dieter Wiemhöfer*[2]
[1]Technical University of Dresden, Institute of Physical Chemistry, Dresden, Germany,
[2]Institute of Inorganic and Analytical Chemistry, University of Münster, Münster, Germany

2.1 Solid-state electrolyte sensors for oxygen-containing gases

2.1.1 Fundamentals of oxygen ion conductivity in oxides

Solid oxide electrolytes with high oxygen ion conductivity and negligible electronic conduction are excellent materials for solid-state electrolyte oxygen sensors at medium and high temperatures. A prerequisite is, however, a sufficient thermodynamic stability over a large range of oxygen activities without significant compositional changes or degradation by redox reactions. The first publications which suggest the use of oxygen ion conducting cubic zirconia (which can be stabilized by substitution of up to 20 mol.% of zirconium ions with yttrium ions) for measuring oxygen activities in gases are due to independent pioneering work of Peters and Möbius [1] and Weissbart and Ruka [2] in 1961. Earlier work by Kiukkola and Wagner had already shown that yttria-stabilized zirconia (YSZ) (mostly denoted simply as YSZ) could be applied as exclusively oxygen ion conducting electrolyte in galvanic cells to measure oxygen activities of solid oxide containing materials [3]. Peters and Möbius [4] were the first to measure the oxygen concentration in gases and gas equilibria. It is now well-established that YSZ exhibits a predominant oxygen ion conductivity higher than 10^{-4} S/cm already at 500°C increasing up to 10^{-1} S/cm at temperatures higher than 1100°C. Due to its large band gap of around 5.1 eV, the partial electronic conductivity remains negligible as compared to the oxygen ion conductivity in the above-mentioned temperature range.

Fig. 2.1 gives an overview on oxides with remarkable oxygen ion conductivities. Among them, YSZ is still the preferred electrolyte material, because it combines good oxygen ion conductivity, excellent chemical stability and at the same time a negligible electronic conductivity in the most interesting oxygen partial pressure range between 10^{-20} bar and 10^5 bar at 1000°C.

This is a precondition to apply an oxygen ion conductor in corresponding solid-state electrolyte gas sensors working in a potentiometric or an amperometric mode

Gas Sensors Based on Conducting Metal Oxides. DOI: https://doi.org/10.1016/B978-0-12-811224-3.00002-0

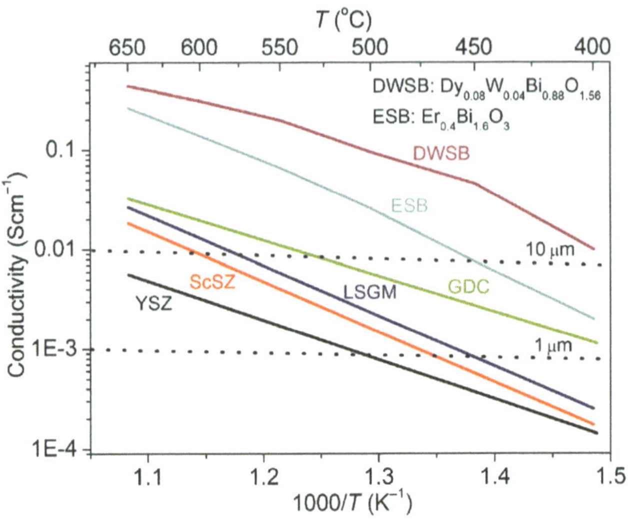

Figure 2.1 Electrical conductivity of the most popular oxygen ion conducting solid-state electrolyte materials. Acronyms: GDC = gadolinia doped ceria, e.g. $Ce_{0.8}Gd_{0.2}O_{1.9}$; LSGM = Sr- and Mg-doped lanthanum gallate, e.g. $La_{0.9}Sr_{0.1}Ga_{0.8}Mg_{0.2}O_{2.85}$; ScSZ = Scandia stabilized zirconia, $Zr_{0.8}Sc_{0.2}O_{1.9}$; YSZ = $Zr_{0.8}Y_{0.2}O_{1.9}$. Gas sensor applications, in particular in an amperometric operation mode, should rely on an area-specific resistance < 0.1 W cm^2. The dotted lines show the corresponding minimum conductivities for the specified electrolyte thickness values.
Source: Z. Gao, L.V. Mogni, E.C. Miller, J.G. Railsback, S.A. Barnett, A perspective on low-temperature solid oxide fuel cells, Energy Environ. Sci. 9 (2016) 1602–1644.

for measuring oxygen activities. Useful oxygen ion conductivities should be equal or higher than 10^{-4} S/cm for an electrolyte thickness of 10 μm (cf. Fig. 2.2). The effect of the much larger band gap of YSZ (5.2 eV) is well recognizable in comparison to GDC. It should be noted that electronic conductivities increase strongly increase with increasing temperature. This limits applications of this solid-state electrolyte in the range of lower temperatures.

The second condition for the use of any solid electrolyte in solid-state electrolyte sensors is a negligible electronic conductivity. This can be expressed by demanding that the oxygen ion conductivity should always be distinctly higher than the electronic conductivity by at least one or better two orders of magnitude. But as the minority conductivities of electrons and holes depend exponentially on the oxygen partial pressure, the condition means that potentiometric and amperometric sensor applications are restricted to a so-called *electrolytic domain*. It is usually centered at the oxygen partial pressure where the electronic conductivity reaches a minimum. Fig. 2.2 compares the useful electrolytic domains for YSZ (with 10 mol.% yttria) and GDC (i.e., gadolinia-doped ceria) in a plot oxygen partial pressure (logarithmic scale) versus temperature. For YSZ at 800°C, the useful electrolyte domain with predominant oxygen ion conduction extends even to more than 40 decades from 10^{-30} bar to 10^{10} bar whereas the range is much narrower for GDC, due to the

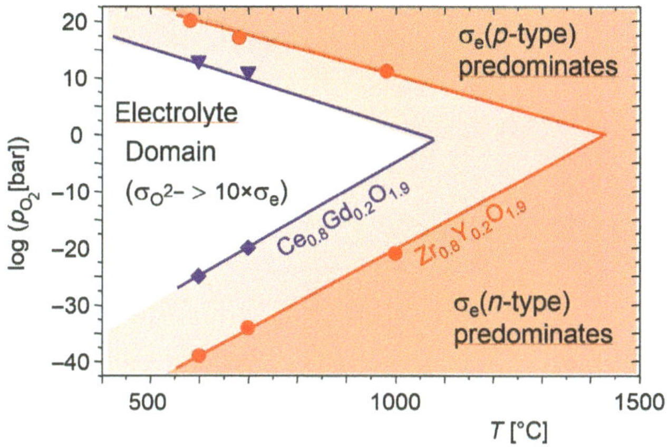

Figure 2.2 Electrolyte domain determined for the oxygen ion conductors YSZ and GDC from determination of the partial conductivities of oxygen ions and of electrons by suitable electrochemical polarization methods.
Source: S. Lübke, H.-D. Wiemhöfer, Electronic conductivity of Gd-doped ceria with additional Pr-doping, Solid State Ionics 117 (1999) 229–243.

lower effective band gap (about 3.2 eV for the transition from the mainly O_2 *p*-derived valence band to the mainly Ce4f-derived conduction band). This leads to significant *p*-type conductivity or n-type conductivity if the oxygen partial pressure is reduced from high to low values. Hence, it is basically the *band gap* which determines the useful width of the electrolytic domain. This is indeed the reason why oxygen ion conductors based on doped Bi_2O_3 or doped CeO_2 have not found widespread applications, as the p-type or n-type conductivity limits the signal accuracy at low oxygen activities. For Sr- and Mg-doped lanthanum gallates (see Fig. 2.1), limitations of the electrolytic domain arise from an increasing p-type conductivity at higher oxygen partial pressures. In addition, LSGM suffers from the loss of gallium under these conditions due to volatile Ga_2O. This effect occurs, in particular, at high temperatures.

2.1.2 Basic operation principles of solid-state electrolyte sensors with ion conductors

Most high temperature solid-state electrolyte sensors are based on an electrochemical cell which is shown schematically in Fig. 2.3. It consists of gas-tight ceramic electrolyte in the form of tubes, discs, or planar substrates which is sandwiched by precious metal electrodes, preferably porous platinum or sometimes gold. Furthermore, metal-oxide composites are used for solid-state electrolyte sensors. Both electrodes are in close contact with two separate gas compartments which contains different gas concentrations. YSZ ($Zr_{0.84}Y_{0.16}O_{1.92}$) is widely used as an oxide

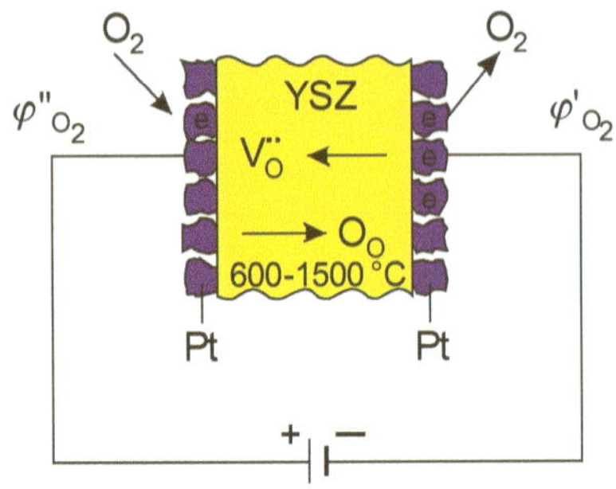

Figure 2.3 Oxygen concentration cell at high temperature.
Source: U. Guth, Solid State Electrochemistry, in: G. Kreysa, K.-o. Ota, R. Savinell (Eds.), Encyclopedia of Applied Electrochemistry, Springer, New York, Heidelberg, Dordrecht, London, 2014, pp. 1993–1998 and 2014–2041.

ionic conductor. In most cases, the electrochemical cell forms a so-called oxygen concentration cell. For YSZ, for example, its conventional notation is

$$O_2\left(p'_{O_2}\right), Pt\left|(ZrO_2)_{0.34}(Y_2O_3)_{0.08}\right|Pt, O_2\left(p''_{O_2}\right) \quad (2.1)$$

in which | means a phase boundary with a charge transfer. Two half-spaces with different oxygen partial pressures and are separated by a separating wall, composed of the ion conductor YSZ and coated with two different porous Pt electrodes.

Let us assume an electrically short-circuited cell or a zero-voltage of the external source (that has a nonnegligible internal resistance). For different oxygen partial pressures in both compartments (i.e., on both sides in Fig. 2.3, $p'_{O_2} > p''_{O_2}$), oxygen is transferred from the compartment of higher concentration to the half-space of lower concentration in order to balance the partial pressure difference. However, a transport of *molecular oxygen* O_2 through the YSZ is not possible since the solid electrolyte is gas-tight. Instead, the balance takes place by dissociation of O_2 and incorporation of *oxygen ions* into the solid electrolyte by reaction (Eq. 2.2) and the reverse reaction (Eq. 2.3) at the opposite electrode

$$O_2(g) + 2V_O^{\bullet\bullet}(YSZ) + 4e^-(Pt) \rightleftarrows 2O_O^{\times}(YSZ) \quad (2.2)$$

$$2O_O^{\times}(YSZ) \rightleftarrows O_2(g) + 2V_O^{\bullet\bullet}(YSZ) + 4e^-(Pt) \quad (2.3)$$

$$O_2\left(p'_{O_2}\right) \rightleftarrows O_2\left(p''_{O_2}\right) \quad (2.4)$$

This process requires the availability of doubly positively charged lattice vacancies in the oxygen sublattice of YSZ (written here as $V_O^{\bullet\bullet}$ according to Kröger-Vink notation and electrons e^- that originates from Pt). The regular oxygen ions in the YSZ are treated as formally uncharged $O_O^{\times}$. In summary, oxygen is transferred from compartment I to II (Eq. 2.4), accompanied by an electron flow through the outer short-circuit or voltage source. This electron flow leads to a measurable current. Under open-circuit condition, no current flows, however, and an electrical potential difference (electromotive force or *emf*) between the two Pt electrodes is generated as the result of the partial pressure difference. The emf can be accurately determined with a voltmeter of infinitely large internal resistance. Fig. 2.3 shows also the case, in which an adjustable voltage source is connected to both Pt contacts. If the external voltage has the same value and polarity as the voltage, generated under open-circuit condition, no electron and ion flow take place; then the adjusted voltage is the same as the ideal-voltmeter reading under the open-circuit condition. This corresponds to a situation in which an *electrochemical equilibrium* is established.

In this potentiometric mode (i.e., at equilibrium), the turnover of oxygen molecules is consequently zero and a voltage U_c is measured (Fig. 2.4). If no equilibrium is achieved (e.g., in an exhaust gas with a mixture of hydrocarbons (HC), nitrogen oxides (NO_x), and oxygen (O_2) interacting with a porous Au electrode), a mixed-potential U_m is generated between the two electrodes; typically, a small current flows in this so-called mixed-potential mode under short-circuit condition (not shown). In coulometric sensors, the flow of electrons is connected with a transport of oxygen according to Faraday's law. This transport is described by the amount of oxygen $\dot{n}_{O_2}$ that passes the electrode per time. It leads to a current I_F. In amperometric sensors, the mass transport is the rate limiting step, typically achieved by means of a diffusion barrier. If all components, which reach the electrode by diffusion, react electrochemically the resulting current I_d depends linearly on the

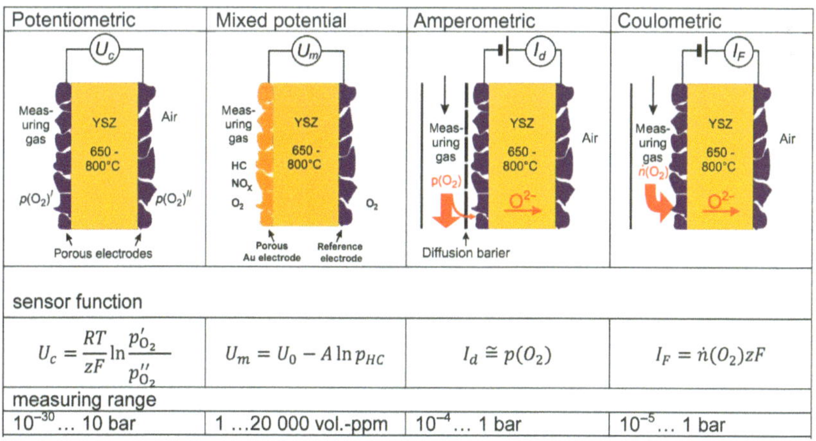

Figure 2.4 Working principles of solid electrolyte sensors. The equations refer to the measured signals. The typical measuring devices of practically applied devices are indicated. *Source:* U. Guth, Gas sensors, in: A. Bard, G. Inzelt, F. Scholz (Eds.), Electrochemical Dictionary, second extended ed., Springer, Heidelberg, Dordrecht, London, New York, 2012, pp. 400–406.

concentration of reacting species. All widely applied solid-state electrolyte sensors fall into one of the operation modes shown in Fig. 2.4 [5]. For amperometric and coulometric modes, external voltage sources are used.

2.2 Potentiometric sensors

2.2.1 Fundamentals

Widespread and most commonly used sensors are potentiometric sensors by which thermodynamically determined voltages according to the Nernst equation can be measured in a wide range of temperature (400–1600°C) and oxygen partial pressure (10^{-20} to 10 bar). The sensor signal is always determined thermodynamically, as derived in the following.

The Gibbs enthalpy of reaction (Eq. 2.4) is negative (i.e., $\Delta_R G < 0$) since the oxygen transport proceeds spontaneously ($p'_{O_2} > p''_{O_2}$). On the other hand, $\Delta_R G$ can be expressed as the difference of the chemical potentials of oxygen at the pressures p''_{O_2} and p'_{O_2}

$$\Delta_R G = \mu''_{O_2} - \mu'_{O_2} \tag{2.5}$$

The "chemical work" is also related to the electrical work for the transport of electrons from the Pt electrode in half-space I to the Pt electrode in half-space II. This work is expressed in terms of the Galvani potentials φ' and equals $\Delta_R G$

$$\Delta_R G = -zF(\varphi' - \varphi'') \tag{2.6}$$

Here, z is the total number of elementary charges of the involved ions per oxygen molecule (here: $z = 4$) and F is the Faraday constant. Combining Eqs. (2.5) and (2.6) and rearranging yields

$$\left\{\mu'_{O_2} + zF\varphi'\right\} = \left\{\mu''_{O_2} + zF\varphi''\right\} \tag{2.7}$$

The sums in the brackets correspond to the electrochemical potentials $\tilde{\mu}'_{O_2}$ and $\tilde{\mu}''_{O_2}$ of oxygen on each electrode of the concentration cell shown in Fig. 2.3. Eq. (2.7) confirms that the electrochemical potentials are the same at each electrode for an equilibrium. Furthermore, also the electrochemical potential $\tilde{\mu}_{O^{--}}$ of oxygen ions does not vary in the ion conductor. It should be noted that the electrochemical potential $\tilde{\mu}_{e^-}$ of the electrons is, however, different at both electrodes. As a consequence, $\tilde{\mu}_{e^-}$ vary within the ion conductor between both levels. Since $\tilde{\mu}_{e^-}$ equals the Fermi energy, $\varepsilon'_F < \varepsilon''_F$ holds for $p'_{O_2} > p''_{O_2}$. More details to follow.

Taking into account the partial pressure dependence of the chemical potentials, that is,

$$\mu_{O_2}^{'('')} = \mu_{O_2}^{o'('')} + RTlnp_{O_2}^{'('')} \tag{2.8}$$

and identical standard chemical potentials $\mu_{O_2}^{o'} = \mu_{O_2}^{o''}$ (since identical reference states can be chosen as standard state in I and II) one obtains the Nernst equation for the oxygen concentration cell with the electrode potential E

$$E = \left(\varphi' - \varphi''\right)_{eq} = \frac{RT}{zF}\ln\frac{p_{O_2}''}{p_{O_2}'} \tag{2.9}$$

The suffix "eq" emphasizes again that we have derived Eq. (2.9) for electrochemical equilibrium. R denotes the gas constant and T the temperature, respectively. $\frac{RT}{zF}$ is the so-called Nernst factor. For $p_{O_2}' = p_{O_2}''$ the electrode potential is zero, $\Delta E = 0$. A positive value of ΔE corresponds to a thermodynamically favored reaction as in this case in which oxygen transport proceeds spontaneously.

Assuming ideal behavior of oxygen and the residual gas components and taking $\phi_{O_2}^{'('')} = \frac{n_{O_2}^{'('')}}{\sum n_i^{'('')}}$ as volume fractions of oxygen in the gas phases I and II (n denotes the amount of oxygen and the residual components i), it holds:

$$E_{eq} = \frac{RT}{4F}\ln\frac{p_{O_2}''}{p_{O_2}'} = \frac{RT}{4F}\ln\frac{\phi_{O_2}'' p''}{\phi_{O_2}' p'} \tag{2.10}$$

and

$$\Delta E_{eq} = \frac{RT}{4F}ln\frac{\phi_{O_2}''}{\phi_{O_2}'} \tag{2.11}$$

for identical total pressures in both gas compartments of the concentration cell, that is, $p'' = p'$.

It is useful to relate this thermodynamic approach with the electronic properties of a solid electrolyte in more detail. This allows to get a deeper understanding of: (1) the extension of the electrolytic domain and the stability limits as function of oxygen activities; and (2) as well on the role of the applied electrode potentials referred to a suitable internal energy scale of the solid electrolyte (Fig. 2.5). Note that the electrochemical potential of electrons (unit J/mol) is equivalent to the Fermi level (unit eV) and, in a general view, the cell voltage is given by the difference of the Fermi levels of both electrodes regardless of the type of electrode (whether a metal electrode, an electronically conducting oxide or a mixed oxygen ion and electron conducting compound). If there is no current flow, the cell voltage is only given by the Fermi level difference throughout the solid electrolyte. At the same time, there is no electrical potential difference within the solid electrolyte. Accordingly, in a good solid electrolyte there is practically no band bending. Therefore, a corresponding cell between

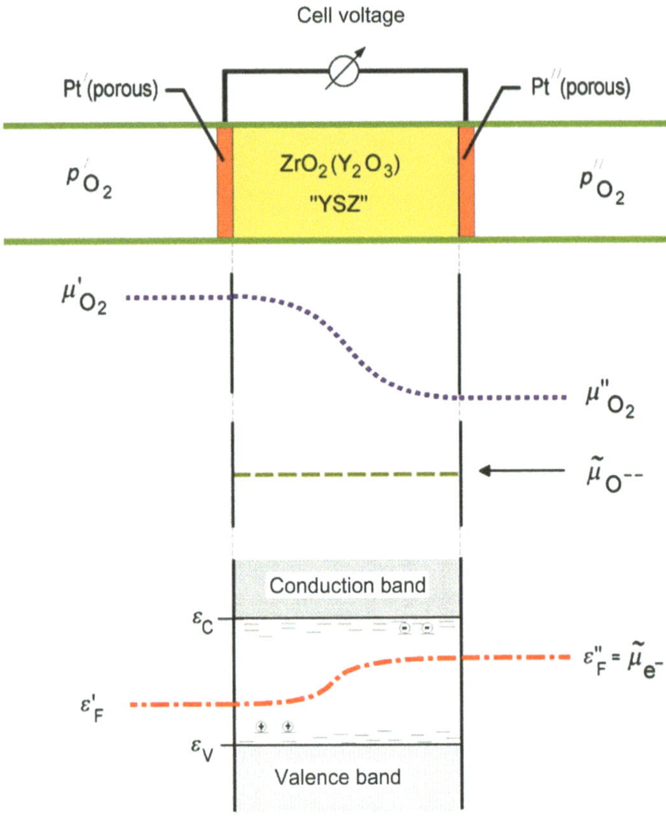

Figure 2.5 Electrolyte domain determined for the oxygen ion conductors YSZ and GDC. $\mu(O_2)$ = chemical potential of oxygen gas; ~ {μ}(O^{2-}) = electrochemical potential of oxygen ions; ~ {μ}(e^-) = electrochemical potential of electrons.

two electrodes in different gas environment with different oxygen partial pressures shows a steady state concentration gradient of electrons and holes reflected by the Fermi level gradient. For very low electronic conductivities as compared to the ionic conductivity, any electronic leakage current will be negligible in good approximation. One can even define an absolute electrode potential scale referring to the energies of the valence band or conduction band edges. This is demonstrated for the system Pt/YSZ in Fig. 2.6 by using the band scheme of a solid oxygen ion conductor between two electrodes and plotting the change of the Fermi level of electrons ε_F throughout the electrolyte between the electrodes, if they are in equilibrium with different oxygen partial pressures of two well separated gas chambers (the solid electrolyte acting as an impermeable separating membrane). The gradient of μ within the solid electrolyte shows that there is a well-defined oxygen activity at all locations within the bulk of the solid electrolyte, too. Using solid metal, metal oxide mixtures, the logarithmic oxygen partial pressure scale in b) was determined using surface spectroscopic techniques.

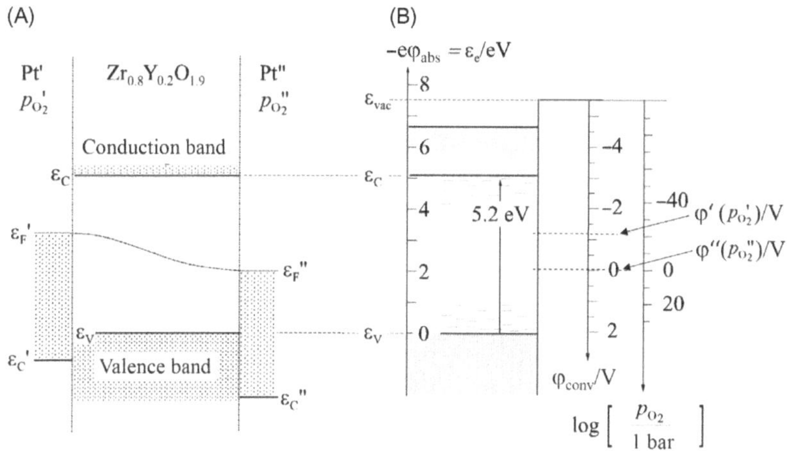

Figure 2.6 Results from surface analytical techniques: The position of the Fermi level with respect to the band edges is an unambiguous function of the local oxygen activity (i.e. the oxygen partial pressure) both at the interfaces to the electrodes.
Source: H.-D. Wiemhöfer, U. Vohrer, Spectroscopy and thermodynamics of electrons in yttria-stabilized zirconia, Ber. Bunsenges. Phys. Chem. 96 (1992) 1646–1651.

For the cell voltage U, determined by the Fermi levels $\varepsilon_F^{'}$ and $\varepsilon_F^{''}$ at oxygen partial pressures $p_{O_2}^{'}$ and $p_{O_2}^{''}$, respecitively, the following equation holds

$$U = -\frac{1}{e}\left(\varepsilon_F^{''} - \varepsilon_F^{'}\right) = \frac{1}{4F}\left(\mu_{O_2}^{''} - \mu_{O_2}^{'}\right) = \frac{RT}{4F}\ln\frac{p_{O_2}^{''}}{p_{O_2}^{'}} \tag{2.12}$$

Eq. 2.12 can be also applied to gas mixtures as well as in molten materials such as liquid metals and molten glasses, standard states must be properly chosen.

Not only oxygen can be measured in the potentiometric mode but all other gases which are in equilibrium with oxygen are also detectable. For electrode reactions of other systems like

$$\frac{1}{2}O_2(g) + V_O^{\bullet\bullet}(YSZ) + 2e^-(Pt) \rightleftarrows O_O^{\bullet\bullet}(YSZ) \tag{2.13}$$

$$H_2O(g) + V_O^{\bullet\bullet}(YSZ) + 2e^-(Pt) \rightleftarrows O_O^{\bullet\bullet}(YSZ) + H_2(g) \tag{2.14}$$

$$CO_2(g) + V_O^{\bullet\bullet}(YSZ) + 2e^-(Pt) \rightleftarrows O_O^{\bullet\bullet}(YSZ) + CO(g) \tag{2.15}$$

$$\frac{1}{2}O_2(\text{dissolv}, Fe) + V_O^{\bullet\bullet}(YSZ) + 2e^-(Fe) \rightleftarrows O_O^{\bullet\bullet}(YSZ) \tag{2.16}$$

$$NiO(s) + V_O^{\bullet\bullet}(YSZ) + 2e^-(Pt)) \rightleftarrows O_O^{\bullet\bullet}(YSZ) + Ni(s) \tag{2.17}$$

Nernst equation must be modified correspondingly. The electrochemical reaction (Eq. 2.17) relates to solid reference systems, an important case, in which the oxygen partial pressure is determined by a chemical equilibrium between metal and metal oxides. Such reference systems are necessary for oxygen determinations in very low pressure ranges and high temperatures, for example, in metal melts. Here, such systems are applied in order to avoid cross-over of oxygen through permeation.

Furthermore, such equilibria can be potentiometrically evaluated and used for measurements of gas components whose concentrations depend on oxygen concentration. Examples are:

$$CO_2 + H_2 \rightleftarrows CO + H_2O \tag{2.18}$$

$$C + CO_2 \rightleftarrows 2CO \tag{2.19}$$

$$2NH_3 \rightleftarrows N_2 + 3H_2 \tag{2.20}$$

$$CH_3OH \rightleftarrows CO + 2H_2 \tag{2.21}$$

$$NO \rightleftarrows N_2 + O_2 \tag{2.22}$$

These equilibria are, for example, relevant for applications of potentiometric solid-state electrolyte sensors in porcelain kilns and for steel hardening by nitridation. For the H_2/H_2O couple and the corresponding resulting emf of such an electrode the following equation is obtained [6]

$$-E/\mathrm{mV} = U_{eq}/\mathrm{mV} = -1290.6 + \left[0.2924 - 0.0992 \log\left(\frac{p_{H_2O}}{p_{H_2}}\right)\right] T/\mathrm{K} \tag{2.23}$$

To derive this relationship, the temperature dependence of the mass-action law (i.e., the equilibrium constant K_p) can be introduced in the Nernst equation.

2.2.2 Application of potentiometric sensors

An overview on different applications of potentiometric sensors in a voltage-temperature diagram is presented in Fig. 2.7. Main applications are in situ measurements in high-temperature processes such combustion control [7], lambda probes [8], steel hardening [9], steel preparation [10], porcelain, brick and glass production [11,12]. Furthermore, potentiometric sensors can be used in trace gas analysis, for example, water in hydrogen, and oxygen in nitrogen, for monitoring the protecting gas in solder processes. From the scientific point of view, the interest in galvanic cells using solid electrolytes also results from the determination of thermodynamic data [13].

For a variety of applications, oxygen probes are nowadays available from many companies worldwide. Figs. 2.8–2.13 show some examples of commercial oxygen sensors, also called oxygen probes. They differ in respect of how the measuring and reference compartments are separated which ensures gas tightness. Depending on

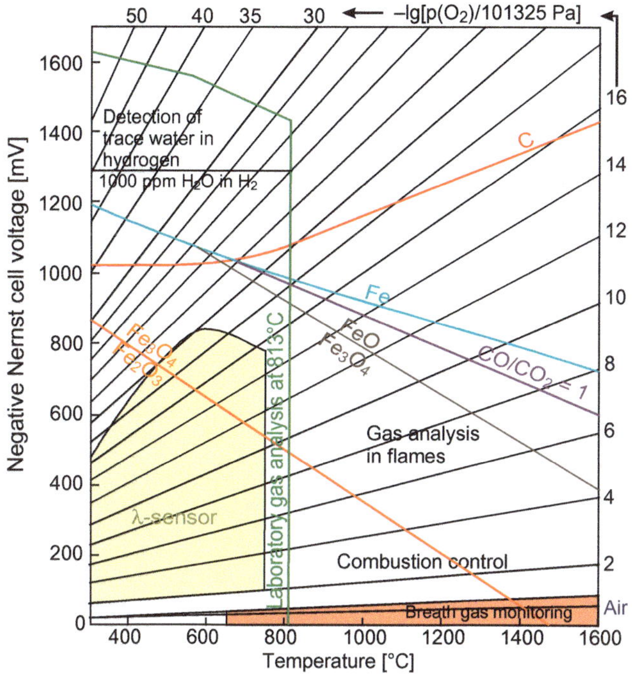

Figure 2.7 Nernst cell voltage (emf) versus temperature for galvanic cells using solid electrolyte.
Source: U. Guth, Gas sensors, in: A. Bard, G. Inzelt, F. Scholz (Eds.), Electrochemical Dictionary, second extended ed., Springer, Heidelberg, Dordrecht, London, New York, 2012, pp. 400–406.

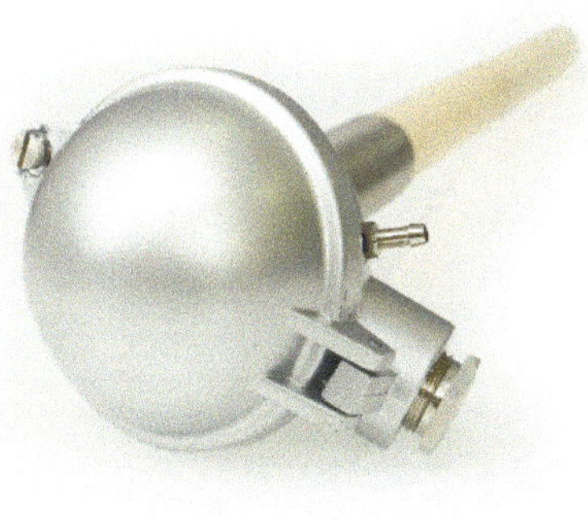

Figure 2.8 High temperature oxygen probes for process control in porcelain industry [14].
Source: With Courtesy of Zirox GmbH, Greifswald, Germany, www.zirox.de, www.stg-cottbus.de.

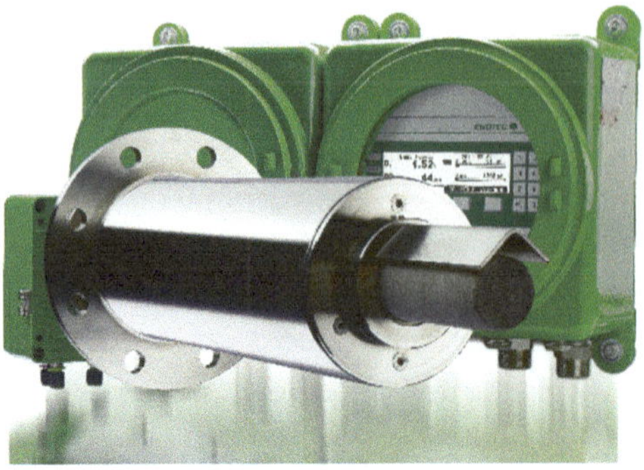

Figure 2.9 Oxygen probes for combustion control [15].
Source: With Courtesy of ENOTEC GmbH, Marienheide, Germany, www.enotec.de.

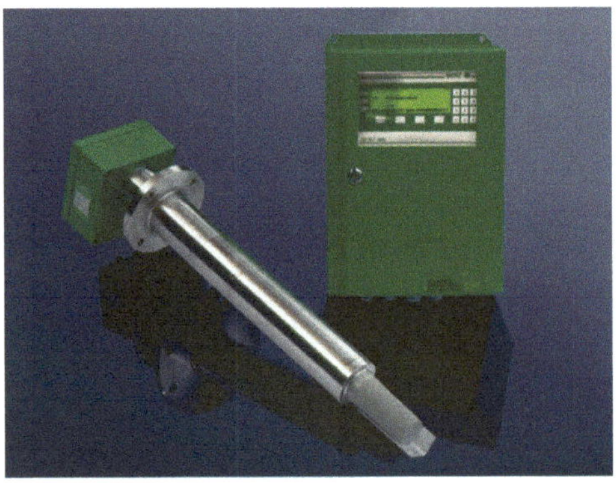

Figure 2.10 Oxygen probes for combustion control [16]
Source: Courtesy of Emerson, USA, www.emerson.com.

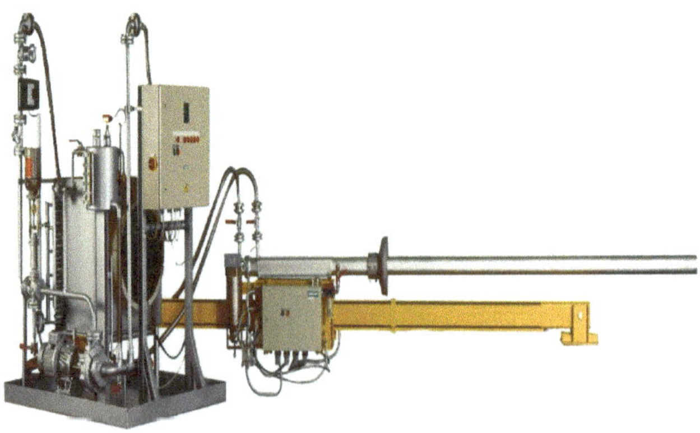

Figure 2.11 Measuring equipment for the application in a cement kiln [15].
Source: Courtesy of ENOTEC GmbH, Marienheide, Germany, www.enotec.de.

Figure 2.12 Oxygen probe for vacuum chambers [14].
Source: Zirox GmbH, Greifswald, Germany, www.zirox.de, www.stg-cottbus.de.

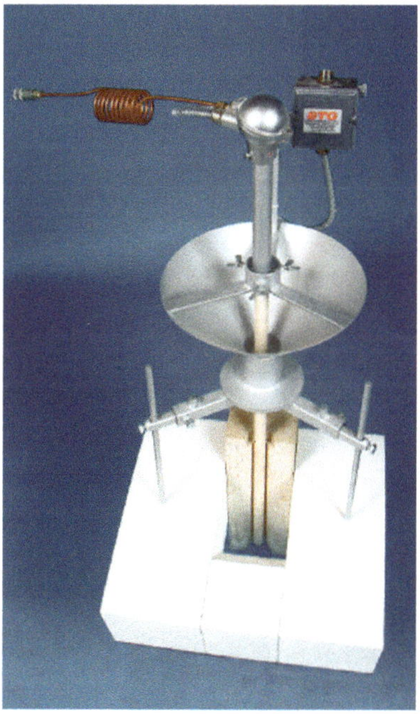

Figure 2.13 STG Oxygen sensor for process control of glass melting furnaces [14].
Source: STG GmbH &Co KG, Cottbus Germany, www.zirox.de, www.stg-cottbus.de.

the operation temperature, an additional external heater can be required; it heats the cell to constant temperatures between 700 and 800°C. In high temperature processes (i.e., above 700°C), an additional heater is, however, not common. Furthermore, the protection tube has to withstand operation temperature and

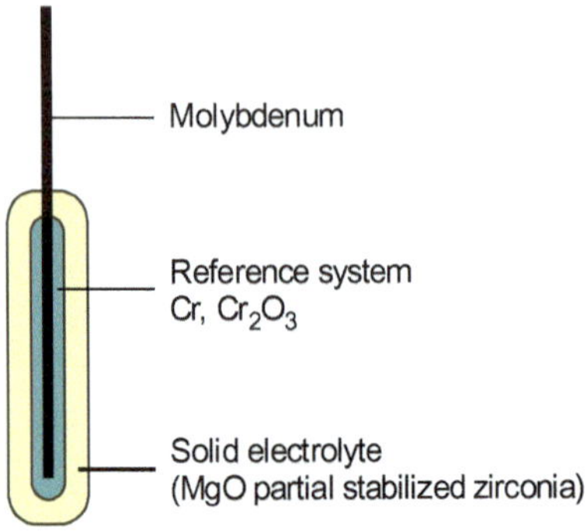

Figure 2.14 Cross section of a steel probe for dissolved oxygen (schematically).
Source: G. Kreysa, K.-o. Ota, R. Savinell (Eds.), Encyclopedia of Applied Electrochemistry. Springer Reference, 2014, pp. 1318–1320.

influence of dust and chemically aggressive substances. Sometimes, it is necessary to cool the outer shielding of the probe, for example, with a water flow [14]. Such sensors are, for example, applied in vacuum chamber to control plasma processes [15]. The life time of oxygen probes varies typically between more than 5 years (e.g., for combustion control for coal fired power plants) and several months for application at temperatures $>1300°C$ in glass kilns (Fig. 2.14).

For measuring oxygen in molten steel, probes for dissolved oxygen are utilized (Fig. 2.13). The measurements are based on the following cell:

$$\mathrm{Fe, O_2}(p_{\mathrm{melt}})\big|\mathrm{MgO - ZrO_2}\big|\mathrm{Me, O_2}(p_{\mathrm{ref}}) \tag{2.24}$$

For the cell voltage, one can use the following equation that also take into account the temperature dependence of the mass action and the corresponding equilibrium constant

$$U_O(O \text{ dissolved in Fe})/\mathrm{mV} = -818 - \left[0.105 - 0.0992 \log\left(c/\mathrm{wt.}\ \%\mathrm{O}\right)\right]T/\mathrm{K} \tag{2.25}$$

Here, c denotes the oxygen concentration in weight percent. This principle is also applied for oxygen determination in other metal melts like sodium (in the cooling circuit of fast breeder reactors), lead, silver, copper, aluminum [10,17].

2.3 Amperometric sensors

Amperometric sensors operate with voltages at which the electrode reaction is determined by the rate of the gas transport (Fig. 2.4). This rate is limited by a diffusion barrier which consists typically of a porous ceramic or a small hole. The measured current I_{lim} is monitored as sensor signal. It holds

$$I_{\text{lim}} = -\frac{zFD_{O_2}Ap}{RTL}\ln\left[1 - x_{O_2}\right] \tag{2.26}$$

For a hole-type diffusion barrier, A denotes the area and L the length of the hole. The diffusion coefficient of oxygen is D_{O_2}. If the mole fraction of oxygen is very small ($x_{x_{O_2}} \ll 1$) then Eq. (2.26) simplifies to

$$I_{\text{lim}} = \left(-\frac{zFD_{O_2}Ap}{RTL}\right)x_{x_{O_2}} \tag{2.27}$$

The term in the bracket is taken as a constant and one obtains a linear relationship between the current I_{lim} and the oxygen mole fraction as well as the oxygen partial pressure

$$I_{\text{lim}} = const\ x_{O_2} \propto p_{O_2} \tag{2.28}$$

A well-known application of amperometric solid-state electrolyte sensors is the wideband lambda probe (Fig. 2.15). In the lean-burn regime, oxygen diffuses through the diffusion barrier. By means of the pump cell, oxygen is removed from the internal cavity until a certain value of the Nernst cell voltage is obtained. Then, the current through the pump cell is proportional to the oxygen concentration [18].

By means of the amperometric principle, a simultaneous determination of oxygen and HC or nitric oxides is possible. For such sensors, tube [19] or planar [18] designs were described. Fig. 2.16 shows a planar device. The measured currents I_r and I_p are proportional to the oxygen concentration and the concentration of nitric oxides, respectively, [18,19] or HC [20].

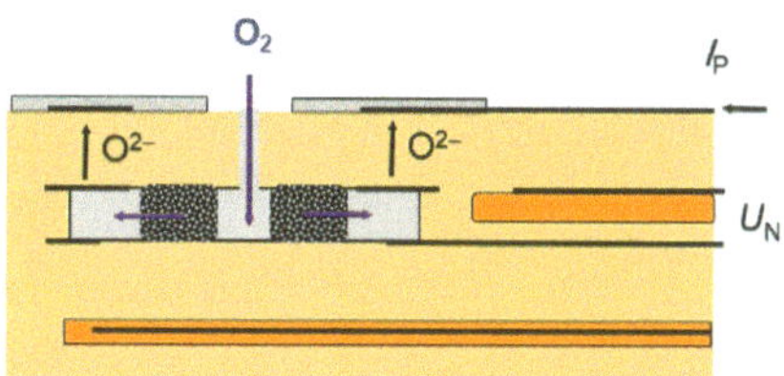

Figure 2.15 Oxygen sensor for automotive application in lean burn regime.
Source: P. Schmidt-Zhang, K.-P. Sandow, F. Adolf, G. Reinhardt, U. Guth, W. Göpel, A novel thick film sensor for simultaneous O_2 and NO monitoring exhaust gases, Sens. Actuators B 70 (2000) 25–29.

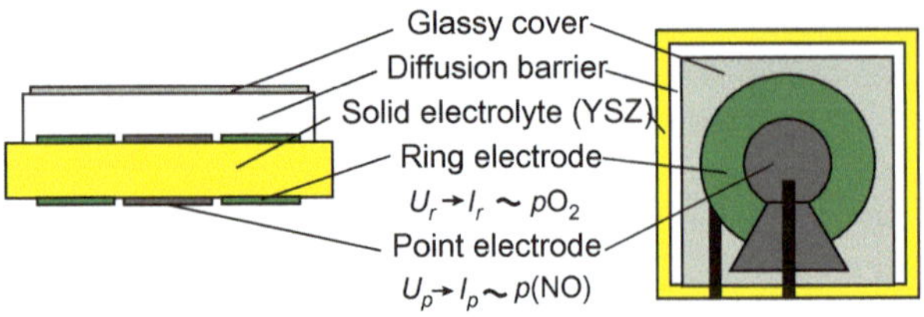

Figure 2.16 Planar design for simultaneous determination of oxygen and nitric oxides. *Source:* S. Somov, G. Reinhardt, U. Guth, W. Göpel, Gas analysis with arrays of solid state electrochemical sensors: implications to monitor HCs and NOx in exhaust, Sens. Actuators B 35–36 (1996), 409–418.

2.4 Coulometric sensors

The coulometric principle is well-known for solid-state electrolyte sensors to measure oxygen, hydrogen, HC as well as other oxidizable and reducible gases [21]. The main advantage is that these devices follow strictly Faraday's law and can operate, consequently, over long periods without calibration. The amount n of the involved electrochemically active gas component (i.e., the analyte) can be determined by

$$n = \frac{Q}{zF} = \frac{It}{zF} \tag{2.29}$$

where Q is the charge, z the number of transferred electrons in the electrochemical reaction, F the Faraday's constant, I the current and t the time. If the sensors are used for gas titration, current peaks are obtained and the amount n results from an integration over the time

$$n = \frac{Q}{zF} = \frac{1}{zF}\int_{t_1}^{t_2} (I - I_b)dt \tag{2.30}$$

Here I_b is the base-line current if no detectible gases are present in the gas mixture. Coulometric solid-state electrolyte sensors for continuous gas flows have been developed as well [22]. Here, the sensor consists of an assembly with pump cell and potentiometric cell; an example is shown in Fig. 2.17. The mole fraction of the analyte, which is transferred at the working electrode, can be obtained by applying Faraday's law. Practically, the residual current I_r due to electronic conductivity of the solid electrolyte must be subtracted from the measured current I. Then, in a gas flow of the analyte-containing gas one obtains the mole fraction

$$x_g = \frac{I - I_r}{zF}\frac{V_m}{dV/dt} \tag{2.31}$$

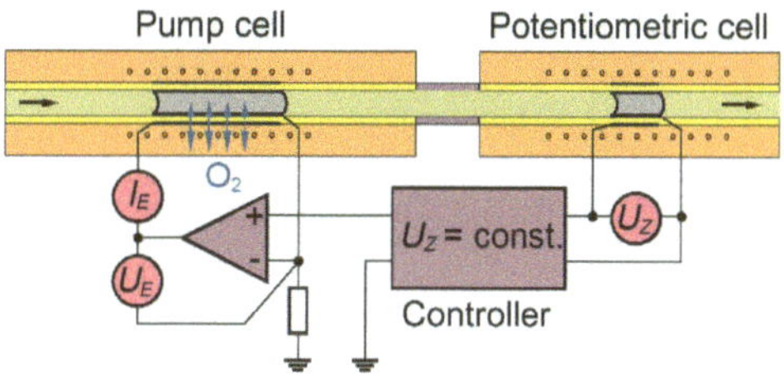

Figure 2.17 Coulometric sensor for determining of gas species in continuous flow.
Source: M. Schelter, J. Zosel, W. Oelßner, U. Guth, M. Mertig, A solid electrolyte sensor for trace gas analysis, Sens. Actuators B Chem. B 187 (2013) 209–214.

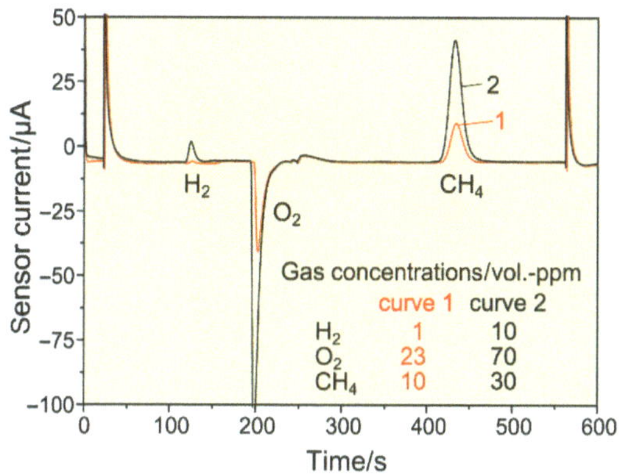

Figure 2.18 Simultaneous determination of different gases by means of coulometric sensor after chromatographic separation with two different concentrations.
Source: M. Schelter, J. Zosel, W. Oelßner, U. Guth, M. Mertig, A solid electrolyte sensor for trace gas analysis, Sens. Actuators B Chem. B 187 (2013) 209–214.

Here, dV/dt denotes the volume flow and V_m the molar volume. According to Faraday's law, the signal of coulometric sensors depends only on the concentration of the analyte component and the flow rate. Again, calibration free operation of these devices is achieved.

The detection limit of coulometric flow-through sensors, applied as detectors for gas chromatography in trace analysis, can be decreased from some 10 vol-ppm level down to the ppb-range. This requires, however, the correction of any electronic noise and also of the electronic contribution in the measured current. As an example, the simultaneous determination of H_2, O_2 and CH_4 after a separation with a gas chromatic column is shown in Fig. 2.18 [22]. According to their elution times, different peaks can be identified and depending on their concentration in the sample gases, peaks of different intensity are collected and evaluated in terms of gas type and concentration.

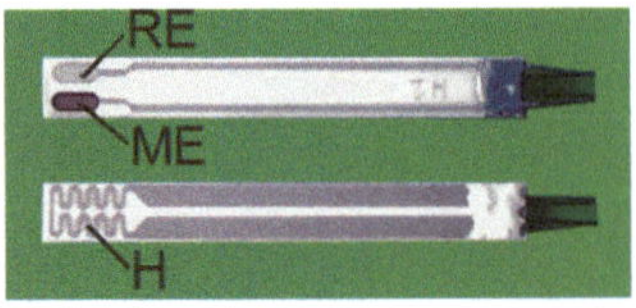

Figure 2.19 Gas symmetric arrangement of mixed-potential sensor (ME, measuring electrode; RE, reference electrode; H, heater).
Source: U. Guth, Gas sensors, in: A. Bard, G. Inzelt, F. Scholz (Eds.), Electrochemical Dictionary, second extended ed., Springer, Heidelberg, Dordrecht, London, New York, 2012, pp. 400–406.

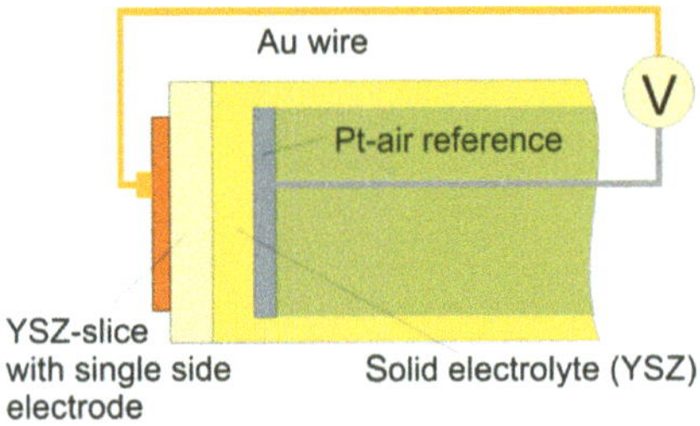

Figure 2.20 Mixed-potential sensor with separate reference electrode.

2.5 Mixed-potential sensors

2.5.1 Principle

In contrast to potentiometric sensors, mixed-potential sensors operate on kinetic principles, that is, that rate by which electrochemical reactions proceed. Typically, mixed-potential sensors consist of an oxygen electrode as a reference electrode and a second (working) electrode, which shows a low oxygen sensitivity but a high sensitivity to the analyte gas (e.g., combustibles like H_2, CO, HC) or nitric oxides (NO_x). Both electrodes can be exposed to the gas to be analyzed in an assembly which is called a gas-symmetric arrangement (Fig. 2.19). Alternatively, the sensor can operate with a platinum electrode exposed to air as an external reference (Fig. 2.20) [23]. The key component is here the choice of suitable electrode material on which the nonequilibrated gas mixture (which contains combustibles and oxygen) do not react chemically prior to its interaction with the sensor.

2.5.2 Catalytic behavior of electrode materials

The observation that the behavior of metals is different when used as catalysts or as electrode materials for solid electrolyte cells goes back to the early 1970s. Cell voltages were found to deviate from thermodynamically expected values, depending on the material and the operation temperature of the cells. One of the first papers on this subject by Sandler described the determination of oxygen in a nonequilibrated methane-oxygen mixture by means of silver electrodes [24]. Another example was a concentration cell with platinum or nickel electrodes, operating with dry CO_2–CO mixtures. On platinum electrodes, a Nernstian behavior of the voltage was only found at temperatures higher than 950°C, whereas on nickel electrodes Nernstian voltages could be measured at temperature lower than 500°C (Fig. 2.21) [25].

Pebler reported measurements of combustibles gases in oxygen-containing gas mixtures by means of a gold electrode on YSZ and with a platinum air reference electrode [26]. Hartung et al. investigated the sensitivity of different precious metal electrodes in nonequilibrated gas mixtures [27]. They found that silver and gold electrodes showed a remarkable deviation from the Nernst potential in air-amyl alcohol mixtures below 700°C under slow cooling conditions (7 K/min), as revealed in Fig. 2.22. Such observations have led to a new category of solid-state electrolyte sensors, the so-called *mixed-potential sensors*.

It is evident from Fig. 2.22 that the gas components do not equilibrate on the different measuring electrodes at temperatures below 750°C and, furthermore, that the deviation depends on the electrode material.

To explain these observations, we consider a gas mixture with HC and O_2. Here, at least two electrode reactions can take place. The first is the electrochemical reduction of oxygen and the second is the electrochemical oxidation of the hydrocarbon. To give an example, the following reactions in Kröger-Vink notation are obtained for an oxygen-propene mixture

$$9/2O_2(g) + 9V_O^{\bullet\bullet} + 18e^- \rightleftharpoons 9O_o(YSZ) \quad (2.32)$$

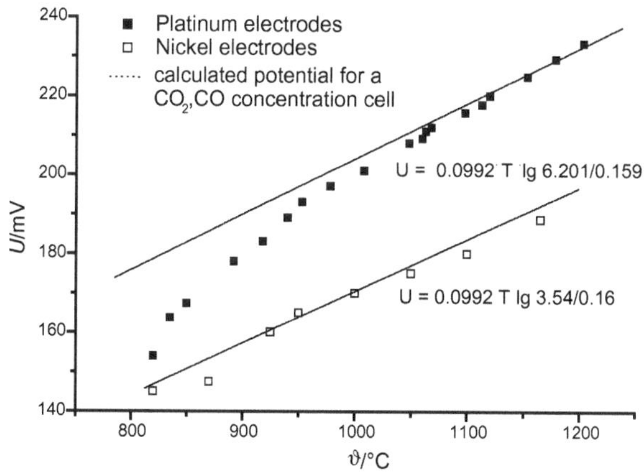

Figure 2.21 Voltage response of CO_2, CO-concentration cell with platinum and nickel electrodes.
Source: R. Moos, K. Sahner, M. Fleischer, U. Guth, N. Barsan, U. Weimar, Solid state gas sensor research in Germany - a status report, Sensors, 9 (2009) 4323–4365.

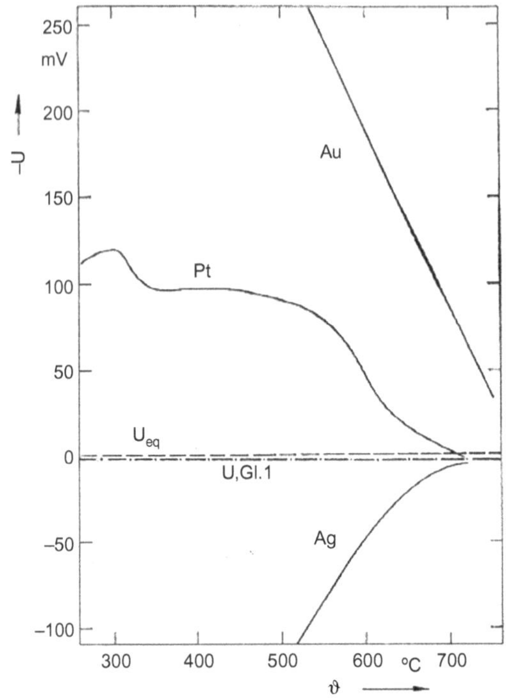

Figure 2.22 M/solid electrolyte/Pt cell (M = Pt, Au, Ag), gas: 0.2 mol% O_2 and 0.06 mol% CO in argon, cooling rate 7 K/min.
Source: R. Hartung, R. Schröder, H.-H. Möbius, Combustible gas sensitive gas symmetric galvanic cell with oxide ion conducting electrolyte, Z. Physik. Chem. 262 (1981) 961–966.

$$C_3H_6(g) + 9O_O(YSZ) \rightleftharpoons 3CO_2(g) + 3H_2O(g) + 9V_O^{\bullet\bullet} + 18e^- \tag{2.33}$$

$$9/2O_2(g) + C_3H_6(g) \rightleftharpoons 3CO_2(g) + 3H_2O(g) \tag{2.34}$$

Eq. (2.34) describes the overall reaction, that is, the stoichiometric oxidation of propane towards CO_2 and H_2O. The Eqs. (2.32) and (2.33) show the two possible electrode reactions with YSZ as solid-state electrolyte. Since the measured open-circuit voltage does not obey the Nernst equation, such electrode behavior is often referred to as *non-Nernstian electrodes* or mixed-potential electrodes. The reason for this behavior is the reduced rate of the reduction of oxygen according to Eq. (2.32), that is, the rate is controlled kinetically. Then, the electrochemical oxidation of the hydrocarbon becomes possible, as demonstrated for propane in Eq. (2.33). The mixed potential can be determined by means of polarization measurements that are performed both, in anodic and cathodic direction. The potential at which the absolute values of the anodic and cathodic current densities are equal ($|i_c| = |i_a|$) is called the mixed-potential U_{mix}. It is obvious from Eqs. (2.32) to (2.34) that a certain amount of oxygen is needed for such sensors. Details of the underlying theory, which is based on the Butler-Volmer equation, is given by Miura et al. [28]. In the case of a system with O_2 and CO electrodes, the following equation was derived for the two current densities corresponding to the O_2 reduction and the CO oxidation, respectively

$$i_{O_2} = i^o_{O_2}\exp\{-4\alpha_1 F(E - E^o_{O_2})/RT \tag{2.35}$$

$$i_{CO} = i^o_{CO}\exp\{-4\alpha_2 F(E - E^o_{CO})/RT \tag{2.36}$$

Here, E is the electrode potential, F the Faraday constant, R the gas constant, and T the temperature. $E^o_{O_2(CO)}$, $i^o_{O_2(CO)}$ and $\alpha_{1(2)}$ are the standard electrode potentials at equilibrium, exchange current density and transfer coefficient of the single electrode reactions. For the exchange current densities, the following equations result from the theoretical treatment

$$i^o_{O_2} = B_1\phi^m_{O_2} \tag{2.37}$$

$$i^o_{CO} = B_2\phi^n_{CO} \tag{2.38}$$

Here, $\phi_{O_{2(CO)}}$ denotes the dimensionless volume fraction of O_2 and CO and B_1, B_1, m, and n are constants. $i^o_{O_2}$ and i^o_{CO} are opposite in sign. The mixed-potential E_m is obtained if the absolute exchange currencies $|i^o_{O_2}| = |i^o_{CO}|$ are equal

$$E_m = E_o + mA\ ln\phi_{O_2} - nA\ ln\phi_{CO} \tag{2.39}$$

E_o and A can be expressed by

$$E_o = \frac{RT}{(4\alpha_1 + 2\alpha_2)F} ln\frac{B_1}{B_2} + \frac{2\alpha_1 E^o_{O_2} + 2\alpha_2 E^o_{CO}}{2\alpha_1 + \alpha_2} \tag{2.40}$$

$$A = \frac{RT}{(4\alpha_1 + 2\alpha_2)F} \tag{2.41}$$

If the oxygen volume fraction is constant, Eq. (2.41) can be simplified to:

$$E_M = E_o - nA\ln\phi_{CO} \tag{2.42}$$

The formation of a mixed potential is schematically presented in Fig. 2.23. Both, the cathodic and anodic polarization curves are experimentally accessible and can be measured by means of a polarization technique. The different electrode behavior is clearly demonstrated by the voltage-*versus*-composition data in Fig. 2.24. In the vicinity of porous platinum electrodes nonequilibrated gas mixtures react completely to H_2O and CO_2. A Nernst potential corresponding to the residual oxygen partial pressure can be measured at least down to 300°C [23] (black curve) and

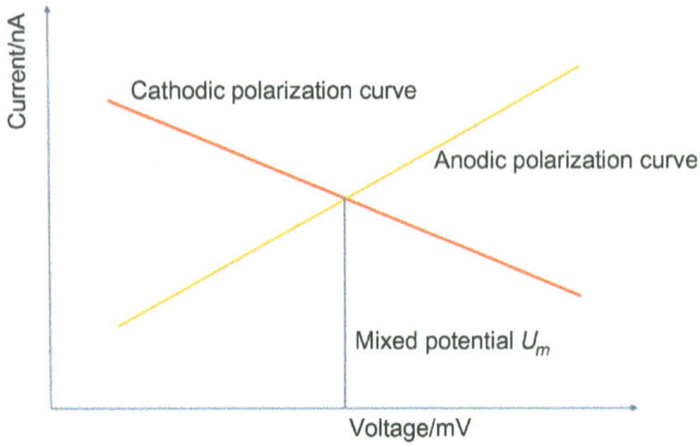

Figure 2.23 Schematic drawing for explanation of a mixed potential.

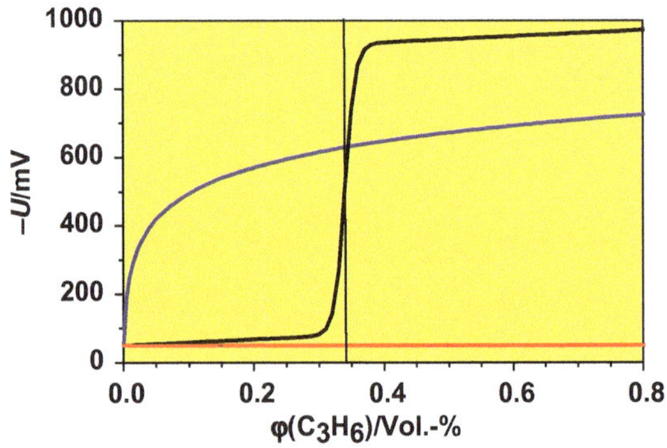

Figure 2.24 Cell voltage in 1.5% oxygen with increasing amount of combustibles (propene) versus platinum reference.
Source: J. Zosel, U. Guth Electrochemical solid electrolyte gas sensors – hydrocarbon and NO_x analysis in exhaust gases, Ionics 10 (2004) 366–377.

a lambda leap ($\lambda = 1$) is observed. Characteristic for the mixed-potential behavior is the logarithmic curve (blue) without a lambda leap in at the dashed line in the middle. On the other hand, free oxygen can be determined by means of electrodes that consist of dense (i.e., nonporous) platinum or catalytically inactive $LaCrO_3$ even in presence of combustibles (red curve). The quantitative relations of the different electrodes are given on the right of Fig. 2.24.

Different equations hold for the different systems. The finally determine the signal-composition relations (i.e., calibration curves) of potentiometric and mixed-potential solid-state electrolyte sensors.

2.5.3 Electrode materials for mixed potential sensors

Binary and ternary oxide compounds are promising electrode materials for mixed-potential sensors. Compared to precious metals, oxides have the advantage of stable morphology and, consequently, does not suffer from grain coarsening. The search for such highly sensitive and selective materials (with respect to a specific gas components) is closely connected with the knowledge on their electro-catalytic behavior and the following examples demonstrate different experimental approaches how suitable electrode materials have been studied up to now.

Fig. 2.25 shows a simple setup, which is suitable for testing materials with respect to catalytic behavior in solid electrolyte cells [19]. The solid electrolyte tube, closed on one side and made of YSZ, is equipped with an inner electrode that consists of the oxide material to be investigated. The outer electrode is formed by a sintered silver layer. The arrangement, surrounded by air, is heated up to 900°C and then cooled down with a certain rate, for example, 15 K/min. At 900°C the chemical potential of oxygen is found to be equal on both electrodes. On the silver electrode the oxygen is in electrochemical equilibrium. During cooling the chemical potential is shifted, according to the catalytic activity of oxides with respect to oxygen incorporation into the solid electrolyte lattice.

Examples for results of such experiments are shown in Fig. 2.26. Cells with oxide electrodes, which are active for the electrochemical incorporation of oxygen into the solid electrolyte, show a cell voltage near zero (e.g., for NiO and Mn_2O_3). With catalytically inactive electrodes like Fe_2O_3 and Cr_2O_3 remarkable deviations are, however, obtained.

The electrode material for nonequilibrium gas mixtures are expected to show two different functions. On the one hand, the material must ensure that the composition of the gas mixture does not change. This means it must not be active with respect to the chemical reaction of the combustible component with oxygen or of the nitrogen oxide (e.g., NO or NO_2) with oxygen. On the other hand, the material

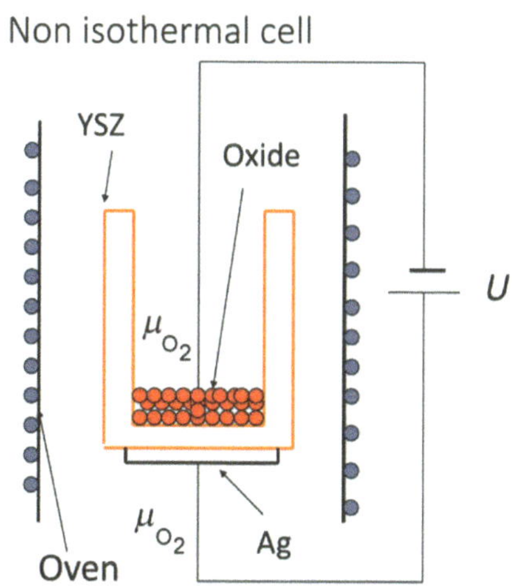

Figure 2.25 Experimental setup for evaluation of catalytic activity of oxygen incorporation. *Source:* J. Zosel, U. Guth Electrochemical solid electrolyte gas sensors – hydrocarbon and NO_x analysis in exhaust gases, Ionics 10 (2004) 366–377.

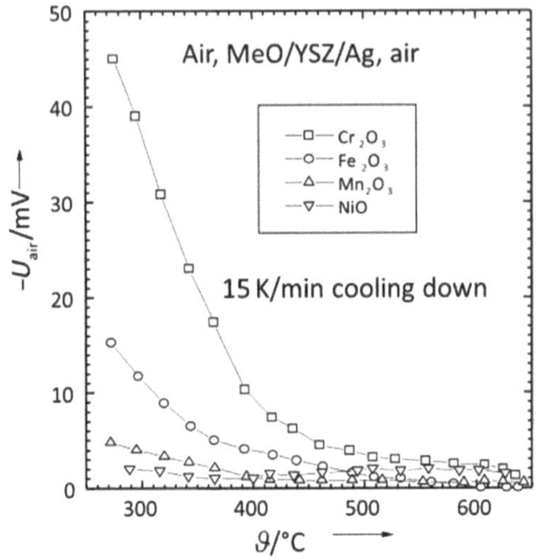

Figure 2.26 Results of experiments obtained using setup described in Fig. 2.25.
Source: J. Zosel, U. Guth Electrochemical solid electrolyte gas sensors – hydrocarbon and NO_x analysis in exhaust gases, Ionics 10 (2004) 366–377.

must catalyze the electrochemical reaction with the solid electrolyte, preferably with the lattice oxide ions and with oxide ion vacancies. Both requirements are opposed to each other with the result that the search for suitable oxide electrodes is always difficult. Furthermore, the catalytic properties depend on both, the type of the oxide and its morphology, especially on grain size and the grain shape.

More detailed information about suitability of an electrode material can be gained with an electrochemical cell in which both gas compartments are separated from each other (Fig. 2.27). The test electrode consists of an YSZ disk with the electrode layer to be investigated. It is pressed on an YSZ tube by a spring mechanism and can be easily exchanged. The inner platinum electrode with air as reference gas is in electrochemical equilibrium so that all effects can be assigned to the test electrode.

In practical applications, mixed-potential sensors like it is shown in Fig. 2.28, are common. On an alumina substrate, single layers which consist of YSZ, reference and measuring electrode, as well as platinum for the back side heater are printed and fired.

Up to now, more or less empirical approaches have been followed in the search for suitable electrode materials for mixed-potential sensors. The mechanism of their electrode reactions is widely unknown. But it is generally believed that ion vacancies and electronic defects (holes or excess electrons) in the oxide play the key role for catalysis on the oxide surface as well as for the electrochemical reaction (Fig. 2.29).

Therefore, measurements of the catalytic activity for the combustion of HC in nonequilibrated gas mixtures may indicate in some cases whether a material is

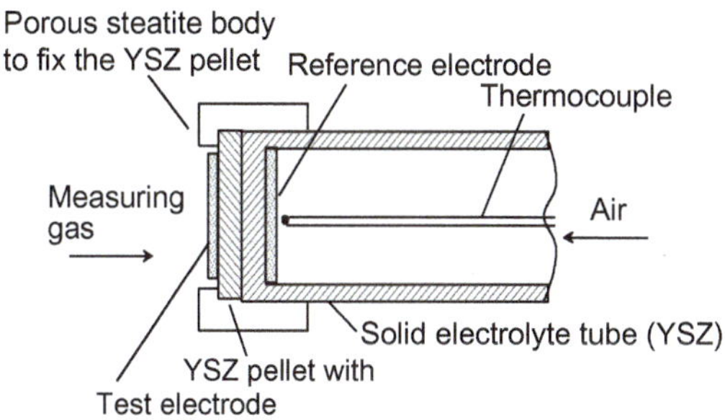

Figure 2.27 Setup of YSZ cell with interchangeable electrodes for potentiometric measurements.

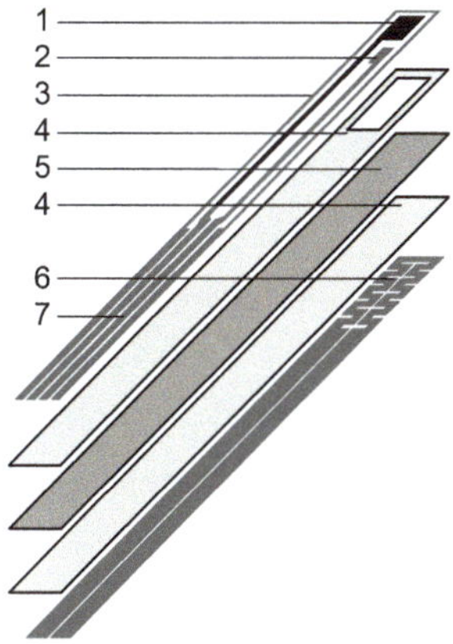

Figure 2.28 Exploded view of a planar solid electrolyte mixed potential sensor with dimensions 4.5 × 66 mm^2: 1 mixed potential electrode, 2 reference electrode, 3 Pt temperature resistor, 4 insulating layers, 5 solid electrolyte YSZ, 6 Pt heater, 7 Pt connecting pads. *Source:* J. Zosel, G. Schiffel, F. Gerlach, K. Ahlborn, U. Sasum, V. Vashook, et al. Electrode materials for potentiometric hydrogen sensors, Solid State Ionics 177 (2006) 2301–2304.

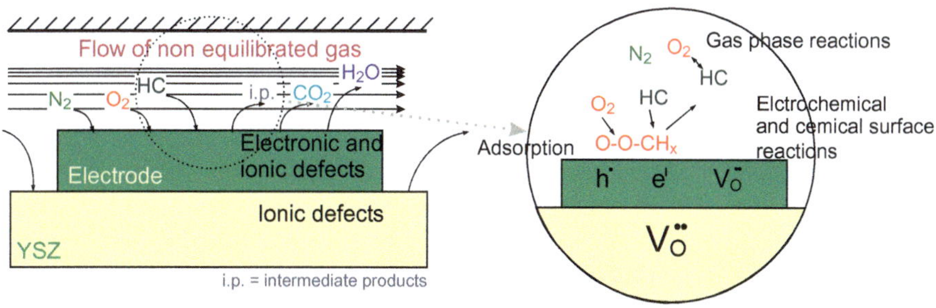

Figure 2.29 Schematic illustration of interactions between gas species and defects in oxide electrodes.

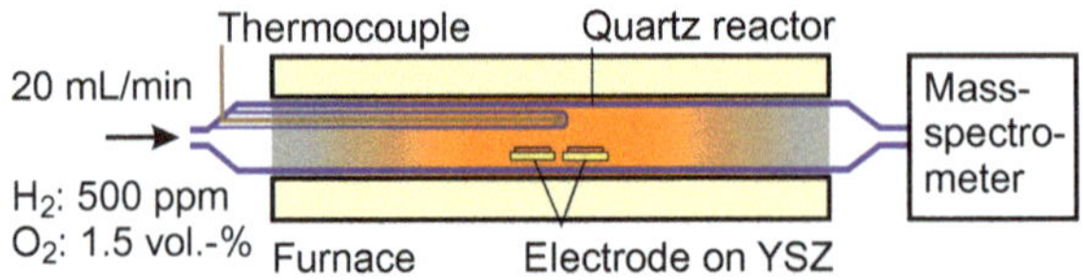

Figure 2.30 Experimental setup for measurement of catalytic activity of the combustion. *Source:* J. Zosel, D. Westphal, S. Jakobs, R. Müller, U. Guth, Au–oxide composites as HC-sensitive electrode material for mixed potential gas sensors, Solid State Ionics 152–153 (2002) 525–529.

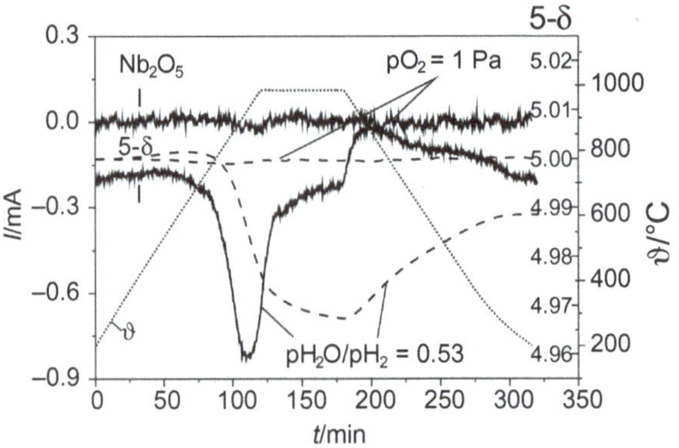

Figure 2.31 Results of Oxygen Solid Electrolyte Coulometry for Nb_2O_5 heated up to 1000°C Under 1 Pa oxygen the stoichiometry is nearly 5. But in reducing gases a remarkable oxygen release can be detected (dashed lines).
Source: J. Zosel, U. Guth Electrochemical solid electrolyte gas sensors – hydrocarbon and NOx analysis in exhaust gases, Ionics 10 (2004) 366–377.

suitable or not. For that purpose, a quartz-tube reactor is used in which the sample can be heated continuously or step wisely. The analytical determination of gas species is carried out by means of a mass spectrometer. Fig. 2.30 reveals such a setup schematically which is used in sensor research.

2.5.3.1 *Binary oxides*

In simple oxides, the defect structure is determined thermodynamically by the temperature and the oxygen partial pressure. According to these parameters, a certain oxygen stoichiometry ($O \pm \delta$) is adjusted. The oxygen stoichiometry, which is connected with the electronic defects, can be determined by the so-called oxygen-solid-electrolyte coulometry [21]. Fig. 2.31 shows the results with respect to stoichiometry for Nb_2O_5. Whereas the oxygen stoichiometry is nearly 5 below an oxygen pressure of 1 Pa, a substoichiometry of 4.97 is found at 1000°C in reducing atmosphere (H_2, H_2O). For some oxides, the electrical conductivity in dependence on the oxygen partial pressure is presented in Fig. 2.32. For the most oxides, the

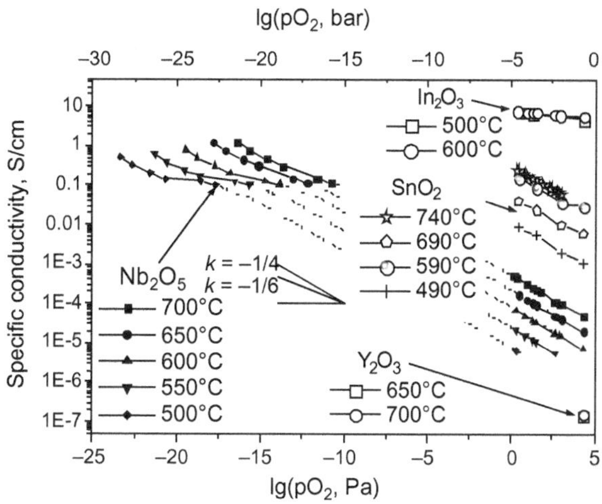

Figure 2.32 Oxygen partial pressure dependence of electrical conductivity for various oxides. *Source:* J. Zosel, K. Ahlborn, R. Müller, D. Westphal, V. Vashook, U. Guth, Selectivity of HC sensitive electrode materials for mixed potential gas sensors, Solid State Ionics 169 (2004) 115–119.

conductivity increases with decreasing oxygen partial pressure. This means that oxygen is released from the lattice and excess electrons e′ are generated according to

$$O_O^{\times} \rightleftarrows \frac{1}{2}O_2(g) + V_O^{\bullet\bullet} + 2e' \tag{2.43}$$

However, a simple relationship between electronic conductivity and catalytic activity does not exist in general. This demonstrates the importance of case studies, a typical selection of them are treated in the following.

2.5.3.2 Electrode materials for hydrogen

Simple oxides exhibit only a very low conductivity in air. To overcome this disadvantage, they must be mixed with gold to enhance the electrical conductivity. In addition, mixing with gold improves the ability for sintering at the surface of a solid-state electrolyte ceramics. Experiments have confirmed that the electrode properties are determined mainly by the oxide and not by the gold. For simple oxides, the relationship between catalytic activity for hydrogen combustion and sensor response have already been demonstrated. All investigated oxide composite electrodes show a non-Nernstian behavior. The corresponding results of catalytic measurements and sensor response behavior confirm that less catalytic oxides provide a high sensitivity to hydrogen. Examples are the systems Au-Nb_2O_5 and Au-Ta_2O_5. On the other hand, composites with SnO_2 and CeO_2 show a much higher conductivity and are more catalytically active. Their voltage response are remarkably lower (Figs. 2.33 and 2.34). Generally, it is assumed that the lower the catalytic activity the higher the sensor response [29].

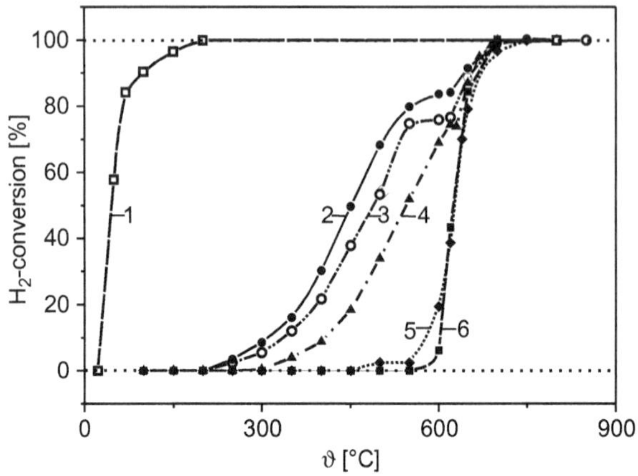

Figure 2.33 Catalytic activity of selected electrode materials on YSZ discs for hydrogen oxidation in comparison with the empty reactor: 1 Pt, 2 Au/SnO_2, 3 Au/CeO_2, 4 Au/Ta_2O_5, 5 Au/Nb_2O_5, 6 empty reactor.
Source: J. Zosel, G. Schiffel, F. Gerlach, K. Ahlborn, U. Sasum, V. Vashook, et al. Electrode materials for potentiometric hydrogen sensors, Solid State Ionics 177 (2006) 2301–2304.

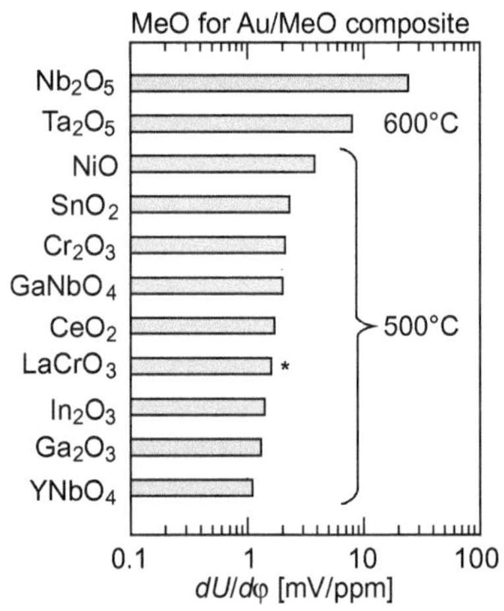

Figure 2.34 Hydrogen sensitivity of composite electrodes with Au and 20 mass-% MeO at $\varphi(H_2) = 10$ vol.-ppm, $\varphi(O_2) = 1.5$ vol.-%, vs. platinum air reference
*$La_{0.95}Sr_{0.02}Cr_{0.8}Ga_{0.2}O_{3-\delta}$ air reference.
Source: J. Zosel, G. Schiffel, F. Gerlach, K. Ahlborn, U. Sasum, V. Vashook, et al. Electrode materials for potentiometric hydrogen sensors, Solid State Ionics 177 (2006) 2301–2304.

The functionality of selected oxide materials for mixed-potential sensors is shown in Fig. 2.35. The shape of the curve measured in a gas-symmetric arrangement is found to be different. Electrodes with Au-Nb_2O_5 are more sensitive in the

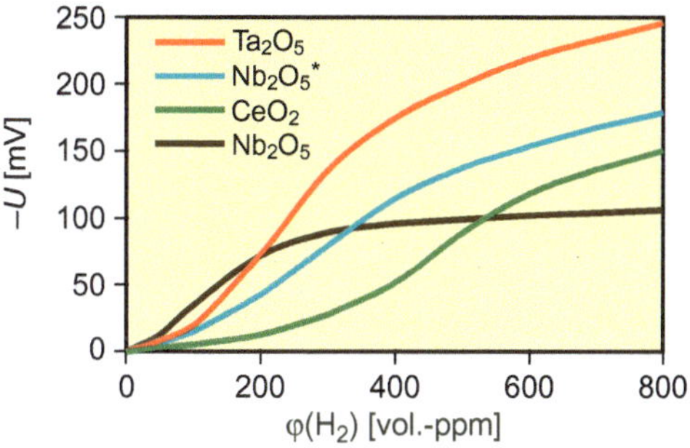

Figure 2.35 Potential response of self- heated gas-symmetrical sensor with measuring electrodes of Au/MeO and Pt reference electrodes in H_2 containing mixtures. $\vartheta = 500°C$. (*sensor with reference electrode of $LaCr_{0.8}Ga_{0.2}O_{3-\delta}$).
Source: J. Zosel, G. Schiffel, F. Gerlach, K. Ahlborn, U. Sasum, V. Vashook, et al. Electrode materials for potentiometric hydrogen sensors, Solid State Ionics 177 (2006) 2301–2304.

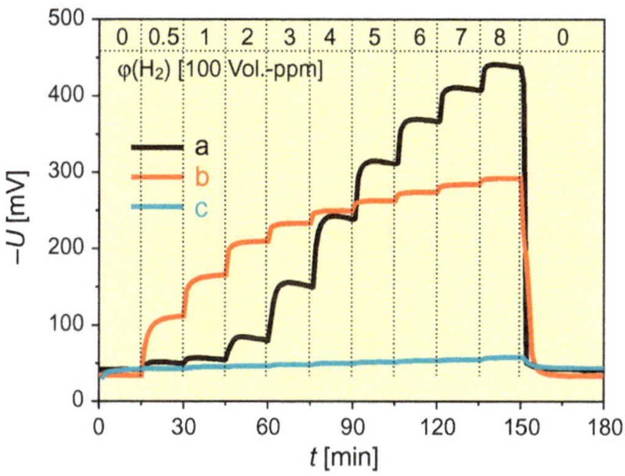

Figure 2.36 Mixed potential response of different electrode materials to H_2 concentration versus Pt-air reference electrode (a) Au/$EuNbO_4$ at 600°C, (b) Au/CeO_2 at 500°C, (c) $LaCr_{0.8}Ga_{0.2}O_{3-\delta}$ at 600°C.
Source: J. Zosel, G. Schiffel, F. Gerlach, K. Ahlborn, U. Sasum, V. Vashook, et al. Electrode materials for potentiometric hydrogen sensors, Solid State Ionics 177 (2006) 2301–2304.

range of low hydrogen concentration whereas electrodes with Au-Ta_2O_5 show higher hydrogen sensitivity for concentrations above 200 ppm. Different results are obtained with mixed-oxide electrodes as compared with Au-CeO_2 electrodes (Fig. 2.36). It is noteworthy that the potential of $LaCr_{0.8}Ga_{0.2}O_{3-\delta}$ electrodes is nearly independent on the hydrogen concentration and, therefore, such compounds are suited for reference electrodes (compare Fig. 2.35).

2.5.3.3 Electrode materials for hydrocarbons

A relationship between catalytic activity for combustion and sensor response was also found for materials that are suitable for hydrocarbon electrodes. The electrode behavior of gold-oxide composites correlates well with the catalytic activity of these materials for hydrocarbon combustion. It was found that the more inactive the material the better is the hydrocarbon sensitivity of electrodes that were prepared with that material. Au-Nb_2O_5 shows the smallest activity for propene conversion and the highest voltage response. The activity of Au-Ga_2O_3 is remarkably higher in comparison with Au-Nb_2O_5, Au-ZrO_2, and Au-Al_2O_3 are, in turn, more active than Au-Ga_2O_3. Among all investigated Au-composites, Au-In_2O_3 exhibits the highest catalytic activity.

Figs. 2.37 and 2.38 confirm that the rate of conversion are different for both HC and hydrogen, as expected for these materials. But the response behavior is similar. Electrodes of Au/Nb_2O_5 and Au/Ta_2O_5 show the lowest catalytic activity and the highest sensitivity with respect to hydrogen and HC.

The catalytic properties for combustion of unsaturated (propene) and saturated HC (propane) are found to be very different. But for both gases, niobium oxide exhibits the lowest activity (Figs. 2.39 and 2.40).

This behavior is confirmed by the sensor response (Fig. 2.41). Electrodes made of Au-Nb_2O_5 show the highest response whereas the sensitivity of Au-In_2O_3-electrodes is much smaller (Fig. 2.42).

The HC sensitivity increases in the direction: Y_2O_3, ZrO_2, In_2O_3, Ga_2O_3, Nb_2O_5 (Fig. 2.41). The cell voltage depends strongly on the structure of chemical compound as shown in Figs. 2.42–2.45. The highest values were found for substituted aromatics (toluene) followed by unsaturated HC (propene) and saturated HC (propane) [20–22].

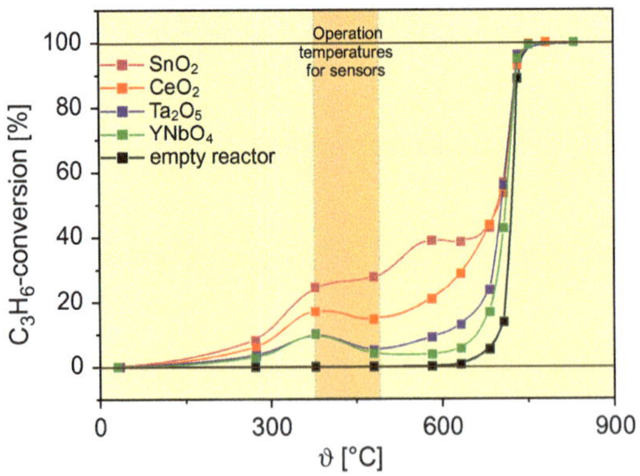

Figure 2.37 Propene conversion on simple oxides.
Source: J. Zosel, R. Müller, V. Vashook, U. Guth, Response behavior of perovskites and Au/oxide composites as HC-electrodes in different combustibles, Solid State Ionics 175 (2004) 531–533.

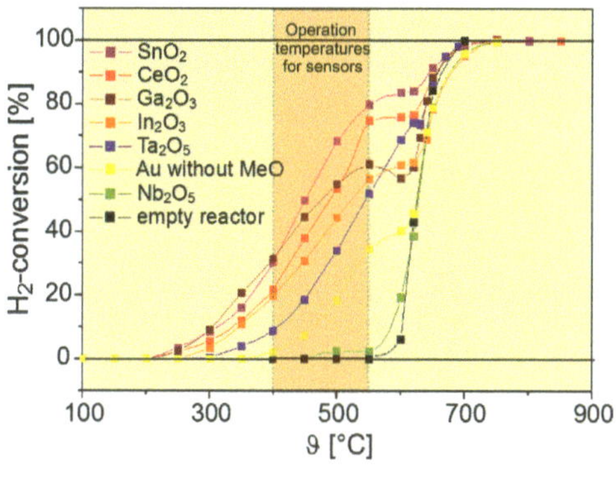

Figure 2.38 Hydrogen conversion on simple oxides.
Source: J. Zosel, G. Schiffel, F. Gerlach, K. Ahlborn, U. Sasum, V. Vashook, et al. Electrode materials for potentiometric hydrogen sensors, Solid State Ionics 177 (2006) 2301–2304.

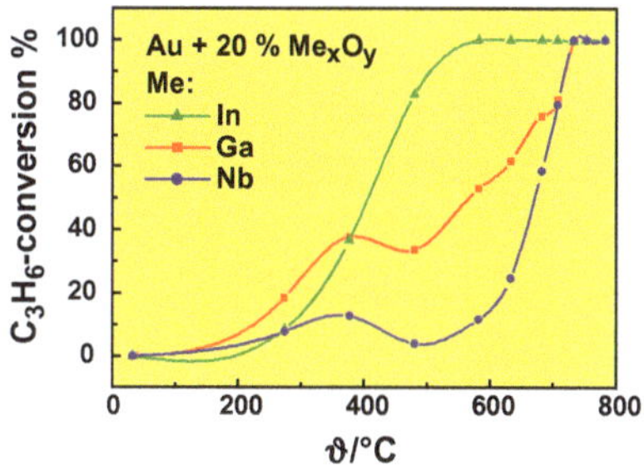

Figure 2.39 Catalytic conversion of propene for different oxides.
Source: J. Zosel, K. Ahlborn, R. Müller, D. Westphal, V. Vashook, U. Guth, Selectivity of HC sensitive electrode materials for mixed potential gas sensors, Solid State Ionics 169 (2004) 115–119.

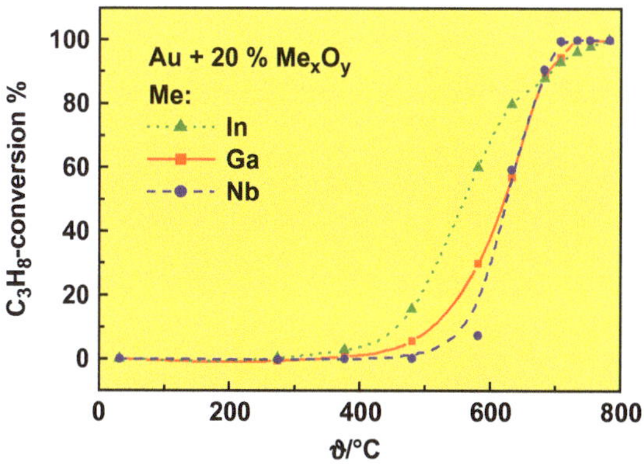

Figure 2.40 Catalytic conversion of propane for different oxides.

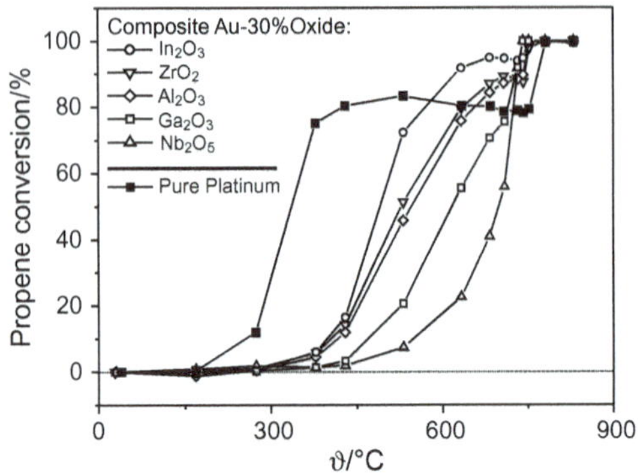

Figure 2.41 Propene (0.5 vol.%) conversation using different oxides in 2.5 vol.% oxygen.

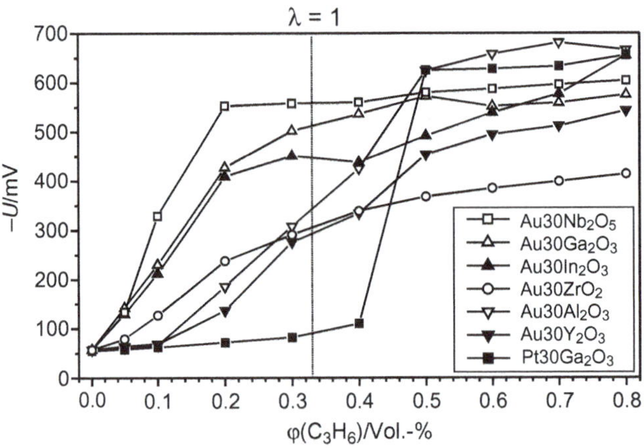

Figure 2.42 Cell voltage of mixed potential sensors at 700°C and 1.5 vol.% O_2.

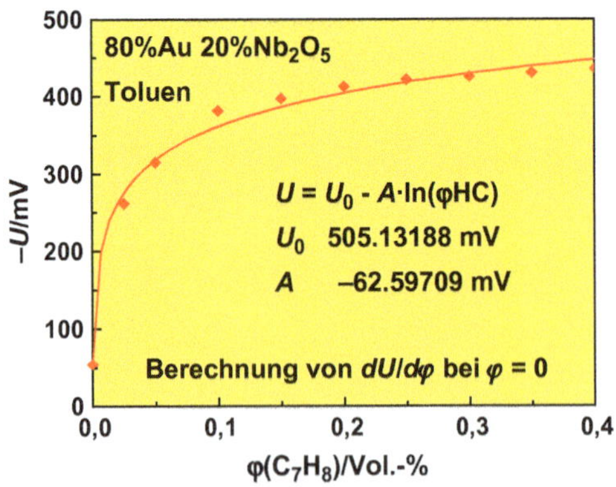

Figure 2.43 Sensitivity to toluene at 800°C in N_2 with 1.5 vol.% O_2.

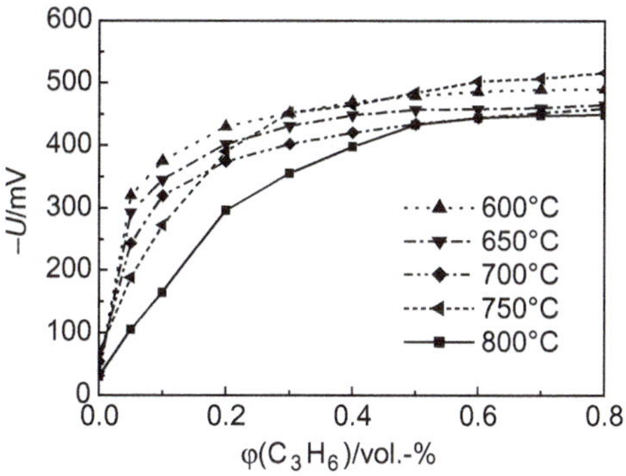

Figure 2.44 Potential dependence of composite electrodes (Au/Nb_2O_5) on propene concentration φ in N_2 with 1.5 vol.% O_2 at different temperatures.
Source: J. Zosel, R. Müller, V. Vashook, U. Guth, Response behavior of perovskites and Au/oxide composites as HC-electrodes in different combustibles, Solid State Ionics 175 (2004) 531–533.

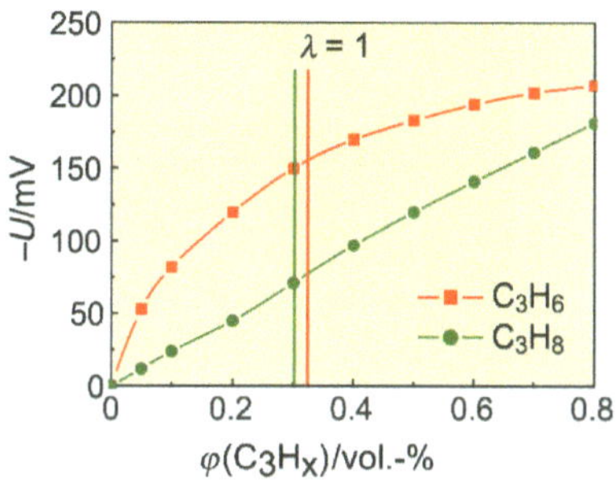

Figure 2.45 Potential response of the sensor in gas symmetric arrangement for different combustibles with Au-Nb_2O_5 measuring electrode versus Pt reference, 9 vol.% O_2 in N_2, $\vartheta = 700°C$.
Source: J. Zosel, U. Guth Electrochemical solid electrolyte gas sensors – hydrocarbon and NOx analysis in exhaust gases, Ionics 10 (2004) 366–377.

Furthermore, the cross-sensitivity to oxygen for propene and propane is shown in Fig. 2.46. For propene, the highest voltage is measured at the lowest oxygen concentration. In propane the sensor behaves reversely: the higher the oxygen concentration the higher the sensor response.

A general problem for the search of suitable electrodes for mixed-potential sensors is to combine a high sensitivity for the desired analyte gas (e.g., CO) with low cross-sensitivities against other gas species which are present in the exhaust gas of the combustion ("raw gas"). Typical interfering gases are NO_x, SO_2, and H_2O. Often all combustibles HC and CO are summarized as CO equivalents and denoted briefly as CO-e.

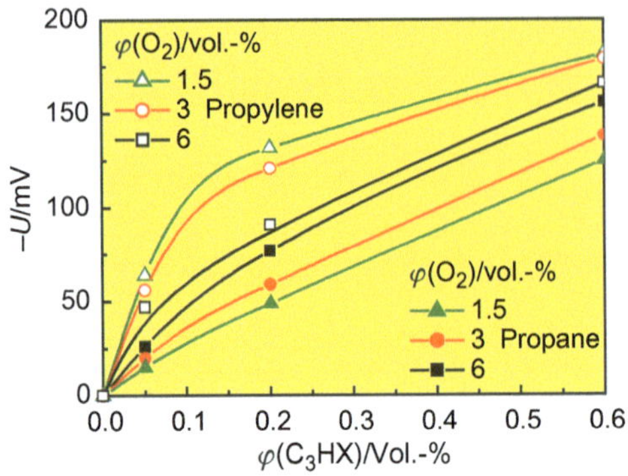

Figure 2.46 Cross sensitivity to oxygen for propan, propen containing gases.
Source: J. Zosel, R. Müller, V. Vashook, U. Guth, Response behavior of perovskites and Au/oxide composites as HC-electrodes in different combustibles, Solid State Ionics 175 (2004) 531–533.

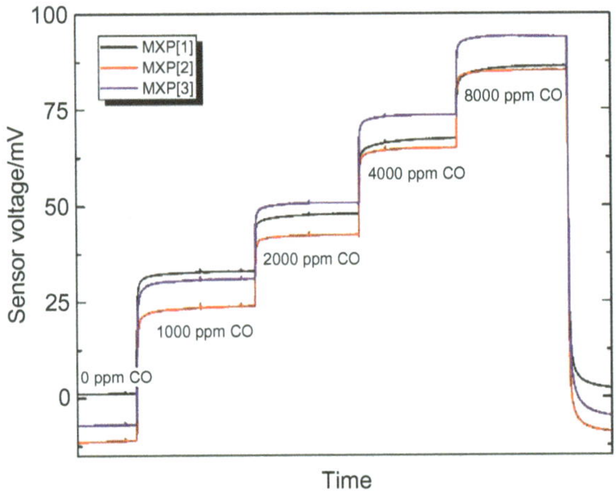

Figure 2.47 Sensor response of a CO sensor with an oxide electrode. MXP [1–3] designated three different sensors of the same setup.
Source: O. Driesner, F. Gumprecht, U. Guth, In situ measurements of O_2 and CO in cement kiln, J. Sens. Sens. Syst. 6 (2017) 327–330.

Up to now, a large number of various substances have been investigated with respect to electrode utilization in mixed-potential sensors [30–32]. As an example, the sensor response for CO-e with a low cross sensitivity to NO_x, SO_2, H_2O. is shown in Fig. 2.47.

2.5.3.4 Mixed oxides

Especially, complex oxides of perovskite-type ABO_3 are considered as suitable electrode materials due to the selection of A and B cations and partial substitution of both, A and B ions, by aliovalent ions. This offers the possibility to create electronic defects (i.e., excess electrons and holes) and oxide ion vacancies as well.

The incorporation of an aliovalent A-cation can be compensated by a corresponding change from a three-valent Cr^{3+} ion into a four-valent Cr^{4+} ion. A Cr^{4+} ion that occupies a Cr^{3+} site results in the formation of an electron hole:

$$LaCrO_3 \underset{xSrO}{\rightarrow} La_{1-x}^{3+}Sr_x^{2+}Cr_{1-x}^{3+}Cr_x^{4+}O_3 \tag{2.44}$$

On the other hand, complex oxides of the perovskite-type can compensate the deficiency on the A site charges by releasing lattice oxygen and formation of an oxygen deficiency.

$$LaMnO_3 \underset{aSrO}{\rightarrow} La_{1-a}^{3+}Sr_a^{2+}Mn_{1-a+2\delta}^{3+}Mn_{a-2\delta}^{4+}\left(V_O^{\bullet\bullet}\right)O_3 \tag{2.45}$$

It is, however, not a straightforward approach to relate such defect-structure phenomena with the catalytic and electro-catalytic behavior of the material. The main factors that influence the electrode activity are electrode morphology and thickness. Fig. 2.48 shows that already the substitution of B cations by isovalent cations changes the catalytic activity significantly. On the other hand, it is likely that the substitution also changes the morphology.

Obviously, the substitution of Cr^{3+} by Ga^{3+} leads to a higher catalytic activity for combustion of propene (Fig. 2.48) and to a small increase of the sensor response (Fig. 2.49). Relevant effects can be seen in Figs. 2.50 and 2.51 for Sr^{2+}-deficiency on A-sites (Figs. 2.52–2.53).

Such materials are also suitable for impedimetric sensors (see Figs. 2.58 and 2.59).

Another possible and suitable class of oxides for mixed-potential sensors have been identified in perovskites with Mn, Co, Ni, and Fe as B-cations. Typical results

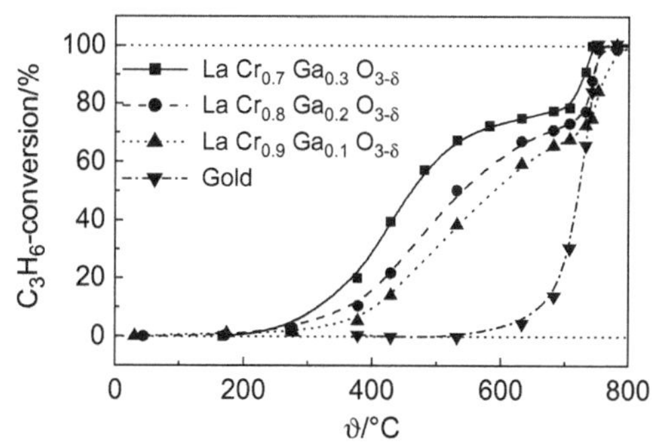

Figure 2.48 Influence of B substitution by an isovalent ion on the amount of conversation. *Source:* J. Zosel, K. Ahlborn, W. Fichtner, R. Müller, U. Guth, Investigations on selected gallium doped lanthanum chromites as electrode materials for HC detection, Ionics 6 (2000) 359–363.

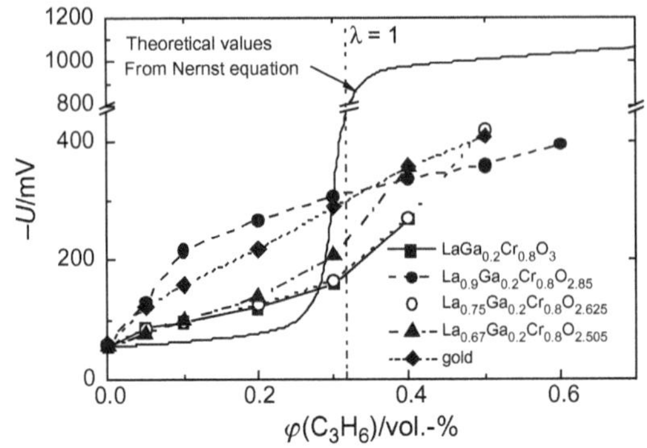

Figure 2.49 Influence of the oxygen deficiency on sensitivity for different mixed oxides electrodes.
Source: J. Zosel, U. Guth Electrochemical solid electrolyte gas sensors – hydrocarbon and NOx analysis in exhaust gases, Ionics 10 (2004) 366–377.

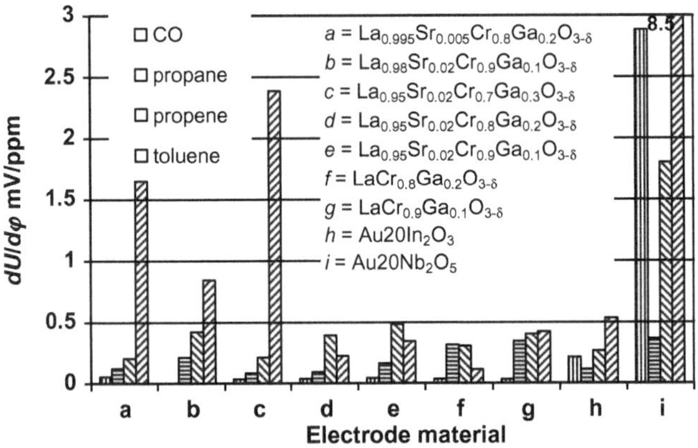

Figure 2.50 Sensitivity at infinite concentrations ($\varphi = 10$ ppm) of various electrode materials in different combustibles diluted in a gaseous mixture of N_2 and 1.5 vol.% O_2.
Source: J. Zosel, K. Ahlborn, R. Müller, D. Westphal, V. Vashook, U. Guth, Selectivity of HC sensitive electrode materials for mixed potential gas sensors, Solid State Ionics 169 (2004) 115–119.

are presented in Figs. 2.50 and 2.51. They show only low sensitivity to combustibles but are obviously very sensitive to nitric oxides.

2.5.3.5 Electrode materials for nitrogen oxides NO_x (Figs. 2.54–2.55)

Double B-site perovskites open new perspectives to tailor materials with respect to electronic excess or holes. A broad investigation was performed in the system

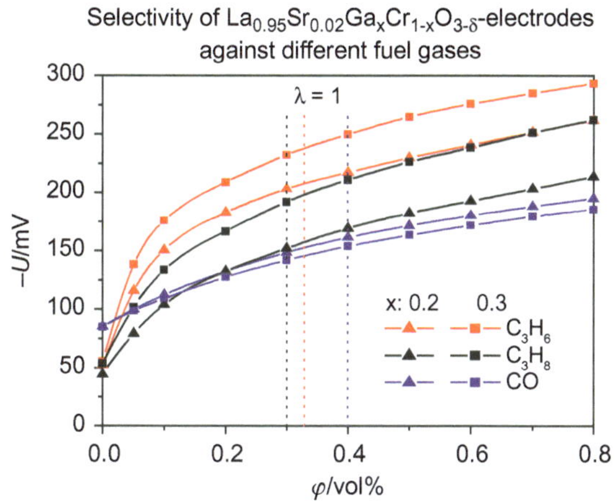

Figure 2.51 Selectivity of $La_{0.95}Sr_{0.02}Ga_xCr_{1-x}O_{3-\delta}$/YSZ-electrodes against different fuel gases.
Source: J. Zosel, U. Guth, Electrochemical solid electrolyte gas sensors – hydrocarbon and NOx analysis in exhaust gases, Ionics 10 (2004) 366–377 [33].

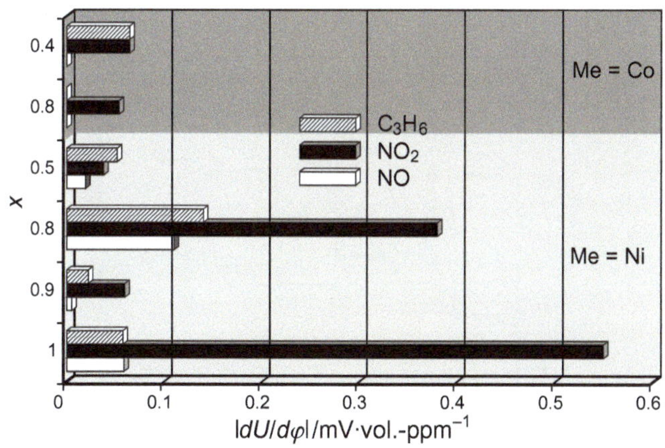

Figure 2.52 Sensitivities for mixed potential cell with electrode materials $La_{0.6}Ca_{0.4}Mn_xMe_{1-x}O_{3-\delta}$, $\varphi = 50$ Vol.-ppm, $\vartheta = 550°C$, flow rate 50 mL/min, measurement against Pt/air reference.
Source: J. Zosel, D. Franke, K. Ahlborn, F. Gerlach, V. Vashook, U. Guth Perovskite related electrode materials with enhanced NO sensitivity for mixed potential sensors, Solid State Ionics 179 (2008) 1628–1631.

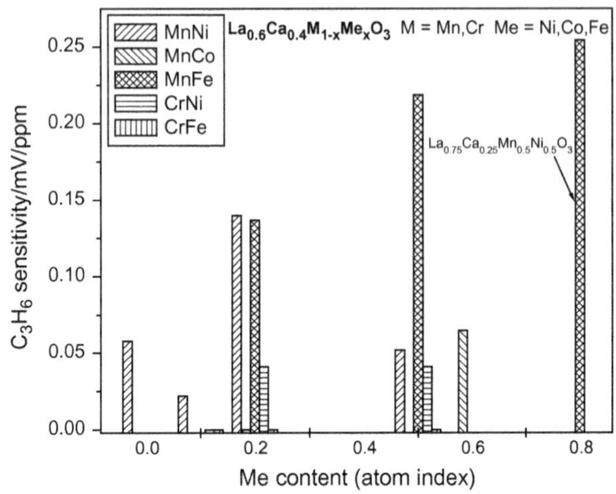

Figure 2.53 Propene sensitivity for different perovskite electrodes at 550°C (under the same conditions as described in Fig. 2.52).
Source: J. Zosel, D. Franke, K. Ahlborn, F. Gerlach, V. Vashook, U. Guth Perovskite related electrode materials with enhanced NO sensitivity for mixed potential sensors, Solid State Ionics 179 (2008) 1628–1631.

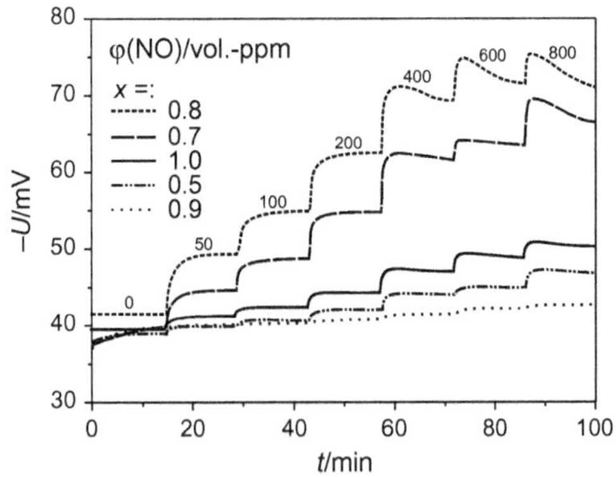

Figure 2.54 Potentiometric response of electrode materials of the group La0.6Ca0.4Mn$_x$Ni$_{1-x}$O$_{3-\delta}$ in a solid electrolyte cell with Pt-air reference, $\vartheta = 550°C$, $\varphi(O_2) = 1.5$ vol.%, flow rate 50 mL/min.
Source: J. Zosel, D. Franke, K. Ahlborn, F. Gerlach, V. Vashook, U. Guth Perovskite related electrode materials with enhanced NO sensitivity for mixed potential sensors, Solid State Ionics 179 (2008) 1628–1631.

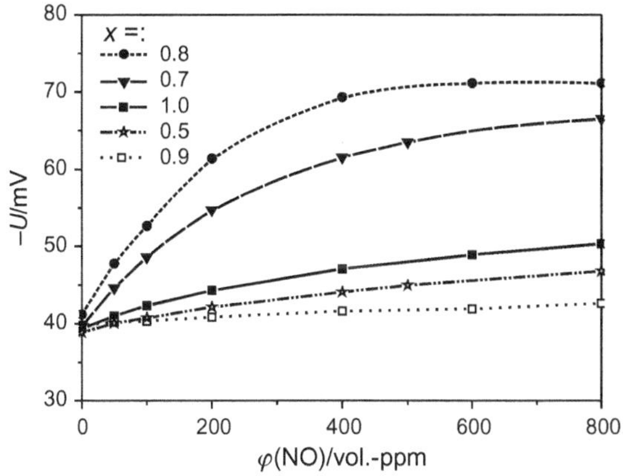

Figure 2.55 Dependence of the potential on the NO concentration of the materials given in Fig. 2.54.
Source: J. Zosel, D. Franke, K. Ahlborn, F. Gerlach, V. Vashook, U. Guth Perovskite related electrode materials with enhanced NO sensitivity for mixed potential sensors, Solid State Ionics 179 (2008) 1628–1631.

$La_{(2+y-2x)/3}Ca_xCr_yTi_{1-y}CrO_3$ which is a material with n- and p-type conductivity. It has been shown previously that substitution of La^{3+} by Ca^{2+} allows for to affect the ratio between Ti^{3+} and Ti^{4+} as well as between Cr^{3+} and Cr^{4+} [34]. Many compounds in this system have been investigated with regard to electrical conductivity and oxygen stoichiometry, but tests of their capability as electrode materials in mixed-potential sensors are still to be done.

2.6 Solid-state electrolyte sensors in dynamic mode

Solid-state electrolyte sensors (Fig. 2.6) can operate in steady-state or in dynamic mode. The latter term refers to all methods in which a time-dependent voltage is applied and a time-dependent current response is measured. It is common to distinguish between periodical and continuous changes of the potential or current in time. Continuous changes are called voltammetry in electrochemistry. The first mode is well-known as impedance measurement. For sensor applications, it is useful to plot the imaginary part versus the real part of the impedance in a Nyquist plot while impedance-*versus*-frequency are less common. Ideally, semicircles are obtained which can be interpreted in terms of single steps of the electrochemical reaction. The intercept at low frequencies, that is, the total cell resistance, is strongly influenced by the electrochemical reaction and this reaction is, in turn, influenced by the involved species. Gas sensors based on this principle are known as impedimetric or impedancemetric sensors [35,36]. Such sensors are often described in literature whereas reports on voltammetric techniques with oxide electrodes are very rare.

Sometimes, the corresponding studies confirm even better results with such electrodes, operating in impedimetric mode, than they are achieved with mixed-potential sensors (Figs. 2.56–2.57).

In some cases, potentiometric response on NO_x is very little. In the case of $La_{0.6}Ca_{0.4}Mn_{0.9}Ni_{0.1}O_{3-\delta}$/YSZ only 0.004 mV/ppm NO has been determined. On the other hand, electrodes with the same material show an exceptionally high sensitivity in impedimetric mode (Figs. 2.58 and 2.59). Even simple oxides like ZnO, NiO, and Cr_2O_3 are found to be sensitive to NO_x [37].

Another approach to receive an NO-sensitive electrode material is to modify the composition of the precious metal. This can be seen in the Nyquist plot of Fig. 2.60 [38]. In comparison with the Pt/YSZ electrode, the cathodic O_2 reduction is strongly inhibited, which is also confirmed by the cell resistance for cathodic O_2 reduction (i.e., R_{cell}(Pt/YSZ) < R_{cell}(PtAu/YSZ) < R_{cell}(RhPtAu/YSZ). The sensitivity to oxygen is high for Pt/YSZ electrodes and is drastically diminished for PtAu/YSZ and RhPtAu/YSZ electrodes. The cathodic NO reduction, however, is promoted by the RhPtAu/YSZ electrode. Using this electrode in a symmetrically digital structure in an amperometric mode, the current response was demonstrated to depend linearly on the NO concentration (Fig. 2.61).

Up to now, only a few papers on voltammetric experiments with solid-state electrolyte sensors have been presented [39]. Commercially available cell setups, for example, in a tube-design (Fig. 2.62) make such experiments possible with the aim to determine low concentrations of hydrogen. Typical experimental data are presented in Fig. 2.63 [38].

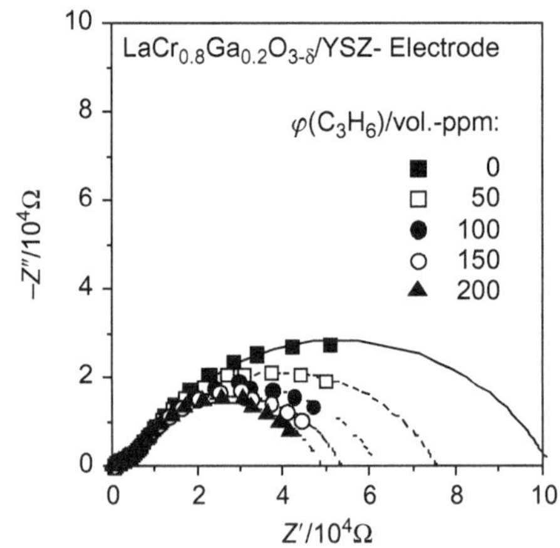

Figure 2.56 Impedance plots at La-cromite-electrodes at different propene concentrations, $\vartheta = 650°C$, $\varphi(O_2) = 1000$ ppm.
Source: J. Zosel, U. Guth Electrochemical solid electrolyte gas sensors – hydrocarbon and NOx analysis in exhaust gases, Ionics 10 (2004) 366–377.

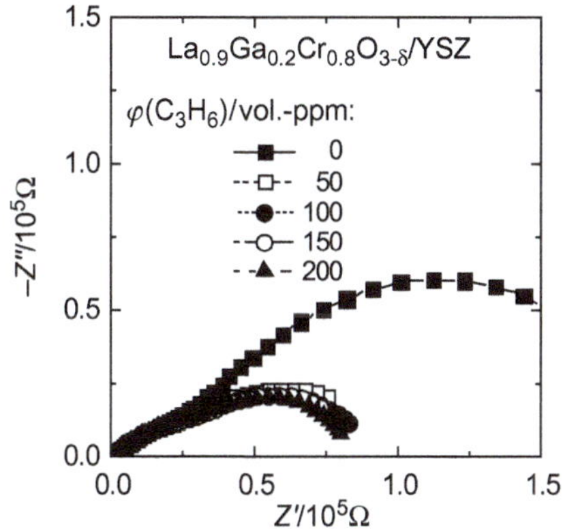

Figure 2.57 Nyquist plots using La-chromite-electrodes with A-side deficiency at different propene concentrations, $\vartheta = 700°C$, $\varphi(O_2) = 1000$ ppm.
Source: J. Zosel, U. Guth Electrochemical solid electrolyte gas sensors – hydrocarbon and NOx analysis in exhaust gases, Ionics 10 (2004) 366–377.

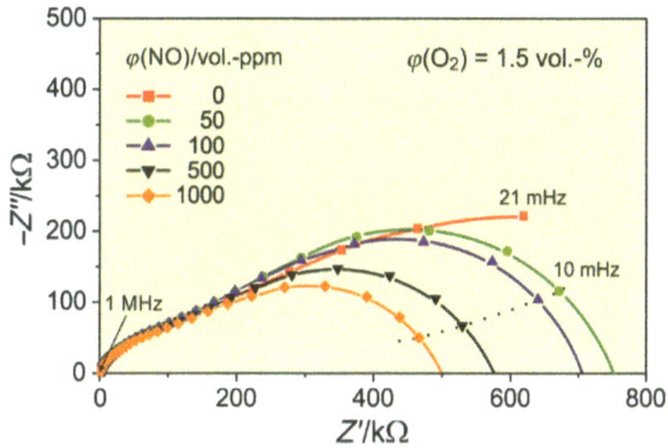

Figure 2.58 Dependence of impedance plots at electrodes of $La_{0.6}Ca_{0.4}Mn_{0.9}Ni_{0.1}O_{3-\delta}$ /YSZ on NO concentration, $\vartheta = 550°C$.
Source: J. Zosel, D. Franke, K. Ahlborn, F. Gerlach, V. Vashook, U. Guth Perovskite related electrode materials with enhanced NO sensitivity for mixed potential sensors, Solid State Ionics 179 (2008) 1628–1631.

The voltammetric curves show that the oxidation peaks depend on the scan rate. The peak height is a function of the square root of the scan rate which is typical for a diffusion process (Fig. 2.63). Voltammetric experiments were also carried out

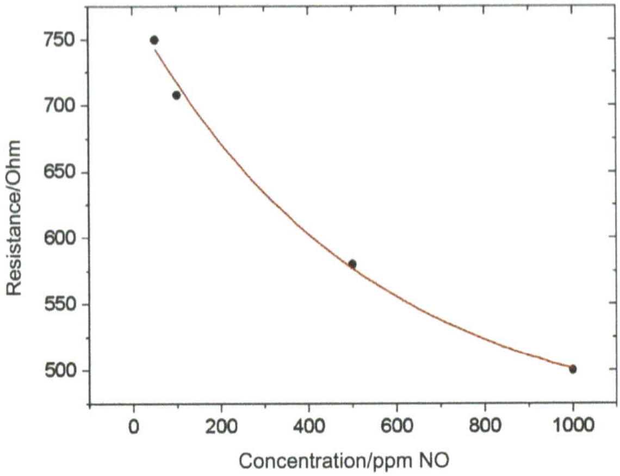

Figure 2.59 Calculated cell resistance from impedance plots according to Fig. 2.56 Potentiometric Response 0.004 mV/ppm NO Impedimetric response (av.) 0.22 Ohm/ppm NO.
Source: J. Zosel, D. Franke, K. Ahlborn, F. Gerlach, V. Vashook, U. Guth Perovskite related electrode materials with enhanced NO sensitivity for mixed potential sensors, Solid State Ionics 179 (2008) 1628–1631, Own results.

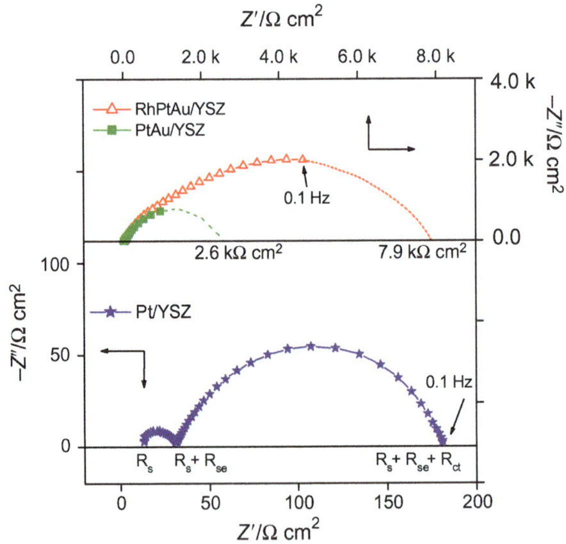

Figure 2.60 Impedance plots for three different electrodes in 1 vol.% O_2, N_2 gas mixture at 600°C.
Source: P. Schmidt-Zhang, W. Zhang, F. Gerlach, K. Ahlborn, U. Guth, Electrochemical investigation on multi-metallic electrodes for amperometric NO gas sensor, Sens. Actuators, B 108 (2005) 797–802.

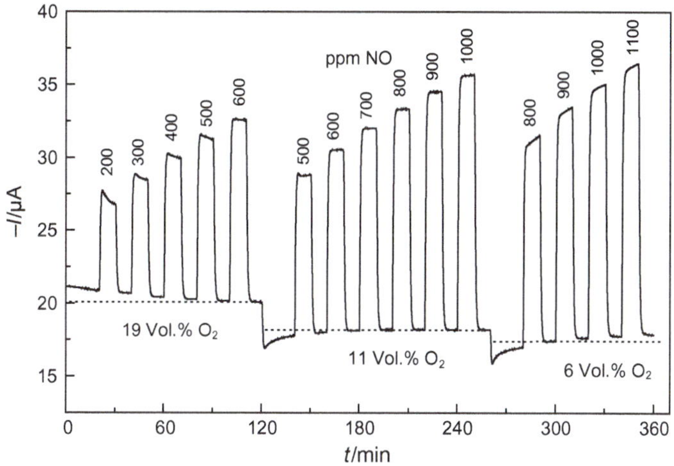

Figure 2.61 Current response of a sensor with RhPtAu/YSZ interdigitated electrodes exposed to gas mixtures with simulated NO, O_2, N_2 concentration patterns of a lean burn engine at 600°C, at potential of—400 mV, gas flow rate 100 mL/min.
Source: P. Schmidt-Zhang, W. Zhang, F. Gerlach, K. Ahlborn, U. Guth, Electrochemical investigation on multi-metallic electrodes for amperometric NO gas sensor, Sens. Actuators, B 108 (2005) 797–802.

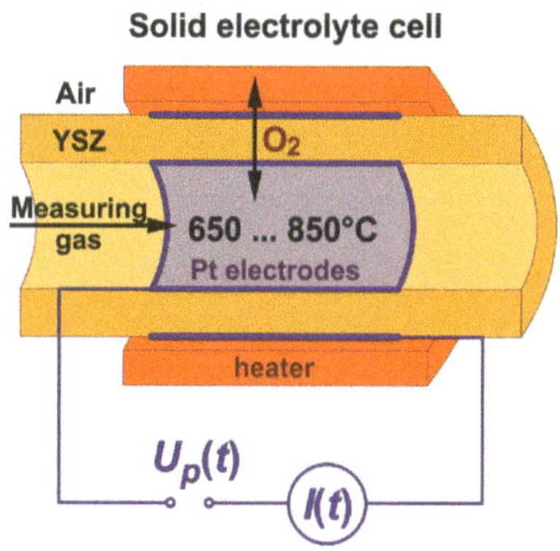

Figure 2.62 Tube designed solid electrolyte cell.
Source: M. Schelter, J. Zosel, W. Oelßner, U. Guth, M. Mertig, Highly selective solid electrolyte sensor for the analysis of gaseous mixtures, J. Sens. Sens. Syst. 5 (2016) 319–324.

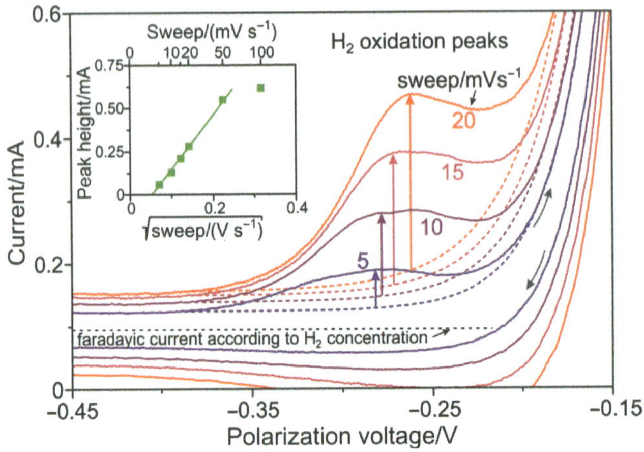

Figure 2.63 Voltammograms measured with tube designed potential sensor (Fig. 2.56) in air, H_2-gas mixture at 750°C, scan rate 10 mV/s.
Source: M. Schelter, J. Zosel, W. Oelßner, U. Guth and M. Mertig, Highly selective solid electrolyte sensor for the analysis of gaseous mixtures, J. Sens. Sens. Syst. 5 (2016) 319–324.

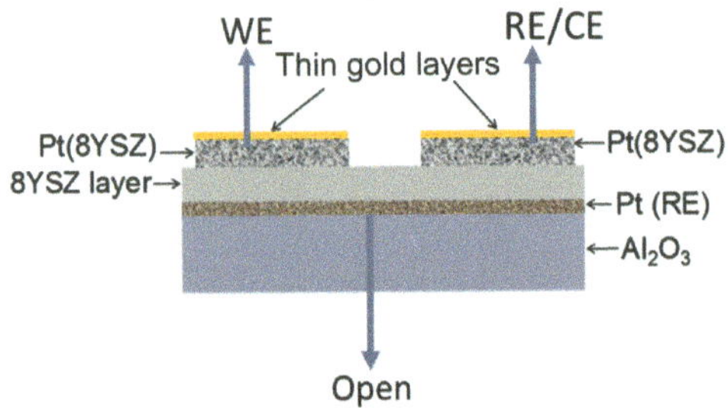

Figure 2.64 Cyclic voltammograms measured in N_2 with different oxygen concentrations, scan rate 50 mV/s.
Source: X. Zhang, H. Kohler, M. Schwotzer, Y. Wu, U Guth, Study of a layered Au, Pt-YSZ mixed-potential sensing electrode by ESEM, XRD and GD-OES with relation to its electrochemical behavior, Eurosensors 2017 Conference, Paris, France, September 3–6, Ext. Abstr., pp. 1–4, 2017.

with mixed-potential sensors. A suitable design of such a sensor is shown in Fig. 2.64. The platinum electrode is covered by a thin layer of gold which is tempered at 1050°C. Typical results, obtained with such a sensor, show peaks which depend on the oxygen concentration (Fig. 2.65) and the CO concentration (Fig. 2.66).

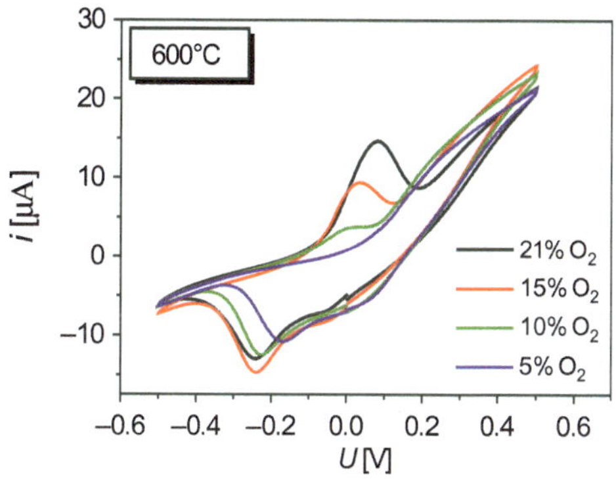

Figure 2.65 Cyclic voltammograms measured in N_2 with different oxygen concentrations, scan rate 50 mV/s.
Source: X. Zhang, H. Kohler, M. Schwotzer, Y. Wu, U Guth (2017) Study of a layered Au, Pt-YSZ mixed-potential sensing electrode by ESEM, XRD and GD-OES with relation to its electrochemical behavior, Eurosensors 2017 Conference, Paris, France, September 3–6, Ext. Abstr. pp. 1–4, 2017.

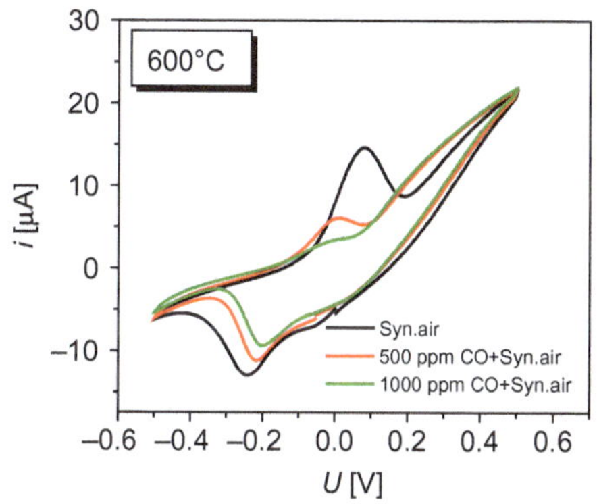

Figure 2.66 Cyclic voltammograms measured with mixed potential sensor (Fig. 2.58) in air, CO-gas mixture.
Source: X. Zhang, H. Kohler, M. Schwotzer, Y. Wu, U Guth (2017) Study of a layered Au, Pt-YSZ mixed-potential sensing electrode by ESEM, XRD and GD-OES with relation to its electrochemical behavior, Eurosensors 2017 Conference, Paris, France, September 3–6, Ext. Abstr. pp. 1–4, 2017.

2.7 Conclusion

High temperature sensors based on solid electrolytes are applied worldwide to measure in situ free oxygen and oxygen in equilibria in high temperature processes. The commercial use of such sensors began in the early 60ies. Over the past 20 years, investigations were performed in order to determine other gas components like CO, HC, and NO_x using the mixed-potential principle. The main challenge is to find electrode materials which are sensitive versus one gas without cross sensitivities to the other species. More recently, development is focused on dynamic methods which exhibit a great potential to determine several components simultaneously.

References

[1] H. Peters, H.-H. Möbius, Procedure for the gas analysis at elevated temperatures using galvanic solid electrolyte elements, DD-Patent 21673, August 1961.

[2] J. Weissbart, R. Ruka, Oxygen gauge, Rev. Sci. Instrum. 32 (1961) 593–595.

[3] (a) K. Kiukkola, C. Wagner, Galvanic cells for the determination of the standard molar free energy of formation of metal halides, oxides, and sulfides at elevated temperatures, J. Electrochem. Soc. 104 (1957) 308–316.
(b) K. Kiukkola, C. Wagner, Measurements on galvanic cells involving solid electrolytes, J. Electrochem. Soc. 104 (1957) 379–387.

[4] H. Peters, H.-H. Möbius, Electrochemical investigation of carbon dioxide dissoziation and Boudouard equilibrium, published in German: Elektrochemische Untersuchung des Kohlensäuredissoziationsgleichgewichtes und des Boudouard-Gleichgewichtes, Naturwissensch. 45 (1958) 309–310.

[5] U. Guth, Gas sensors, in: A. Bard, G. Inzelt, F. Scholz (Eds.), Electrochemical Dictionary, second extended ed., Springer Heidelberg, Dordrecht, London, New York, 2012, pp. 400–406.

[6] H.-H. Möbius, Solid state electrochemical potentiometric sensors for gas analysis, in: W. Göpel, J. Hesse, J.N. Zemel (Eds.), Sensors, VCH, Weinheim, 1991, p. 1118.

[7] P. Shuk, E. Bailey, U. Guth, Zirconia oxygen sensor for the process application: state-of-the-art, Mod. Sens. Technol. 90 (2008) 174–184.

[8] T. Baunach, L. Diehl, K. Schänzlin, Clean flue gas by using of ceramic sensors (in German) Sauberes Abgas durch Keramiksensoren, Phys. J. 5 (2006) 33–38.

[9] H.-H. Möbius, R. Hartung, Potentiometric gas sensors based on ZrO_2 solid electrolytes for gas nitridation and nitrocarbonizing (in German), HTM 53 (1998) 245–254.

[10] W.A. Fischer, D. Jahnke, Metallurgical Electrochemistry (in German) Stahleisen Düsseldorf, Springer, Berlin, Heidelberg, New York, 1975.

[11] K. Kämpfer, E. Prescher, H.-H. Möbius, The use of solid-electrolyte probes in porcelain production kilns, Ceram. Forum Int. 68 (1991) 126–131.

[12] H.-H. Möbius, T. Handler, Control of gas composition induring burning process to garantee the colouring of ceramics (in German), Keram. Z. 50 (1998) 918–933.

[13] R. Hartung, On the analysis of CO_2, H_2 and CO, H_2O-mixtures by water-gas potentiometry with solid electrolyte cells, Fresenius J. Anal. Chem 356 (1996) 228–232.

[14] www.zirox.de, www.stg-cottbus.de

[15] www.enotec.de

[16] www.emerson.com

[17] H. Ullmann, Ceramic Gas Sensors (In German), Akademie Verlag, Berlin, 1993.

[18] P. Schmidt-Zhang, K.-P. Sandow, F. Adolf, G. Reinhardt, U. Guth, W. Göpel, A novel thick film sensor for simultaneous O_2 and NO Monitoring exhaust gases, Sens. Actuat. B 70 (2000) 25–29.

[19] S. Somov, G. Reinhardt, U. Guth, W. Göpel, Gas analysis with arrays of solid state electrochemical sensors: implications to monitor HCs and NOx in exhaust, Sens. Actuat. B 35–36 (1996) 409–418.
[20] S. Somov, U. Guth, A parallel analysis of oxygen and combustibles in solid electrolyte amperometric cells, Sens. Actuat. B 47 (1998) 131–138.
[21] V. Vashook, J. Zosel, U. Guth, Oxygen Solid Electrolyte Coulometry (OSEC), J. Solid State Electrochem. 16 (2012) 3401–3421.
[22] M. Schelter, J. Zosel, W. Oelßner, U. Guth, M. Mertig, A solid electrolyte sensor for trace gas analysis, Sens. Actuat. B: Chem. B 187 (2013) 209–214.
[23] H.-D. Wiemhöfer, U. Vohrer, Spectroscopy and thermodynamics of electrons in yttria-stabilized zirconia, Ber. Bunsenges. Phys. Chem. 96 (1992) 1646–1651.
[24] Y.L. Sandler, The response of the stabilized zirconia galvanic cell to methane-oxygen mixtures, J. Electrochem. Soc. 118 (1971) 1378–1381.
[25] R. Moos, K. Sahner, M. Fleischer, U. Guth, N. Barsan, U. Weimar, Solid state gas sensor research in Germany - a status report, Sensors 9 (2009) 4323–4365.
[26] A.R. Pebler, US. P. 4005001, filed: March 27, 1973, date of patent Jan 25,1977.
[27] R. Hartung, R. Schröder, H.-H. Möbius, Combustible gas sensitive gas symmetric galvanic cell with oxide ion conducting electrolyte Z. physik, Chem. (Leipzig) 262 (1981) 961–966.
[28] N. Miura, T. Raisen, G. Lu, N. Yamazoe, Highly selective CO sensor using stabilized zirconia and a couple of oxide electrodes, Sens. Actuat. B 47 (1998) 84–91.
[29] J. Zosel, G. Schiffel, F. Gerlach, K. Ahlborn, U. Sasum, V. Vashook, et al., Electrode materials for potentiometric hydrogen sensors, Solid State Ionics 177 (2006) 2301–2304.
[30] F.H. Garzon, R. Mukundan, E.L. Brosha, Solid-state mixed potential gas sensors: theory, experiments and challenges, Solid State Ionics 136–137 (2000) 633–638.
[31] N. Miura, T. Sato, S.A. Anggraini, H. Ikeda, S. Zhuiykov, A review of mixed-potential type zirconia-based gas sensors, Ionics 20 (2014) 901–925.
[32] W. Jeffrey, J. Fergus, Sensing mechanism of non-equilibrium solid-electrolyte-based chemical sensors, J. Solid State Electrochem. 15 (2011) 971–984. Available from: https://doi.org/10.1007/s10008-010-1046-4.
[33] J. Zosel, U. Guth, Electrochemical solid electrolyte gas sensors – hydrocarbon and NOx analysis in exhaust gases, Ionics 10 (2004) 366–377.
[34] D. Franke, J. Zosel, U. Guth, NO sensitivity of perovskite-type electrode materials $La_{0.6}Ca_{0.4}B'_{1-x}B''_xO_{3\pm\delta}$ B' = Mn, Cr B" = Ni, Fe, Co x = 0, 0.1,. . ., 0.6 in mixed potential sensors, Sens. Actuat. B 223 (2016) 723–729.
[35] P. Elumalai, J. Zosel, U. Guth, N. Miura, NO_2 sensing properties of YSZ-based sensor using NiO and Cr-doped NiO sensing electrodes at high temperature, Ionics 15 (2009) 405–411.
[36] N. Miura, M. Nakatou, S. Zhuiykov, Impedancemetric gas sensor based on stabilised zirconia solid electrolyte and oxide sensing electrode for detecting total NOx at high temperature, Sens. Actuat. B 93 (2003) 221–228.
[37] N. Miura, J. Wang, M. Nakatou, P. Elumalai, S. Zhuiykov, M. Hasei, High-temperature operating characteristics of mixed-potential-type NO_2 sensor based on stabilized-zirconia tube and NiO sensing electrode, Sens. Actuat. 114 (2006) 903–909.
[38] P. Schmidt-Zhang, W. Zhang, F. Gerlach, K. Ahlborn, U. Guth, Electrochemical investigation on multi-metallic electrodes for amperometric NO gas sensor, Sens. Actuat. B 108 (2005) 797–802. Available from: https://doi.org/10.1016/j.snb.2004.12.074.
[39] J. Yi, A. Kaloyannis, C. Vayenas, High temperature cyclic voltammetry of Pt catalyst-electrodes in solid electrolyte, Cells Electrochim. Acta 38 (1993) 2533–2539.

Further reading

O. Driesner, F. Gumprecht, U. Guth, In situ measurements of O_2 and CO in cement kiln, J. Sens. Sens. Syst. 6 (2017) 327–330.

J.W. Fergus, Electrolytes for solid oxide fuel cells, J. Power Sources 162 (2006) 30–40.

U. Guth Solid state electrochemistry, in: G. Kreysa, K.-o. Ota, R. Savinell (Eds.), Encyclopedia of Applied Electrochemistry, New York, Heidelberg, Dordrecht, London, 2014, pp. 1993–1998 and 2014–2041.

S. Lübke, H.-D. Wiemhöfer, Electronic conductivity of Gd-doped ceria with additional Pr-doping, Solid State Ionics 117 (1999) 229–243.

M. Nakatou, N. Miura, Detection of propene by using new-type impedancemetric zirconia based sensor attached with oxide sensing-electrode, Sens. Actuat. B 120 (2006) 57–62.

K.-P. Sandow, S. Jakobs, R. Hartung, D. Westphal, U. Schönauer, U. Guth, Chemical modification of chromite-YSZ electrode and their influence on the non-NERNST behavior, Ionics 2 (1996) 435–441.

M. Schelter, J. Zosel, W. Oelßner, U. Guth, M. Mertig, Highly selective solid electrolyte sensor for the analysis of gaseous mixtures, J. Sens. Sens. Syst. 5 (2016) 319–324. 2016.

E.D. Wachsman, K.T. Lee, Lowering the temperature of solid oxide fuelc cells, Science 334 (2011) 935–939.

W. Weppner, Electronic transport properties and electrically induced p-n-junction in ZrO_2 + 10mol percent Y_2O_3, J. Solid State Chem. 20 (1977) 305–314.

X. Zhang, H. Kohler, M. Schwotzer, Y. Wu, U Guth, Study of a layered Au, Pt-YSZ mixed-potential sensing electrode by ESEM, XRD and GD-OES with relation to its electrochemical behavior, Eurosensors 2017 Conference, Paris, France, September 3–6, Ext. Abstr., pp. 1–4, 2017.

J. Zosel, K. Ahlborn, R. Müller, D. Westphal, V. Vashook, U. Guth, Selectivity of HC sensitive electrode materials for mixed potential gas sensors, Solid State Ionics 169 (2004) 115–119.

J. Zosel, D. Franke, K. Ahlborn, F. Gerlach, V. Vashook, U. Guth, Perovskite related electrode materials with enhanced NO sensitivity for mixed potential sensors, Solid State Ionics 179 (2008) 1628–1631.

Basics of semiconducting metal oxide–based gas sensors

3

Alexandru Oprea[1], David Degler[2], Nicolae Barsan[3], Anne Hemeryck[4] and Julia Rebholz[5]
[1]Institute for Physical and Theoretical Chemistry, University of Tübingen, Tübingen, Germany, [2]European Synchrotron Radiation Facility, Grenoble, France, [3]Institute for Theoretical and Physical Chemistry, University of Tübingen, Tübingen, Germany, [4]Equipe M3, LAAS-CNRS, Toulouse, France, [5]Innovative Sensor Technology IST AG, Ebnat-Kappel, Switzerland

The operation principle of SMOX-based sensors should be simple: as a result of exposure to a gaseous analyte, the resistance of a heat-sensitive SMOX layer changes. It is important that the parameter that changes is the electrical resistance because it is easy and cheap to measure and there are a lot of electronics available. The reason for the resistance change is the involvement of the material's free charge carriers in various surface reactions that take place with gases present in the ambient atmosphere. This is not unique for SMOXs, but this resistance change is detectable—in the case of metals the detectable changes are minute since there are too many available free-charge carriers. In addition, since SMOXs are already oxides, the irreversible changes observed in other semiconductors like Si or Ga due to oxidation are prevented, and the changes in the electrical resistance are reversible. There are also other oxides that are very surface active, but because they are insulators it is not possible to easily quantify, for example, by measuring the electrical resistance, the effects of exposure to gases. In summary, SMOXs are good sensitive materials because they are semiconductors and oxides. Consequently, one interesting question to ask is what makes some oxides semiconductive.

In spite of extensive research done on SMOXs, the exact origin of conductivity is still a matter of debate. The most generally accepted idea is that the native nonstoichiometry is the main cause. Section 3.1 provides an overview of existing ideas in the context of their experimental foundations and compared to theoretical calculations. It becomes clear from this initial comprehensive and structured overview that no definitive conclusion can yet be drawn. The specific weaknesses of the state of the art are identified and the knowledge gaps which future research must address are highlighted.

Even if the origin of their conductivity remains elusive, the application of SMOX-based sensors has flourished in the past few decades. A great deal of research has been done on how the sensing characteristics can be tuned, which is based on the basic understanding of sensing. For most of the almost 60 years of SMOX-based gas sensing, this was done by combining simple sensor resistance measurements with results from basic surface science investigations performed on model systems that are

Gas Sensors Based on Conducting Metal Oxides. DOI: https://doi.org/10.1016/B978-0-12-811224-3.00003-2

different from the actual sensors and under conditions that drastically vary from the application ones. This approach was largely unsuccessful, so most of the developments were achieved on a try-and-see basis and crude modeling.

For the initial basic advancements, this empirical process was successful, but as the sensors become sophisticated and the application criteria more stringent; a knowledge-based approach is necessary. This requires a fundamental understanding of gas sensing which will allow modeling. The final goal is to develop models that will enable the engineering of gas sensors for specific applications. Experimental insights obtained in application relevant conditions are the necessary foundation of such models and over the past few decades, in situ and operando methods have been developed. Section 3.2 provides an overview of both classical analytical methods and operando techniques, as well as their most important results. Special emphasis is put on SnO_2, an exemplary and well-researched SMOX. Operando measurements are done on sensors which are poor model systems because of their high heterogeneity—various crystalline faces, structural phases, and morphologies, etc. Consequently, the interpretation of the results is challenging. DFT calculations, presented in Section 3.3, can simplify the interpretation by providing insights on a microscopic level about the elementary steps of the reaction processes. It is discussed how the calculations are performed and the limits of their validity are examined. In addition, the role of kinetics in establishing practical sensor models is presented. From these calculations, in combination with the results from experimental techniques, it will be possible to identify which reactions play the dominant role in sensing. This input is used to establish practical sensor models, in which target gas concentrations and sensor response are correlated. Ideally, the effects of interfering gases and variation in the sample preparation will also be correctly quantified in the model. As the experimental inputs become more precise and the theoretical possibilities more sophisticated, the quality of the models will increase. Section 3.4 begins with an historical perspective on the efforts made on sensor modeling. It continues with an example based on state-of-the-art operando experimental inputs and concludes with an outlook.

3.1 Origin of conduction in SMOX

3.1.1 Preliminary considerations on the origin of conduction in SMOXs

The SMOXs employed in gas sensing are—with the exception of SnO_2, In_2O_3, and Ga_2O_3—compounds of transition metals. Analogous with oxides of the metals from the main groups, SMOXs should be insulators at least in their pure, single-crystal form. However, as the name indicates, SMOXs are semiconductors with a rather large charge carrier concentration and mobility. This surprising feature has inspired both scientific and practical efforts to find explanations. The most obvious reasons for SMOX's conduction are structural and stoichiometric defects of the real materials, for example, point defects like missing ions (vacancies), additional compound ions between the regular ones (interstitials, or, more precisely, self-interstitials), impurity ions instead of compound ions (substitutional impurities), or in interstitial positions (interstitial impurities). More exhaustive approaches are given in [1,2]. It is important to specify that the defect formation is not a technological issue, but a principle of thermodynamic consequence above 0K. Indeed, under constant temperature and pressure (as the gas sensors are roughly operated), the free enthalpy G of the material should reach a minimum when the structural equilibrium is attained [1,3]. The defect formation raises the crystal enthalpy H, thus raising the free enthalpy G. The entropy S also increases as the crystal disorder increases due to defect formation. However, the entropic contribution to the free enthalpy produces a decrease of G because it is contained in the negative term $-TS$ as a product with the temperature T. Consequently, the free enthalpy attains a minimum for a given defect concentration (as suggested in Fig. 3.1).

$$\delta_f^{(d)}G = \delta_f^{(d)}H - T\delta_f^{(d)}S$$

For a relatively low concentration of point defects (which is often the case for sensor materials) the evaluation of the entropic contribution to the free enthalpy is simple and follows from the statistical definition of configurational entropy [1]. Under equilibrium conditions at constant pressure and temperature—that is, at the minimum of G in respect with the defect number N_d, at a constant number of sites in the lattice N—one obtains:

$$N_d = N^{\bullet} e^{\frac{\delta_f^{(d)}H}{k_B T}} \tag{3.1}$$

where $\delta_f^{(d)}H, S$ is the enthalpy of the defect formation and k_B the Boltzmann constant.

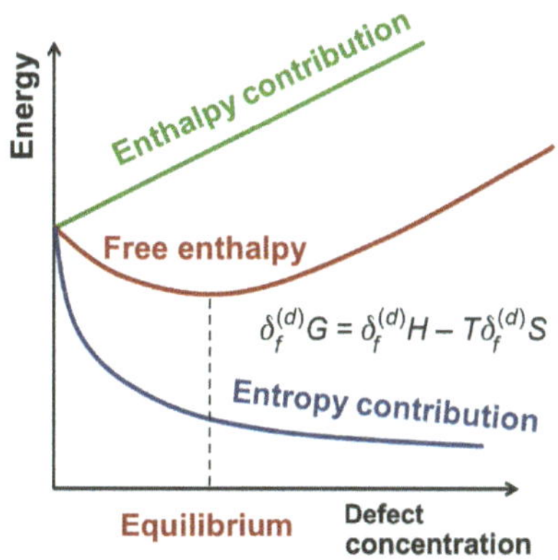

Figure 3.1 The evolution of the enthalpy, entropic contribution, and free enthalpy.
Source: After J. Maier, Physical Chemistry of Ionic Materials: Ions and Electrons in Solids, John Wiley & Sons, Ltd, 2004, https://doi.org/10.1002/0470020229.

The thermal generation of defects is very important because it affects both the ionic and electronic conductivity of SMOXs. The weight of each conduction type is expressed by transference (transport) numbers [4]. For most applications, only one type of mobile charge carriers (one type of ions or electrons, respectively, holes) is desirable. For ionic devices, a transference of one for a certain ion is needed. SMOX conductometric gas sensors require a transference of 1 for electrons or holes.

3.1.2 The role of defects in the electronic conduction of SMOXs

The existence of defects under thermodynamic equilibrium conditions leads to a significant perturbation of the electronic structure of solids. In the first approximation this perturbation can be regarded as a superposition of the defect electronic states with those belonging to the ideal crystal.[1] Because the defects do not respect the crystal symmetry and are randomly distributed, their states are more or less localized in space and can have energy levels spread over the gap. In certain cases, the defect energy spectrum may include resonances (localized states with the energy levels overlapping the valence and conduction bands (CBs), but with reduced interaction with these bands). The charge carriers from the shallow levels (defect states with the energy only a few k_BT away from the band edges) play a chief role in the electrical conduction since they can be thermally promoted into the bands and participate in transport phenomena. Conversely, the deep levels (local energy levels further away from the band edges) cannot exchange electrons with the bands under equilibrium conditions and, therefore, they do not directly influence the transport phenomena. Disregarding the depths of their levels, the localized states are donor levels if they "donate" electrons to other states and acceptors if they "accept" electrons from other states. The direct electron exchange between donors and acceptors leads to semiconductor "compensation," hindering the access of charge carriers to the bands as conduction electrons and holes. Although "dead" for conduction, the electrons/holes on the deep levels are important in optical or equivalent processes, giving a defect fingerprint.

3.1.3 Types of defects involved in electronic conduction

3.1.3.1 Intrinsic defects

Intrinsic defects are deviations of the solid from its ideal structure or composition, without adding atoms/ions other than the constitutive ones. In the case of compound semiconductors such as SMOXs, the deviation from stoichiometry seems to be the main defect acting as a dopant because it enables the formation of electronic localized states, which are often in the gap. The electron exchange between the bands

[1] The total number of electronic states in the crystal remains constant for a constant number of atoms. Therefore, the states associated to defects diminish the number of ideal crystal states under such a restriction.

and the defect levels has a large impact on the electrical charge transport. This process can be regarded as an equilibrium, quasi-chemical reaction [1]:

$$D^x \Leftrightarrow D^\bullet + e' \quad A^x \Leftrightarrow A' + h^\bullet \tag{3.2}$$

D designates donors, *A* acceptors, *e* electrons, and *h* holes. Kröger–Vink notations have been used [1,5,6]. This approach implicitly ensures the overall neutrality of the solid [6]. Complementary theoretical perspectives on the free charge carrier generation by defects are discussed in the next sections.

Usually, the stoichiometric defects are linked to structural ones, inducing lattice distortion. Defect formation is more likely to occur if the subsequent lattice distortion is small. Consequently, the point defects should have the largest spontaneous occurrence. In spite of their name, the point defects often exceed one lattice point because of the disturbance of the periodicity around them. A short overview on intrinsic point defects will give brief descriptions of defects relevant for SMOXs and will roughly explain their electronic behavior.

1. *Vacancies* are empty lattice points, which are missing ions/atomic cores otherwise present in the ideal crystal. An anion vacancy V_A (an oxygen one V_O, in the case of oxides) should have excess electrons localized at its boundary, namely the electrons the cation would transfer to the anion to form the nondefective, stoichiometric compound. Thus, in the case of V_O, two electrons are bound to the defect. Depending on the binding strength, they have shallower or deeper associated energy levels in the gap. The shallow ones can be thermally ionized, promoting electrons to the CB. They behave as effective donors. In principle, the deeper levels could have a donor character as well, but they will not provide delocalized electrons. Cation vacancies V_C (metals in SMOXs, V_M) should behave similarly, but symmetrical in respect with the charge sign, that is, producing acceptor levels in the gap, possibly with thermal generation of holes in the valence band (VB).
2. *Interstitials* are ions/atomic cores located between regular positions in the lattice in a metastable equilibrium state. If they appear in an ideal crystal by adding external ions/atoms of the same type as the constitutive ions/atoms (a growth atmosphere rich in the respective element) then the defect remains very simple. If the interstitials are promoted from lattice ions, then a vacancy formation follows resulting in a Frenkel pair defect. Cation interstitials, metals for SMOXs, are donors because the electrons they possess are not required in the crystal bonding, already ensured by the regular lattice ions, and remain available for delocalization in bands through thermal ionization mechanisms. Conversely, anion interstitials (oxygen in this case) could act as acceptors and exchange charge carriers with the VB.
3. *Antisite defects*, that is, anions at cations sites A_C and cations at anions ones C_A are constitutive ions/atomic cores in wrong lattice points, the anions replacing the cations and vice versa. For SMOXs, that means oxygen atoms/ions at metal sites O_{Me} and metal atoms/ions at oxygen sites Me_O. Due to large difference in atomic features, the antisites are expected to produce deep centers reciprocally compensated.

Ionized defects might be instable in applied electrical fields or due to concentration gradients and migrate to the surface outside the bulk, resulting in surface defects not relevant for the bulk conductivity.

3.1.3.2 Extrinsic defects

The extrinsic defect concept encompasses all structural and compositional modifications of the solid that involves foreign elements. They either unintentionally appear during material synthesis/annealing or are intentionally introduced for technological and applicative reasons. In traditional semiconductors (Ge, Si), extrinsic defects are typically substitutional elements with one valence electron more or less than the host atoms leading to *n*- or *p*-conduction type. In compound semiconductors, this simple description still holds true, but several new features/complications exist.

The unintentional extrinsic defects most encountered in SMOXs

1. *Hydrogen* found either intentionally or unintentionally from:
   - compound precursors as a component of solvents (water or organic ones), of organometallic compounds, in combustion gases of pyrolytic methods (like flame spray pyrolysis [7,8]);
   - synthesis in ambient conditions if no special precautions are taken (e.g., as a result of humidity);
   - annealing/measurement atmosphere, introduced to control the oxygen partial pressure.
2. *Cations with similar properties* (transition metals mostly in the case of SMOXs) as the host metals, which cannot be avoided during the synthesis of the compound. Usually, due to cost constraints, the metal sources used in the preparation routes have a limited purity.
3. *Anions other than oxygen*, present in the synthesis precursors or as gases/vapors in the treatment atmosphere. Some are intentionally used to set the desired value of the oxygen partial pressure during preparation or characterization, for example, hydrogen. However, the resulting oxide impurification is often neglected. Besides hydrogen, the most frequently encountered gas is nitrogen.

Intentional extrinsic defects

In order to enhance the material's native conductivity towards values favorable for device manufacturers or to improve other device parameters, like the sensitivity and specificity of gas sensors, different foreign elements are intentionally introduced in minute amounts. The expected effect of the extrinsic defects is often not achieved due to uncontrollable complexation with native defects that results in semiconductor compensation or defect passivation.

3.1.4 Experimental methods for the characterization of localized states

The experimental approaches employed to evaluate the properties of defect states depend on the localization, the type and material morphology. Many are based on transitions between defect electronic states and depend strongly on the defect configuration (roughly the position of atoms/ions around defects). During the transition, the defect configuration changes due to modified charges and/or interactions until it finally reaches a new equilibrium (relaxes). When the employed technique probes the defect only in relaxed configurations, it acquires thermodynamic equilibrium values of the measured parameters. The principle of some useful experimental methods, however, cannot fulfill the equilibrium condition for the final configuration. They make use of fast electronic transitions where the initial electronic state may correspond to an equilibrium defect configuration, but not the final one. Hence, the output transition energy differs from the value recorded for the equilibrium defect configuration. The experimental methods based on temperature sweep measurements are very slow allowing the defect configuration to reach equilibrium. Thus, relevant thermodynamic parameters (as statistical means of microscopic parameters) can be acquired. Contrarily, spectroscopic techniques are constrained to output "instant" values. The defect configuration at equilibrium for the initial electronic state will remain unchanged during very fast electronic transitions, but will become a nonequilibrium one for the final electronic state. The defect configuration relaxation will occur long after the electronic transition already ended accommodating the defect in its new electronic state (More details in Section "Comparing the experimental data from nonequilibrium measurements with the theoretical simulations").

3.1.4.1 Experimental techniques operating under thermodynamic equilibrium of the defect configuration

The methods listed below are based on electronic exchanges between localized electronic states of the defects and the bands promoted by temperature changes.

The activation energy of the DC currents can be determined from the temperature dependence of the sample conductivity, which is mainly provided by the charge carrier concentration (Eq. (3.9) and Section 3.1.4.1). It allows for the evaluation of the local level depth in the gap with respect to the CV and VB edges. This method is also able to identify the onset of the ionic conductivity, so that it is extensively employed to characterize both the electronic and ionic properties of the compounds with stoichiometric defects. Several examples are given later. A more accurate variant is the temperature dependence of the Hall constant, which directly assesses the charge carrier concentration and their polarity [9].

Thermally stimulated currents [10] can be measured in materials with several morphologies. In the first step, the local levels of the sample are filled by electrical or optical electron injection. Afterwards, the sample is heated at a constant rate and the current is measured. Each time the thermal energy k_BT corresponds to the difference between a gap level and the neighboring band edge, the captured charge carriers are released in the band resulting in current peaks. The lower the starting temperature is, the shallower the perceived levels are.

Deep level transient spectroscopy (DLTS) [11] uses capacitance measurements performed during a temperature sweep of Schottky contacts or *pn*-junctions to assess the energy and concentration of gap states in the depletion region. The technique is appropriate for single crystals but inapplicable to the granular materials used for gas sensing. Nevertheless, local levels in SMOX single crystals were identified this way.

3.1.4.2 Experimental techniques operating away from the thermodynamic equilibrium of the defect configuration

The methods considered in this paragraph use electronic or other fast transitions. They do not allow the final defect configuration to relax.

Optical absorption methods evaluate the decrease in the light intensity (photon flux) incident on a solid sample because of transitions induced between electronic states [12]. Since the transitions occur through an electron energy increase, the sample is brought into a nonequilibrium state (the higher energy electronic states are more and the lower energy ones are less populated than the equilibrium distribution requires). The obtained absorption spectrum contains information about the energy of the transitions and the density of the involved states. Levels that are not electrically active (deep donors and acceptors) can also be identified this way.

Photoluminescence is the counterpart of optical absorption. The sample in nonequilibrium after optical absorption relaxes in a radiative manner emitting light [12]. The output in this case is the spectral dependence of the emergent light.

Cathodoluminescence is similar to photoconduction, but occurs after the semiconductor is exposed to an electronic beam [12]. With the exception of cathodoluminescence, which requires a vacuum, the other optical methods are possible in gaseous ambient at different temperatures.

FTIR (Fourier transform infrared) absorption spectra can be used to investigate the vibrational modes of defects [13–15]. Dispersive spectrometers may be used, but result in increased acquisition time.

Raman spectroscopy utilizes inelastic scattering of monochromatic light in NIR (near infrared), vis (visible), and UV (ultraviolet) spectral ranges [16–18]. The frequency shifts are in the range of the roto-vibrational modes of the defects.

XPS, UPS, and ARUPS (X-ray photoelectron spectroscopy, ultraviolet electron spectroscopy and angle resolved ultraviolet photoelectron spectroscopy) are techniques based on the photoelectric effect [19–24]. They are powerful tools for surface states or bulk states at the surface and can assess both the energy level position and the density of states. However, the characteristics of bulk states can be strongly modified in the surface vicinity so that photoelectron methods do not provide reliable data for bulk defect states. They require high/ultrahigh vacuum conditions and are unsuited for the evaluation of gas sensor materials under operation conditions. Strong ultraviolet sources like synchrotron radiation for UPS and ARUPS enable increased performance and versatility.

Electron paramagnetic resonance (EPR), also known as *electron spin resonance* (ESR) is a spectroscopic technique which allows unpaired electrons either from the CB or from localized levels to be detected [25,26]. It uses the microwave resonant transition of electrons between states whose energy spin degeneration was removed by an external magnetic field (Zeeman Effect). The splitting of the degenerated energy levels, small in respect with the depths of most gap states, is proportional with effective gyromagnetic ratio g connecting the electron magnetic momentum to the Bohr magneton μ_B. From the value of g, relevant information about the electron binding to defects and defect characteristics, for example, concentration of spins, resonance line width and asymmetry, can be inferred. The investigation power increases if the EPR is extended towards ENDOR (electron nuclear double resonance) and ODMR (optically detected magnetic resonance). ODMR opens the possibility to use magnetic resonance for localized levels with paired electronic spins (doubly occupied electronic levels) because the involved optical processes (absorption, fluorescence and phosphorescence) yield unpaired electrons

Muonium spin resonance (μSR) is a technique similar to EPR, but it uses positive muons (leptons with positive charge, ½ spin and 1/9 proton mass with a relatively large lifetime of 2.2 μs) implanted in the investigated material [27,28]. When a muon captures an electron, it forms the muonium complex $Mu = \left[\mu^+ e^-\right]$ which behaves like a lighter hydrogen isotope. It may act as a hydrogen tracer and can overcome the limitations of the standard EPR.

3.1.4.3 Experimental techniques not using electronic state transition

Secondary ion mass spectrometry (SIMS) analyzes the composition of the (secondary) ion beam resulting from sputtering of the material surface with a primary ion beam by means of mass spectrometry [29]. It can be used to identify the extrinsic dopants, mainly when unintentionally present in the sample material.

Rutherford backscattering spectrometry (RBS) employs the changes in direction and the energy of an incident ion beam (often alpha particles $^4_2H^{2+2+}$) to infer the composition, concentration and depth profile of impurities in a sample [30].

X-ray absorption spectroscopy (XAS) techniques, like extended X-ray absorption fine structure (EXAFS) or X-ray absorption near edge structure/near edge X-ray absorption fine structure (XANEX/NEXAFS), exploit the X-ray absorption coefficient dependence on energy to provide information about the adsorbing species: the oxidation state and coordination chemistry (XANEX) or bond length, chemical nature of the neighbors and coordination number (EXAFS) [31,32].

McCluskey and Haller provide more extensive information about the experimental methods used to examine the main properties of dopants and defects in semiconductors [33].

3.1.5 Localized levels in the metal oxides utilized for gas sensors—an overview

3.1.5.1 Intrinsic doping[2]/native doping of SMOXs

During the SMOXs synthesis/growth, stoichiometric and/or structural defects appear out of thermodynamic reasons (Section 3.1.2). Several innovative studies on transport phenomena in SMOXs indicated that the shallow donors responsible for the *n*/*p*-type conductivity are native defects [5]. As the experimental and theoretical

[2] The phrase "intrinsic doping" in this context refers to the changes in the band structure and transport properties of a fully symmetrical (perfect) crystal in the presence of defects which do not involve other type of atoms/ions than the ones building the compound.

methods advanced, justified doubts replaced the initial trust in this approach. Therefore, alternative explanations for the origin of donor states, often with rather limited experimental support, have been suggested. The new explanations, however, do not provide better consistency and scientific consent, so that pertinent models for the intrinsic doping of SMOXs are still missing.

Since the nineteen thirties, the most productive approach to relate localized electronic states in solids to the stoichiometric and structural defects was defect chemistry [1,5,6]. The intrinsic *n*-type character of most SMOXs [34,35] and the dependence of their conductivity on the oxygen partial pressure [36–38] at high temperatures has been known for a long time. They have the generic formula $\left(Me_a^{\frac{2b}{a}+} O_b^{2-}\right)$ and, all of them, except WO_3,[3] the ionic configuration $a \cdot [NG(n-1)] : (n-1)d^{10} \otimes b \cdot [He] : 2s^2 2p^6$ which seems to be a requirement for transparency and conduction in large gap oxides [39]. *Me* represents a metal (mostly transition metals), *n* the period of the metal (its main quantum number) and *NG(n)* the noble gas from the *n*th period (having the main quantum number *n*). The *2p* orbitals of oxygen are formally fully filled in ionic O^{2-} using the electrons from the *s*, *p*, and *d* orbitals of the metal.

Generally, the equilibrium of a SMOX surrounded by an atmosphere containing oxygen is described by quasi-chemical reactions like [40,41] in Kröger-Vink notation:

$$\frac{1}{2}O_2 + V_O^{\bullet\bullet} + 2e' \Leftrightarrow O_O^x \tag{3.3}$$

The enthalpy, however, is unknown. Eq. (3.3) can be divided into three quasi-chemical reactions:

$$\frac{1}{2}O_2 + V_O^x \Leftrightarrow O_O^x \tag{3.4}$$

$$V_O^x - e' \Leftrightarrow V_O^\bullet \quad V_O^\bullet - e' \Leftrightarrow V_O^{\bullet\bullet} \tag{3.5}$$

They assess the virtual physisorption of neutral gaseous oxygen in a neutral oxygen vacancy (Eq. (3.4)) and the two-step oxidation (ionization) of the neutral vacancy (Eq. (3.5)).

For a general reaction of the form $\sum_i \nu_i R_i = 0$, whereby the reactants R_i have positive stoichiometric coefficients ν_i as products and negative ones as educts, one obtains a reaction free enthalpy of [3]:

$$\Delta_r G(T,p) = \Delta_r G^0(T,p) + nRTlnQ \quad Q = \prod_i a_{R,i}^{\nu_i} \quad \Delta_r G^0(T,p) = \text{standard value} \tag{3.6}$$

[3] Tungsten trioxide W^{6+} ion has an equivalent $4f^{14}$ configuration.

with the activities a_i almost equal with the molar fractions x_i in the case of high reactant dilutions ($a_i = \gamma_i \cdot x_i$; γ_i = activity coefficients) [3].

Under constant pressure and temperature, the Eq. (3.6) reaches the equilibrium when the reaction free enthalpy vanishes leading to the well-known mass action law:

$$\prod_i a_{R,i}^{\nu_i} = K = e^{-\frac{\Delta_r G^0}{nRT}} = e^{-\frac{\Delta_r G_m^0}{RT}} = e^{\frac{\Delta_r S_m^0}{R}} \cdot e^{-\frac{\Delta_r H_m^0}{RT}} \tag{3.7}$$

whereby the standard enthalpy and entropy of the process are almost constant over a large range of temperature [42]. In the case described by the defect reaction (3.3) one obtains:

$$K = \prod_i a_{R,i}^{\nu_i} = \frac{a_{O_O^x}^1}{a_{O_2}^{1/2} \cdot a_{V_O^{\bullet\bullet}}^1 \cdot x_{e'}^2} \overset{x_{e'} = 2x_{V_O^{\bullet\bullet}}}{\rightarrow} \frac{1}{\left(p_{O_2}/p^0\right)^{1/2} \cdot 2x_{e'}^3} \tag{3.8}$$

so that $x_{e'} = 2^{1/3} K^{-1/3} \left(p^0\right)^{1/6} p_{O_2}^{-1/6} \propto p_{O_2}^{-1/6}$.

The electron molar fraction is proportional to the electron concentration leading to the observed dependence of the conductance on the oxygen partial pressure.

When the partial pressure of the oxygen is kept constant, Eqs. (3.7) and (3.8) result in:

$$\left(\frac{\partial ln\left[e'\right]}{\partial (T^{-1})}\right)_p = \left(\frac{\partial ln(n)}{\partial (T^{-1})}\right)_p \cong -\frac{\Delta_r H_m^0}{2R} \Rightarrow \sigma = e\mu_n n = e\mu_n \cdot n_0 e^{-\frac{\Delta_r H_m^0}{2RT}} \tag{3.9}$$

if the other small temperature dependencies are neglected. The activation of the conductance immediately follows from Eq. (3.9). A comparison of Eq. (3.9) with the solid state results [43,44], shows the meaning of the involved parameters. The molar enthalpy[4] of the defect formation ($V_O^{\bullet\bullet}$ in this case) corresponds to the energy separation of the donor level from the conduction band maximum (CBM). Therefore, the most popular and used experimental technique for the characterization of native defects was the measurement of the activation enthalpy of the conductance in connection with gravimetric methods to assess the mass/phase changes (thermogravimetric analyses [TGA], thermal desorption [TD], differential thermal analysis [DTA]).

Nevertheless, the dependence of the conductance on the oxygen partial pressure, according to the thermodynamic prediction based on the reaction of the defect formation, is not a certain proof of the defect nature. Several coupled defects—by formation process or by material neutrality condition—could lead to the same conductance dependency (Fig. 3.2).

[4] The enthalpy of defect formation is almost equal with the energy of formation because of the small compressibility of solids. This actually assumes that no mechanical work is involved in the process.

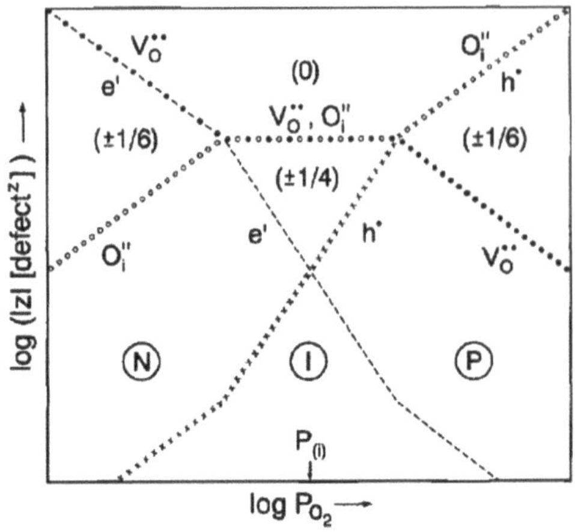

Figure 3.2 The dependence of several defects in a model oxide on the oxygen partial pressure. The standard symbols designate the defects and free carriers in bands. The numbers in round brackets refer to the power of the oxygen partial pressure dependency.
Source: Reprinted with permission from J. Maier, Physical Chemistry of Ionic Materials: Ions and Electrons in Solids, John Wiley & Sons, Ltd, 2004, https://doi.org/10.1002/0470020229.

The choice among possible origins for the shallow donors demands additional criteria. Several representative examples of native defects still accepted as the cause for native conductivity and still employed to model in the field of SMOX gas sensors are given in the following, ordered by materials.

Tin dioxide, SnO_2, $\left([Kr]4d^{10} \otimes 2 \cdot [He]2s^2 2p^6\right)$

SnO_2 remains the most representative gas sensor SMOX

Marley and Dockerty [37] and Samson and Fonstad [45] assessed the enthalpies of the ionization reactions of oxygen vacancies (Eq. (3.5)) from temperature dependent conductance measurements on SnO_2. The obtained values were ~3, 6, and 150 meV. Nagasawa and Shionoya [46] found ~2 meV at 10–130K for the shallow level and Fonstad and Reddiker [47] found 140 meV for the deeper level confirming the previous results. Mizokawa and Nakamura [48] reported activation energies of 100–150 meV and 300 meV, while van Daal [49] only observed a single, much lower level at ~1 meV at 50–700K. He, however, related the result to trace transition metals in his crystals.

These authors considered the V_O as the main free electron source due to the double donor level structure which is in line with the single and double ionization capability of the defect [45]. Additional experimental evidence has reinforced this assumption. For example, a strong EPR signal was recorded on SnO_2 powder even for higher temperatures than that of liquid nitrogen [48]. By comparing the concentration of donors to the one of the spins undergoing the paramagnetic resonance, the authors concluded that the 100–150 meV deep donor is due to a bulk oxygen vacancy while the 300 meV could be an unionized surface oxygen vacancy.

Zinc oxide, ZnO, $([Ar]3d^{10} \otimes [He]2s^2 2p^6)$

Conduction activation energy data for ZnO has been available in literature for a long time [36,50–53]. Different values in the range ~17–45 meV at 100–500K were extracted with the standard Arrhenius function $\sigma = A \cdot e^{(-E_A/k_BT)}$ from the experimental records. The corresponding donor level was interstitial zinc Zn_i and the dopant concentration expressed as cation excess due to lacking oxygen. Bevan et al. [50] observed an increase in the activation energy of the samples as a result of increased temperature and oxygen partial pressure beginning with 38 meV at 470°C at $1.3 \cdot 10^{-4}$ mbar. Later, reports assessed donor levels at ~3 and 61 meV [52,53], in agreement with a hydrogenic model for Zn_i giving ~6 meV below the CBM [53]. The EPR measurements of Kasai [54], conversely, show the deep character of the oxygen vacancy (~0.8 meV from the CBM) and disregard the Zn_i as a shallow donor. From 4K photoluminescence experimental data and some theoretical input (DFT-LDA), Look et al. assigned the ~3 meV/$0.45 \cdot 10^{16}$ cm^{-3} level to Zn_i defect [55]. By annealing ZnO crystals at 1100°C in Zn vapor atmosphere, Halliburton et al. obtained a red ZnO coloration and increased the n-type conductivity [56]. Based on the complete lack of hydrogen during the experiments and EPR data, the authors considered V_O and/or Zn_i as the possible donors and explain the red coloration through the 550 nm blue optical absorption with the participation of the neutral oxygen vacancies near the CBM.

Pöppl and Völkel assigned a small depth to V_O^x (~20–40 meV) and a much larger value to the paramagnetic center $V_O^\bullet$ (~300–450 meV) [57]. They employed EPR and photo EPR in their investigations. Combining EPR, PL and OA, Vanheusden et al. found a strong correlation between the PL peak, the free-carrier concentration, and the concentration of $V_O^\bullet$ in commercial ZnO polycrystalline powders [58]. On the other hand, the difference between the electron concentration $[e'] = 10^{17}$ cm^{-3} and the one of the oxygen single ionized vacancies $[V_O^\bullet] \sim 10^{14}–10^{15}$ cm^{-3} (actually the spin concentration) is significant and the contribution of another shallow donor or paired electron complexes with $V_O^\bullet$ is needed to justify this. Hofmann et al. ascribed the levels at CBM—530 meV and CBM—140 meV to the $(V_O^x \rightarrow V_O^{\bullet\bullet})$ and $(V_O^\bullet \rightarrow V_O^{\bullet\bullet})$ transitions by EPR, ENDOR, DLTS and ODMR [59]. In order to explain the low activation energy (20 meV) in the middle temperature range (~50°C), Miller suggests the formation of zinc interstitial pairs which would also account for the difference between the charge concentration of 10^{15} cm^{-3} and the evaluated interstitial zinc concentration of 10^{18} cm^{-3} [60]. Nevertheless, the model was not proven using adequate experiments. Through EPR measurements a single ionized oxygen vacancy together with a lattice relaxation (Zn_{Zn}^x displaced away from the vacancy) was found [61]. The defect was visible even above room temperature $V_O^\bullet$ [62]. According to Leutwein and Schneider [63], divacancies ($Zn–O$) or even trivacancies ($Zn–O–Zn$) appear under strong irradiation (3 meV electrons or fast neutrons). The antisite defects Zn_O and O_{Zn} produced by Zn and O ion implantation in ZnO appeared to be deep centers involved in photoluminescence [64]. An exhaustive table with references containing EPR data on oxygen vacancies in ZnO can be found in Ref. [65] while Ref. [66] provides a review on all types of defects in ZnO.

Indium oxide, In_2O_3, $(2 \cdot [Kr]4d^{10} \otimes 3 \cdot [He]2s^22p^6)$

In In_2O_3 the generation of ionized oxygen vacancies ($V_O^{\bullet\bullet}$) was seen as a nonstoichiometric dissociation of the semiconductor with electron transfer in the CB (most probably according to the reaction $(2In_{In}^x + 3O_O^x \Leftrightarrow 2In_{In}^x + 3V_O^{\bullet\bullet} + 1.5O_2 + 6e^{'})$ [67–70]. A very small donor activation energy of 10 meV has been observed [67]. Rupprecht, however, had to write a different defect equilibrium equation $(In_2O_3 \Leftrightarrow 2In_i^x + 1.5O_2 + 6e^{'})$ which gives the role of the donor to the interstitial indium ion in order to explain the partial pressure dependence of the conductance he observed [71]. The findings of Rosenberg were in accordance [72].

Gallium oxide, Ga_2O_3, $(2 \cdot [Ar]4d^{10} \otimes 3 \cdot [He]2s^22p^6)$

Several articles refer to the dissociative oxygen loss from Ga_2O_3 with the related stoichiometric defects ($V_O^{\bullet\bullet}$, $Ga_i^{3\bullet}$) and donor level formation [73–75]. The reported activation energy of the donor depends on the temperature with reported values between 0.28 and 1.3 eV (mainly in middle and high temperature ranges). Irmscher et al. found evidence of unintentionally doped (UID) Ga_2O_3 single crystals for a shallow level whose depths depends on the electron concentration ($\sim$17 meV/ $6 \cdot 10^{17}$ cm^{-3} and $\sim$36 meV for the concentration extrapolated to 0) [76]. An accidental contamination of the samples with Si, detected by SIMS, could be responsible for a residual electron concentration of $7 \cdot 10^{16}$ cm^{-3}. EPR investigations also revealed an unintentional impurification with Fe^{3+} and Co^{3+} which might be partly responsible for the three deep electron traps visible in DLTS spectra. By means of the same technique (EPR), Kümmerer et al. identified oxygen vacancies in Ga_2O_3 and provided evidence for a nonuniform defect distribution or clustering [77].

Tungsten trioxide, WO_3, $([Xe]4f^{14} \otimes 3 \cdot [He]2s^2p^6)$

WO_3 shows a large spectrum of activation energies, mainly depending on stoichiometry but also on material structure/morphology. The values recorded for single crystals in monoclinic phase γ, stable at and above room temperature (17–350°C) are spread between 10 and 50 meV [78,79] while for thin films annealed at different oxygen partial pressures even more $\sim$0–480 meV, [80,81]. The origin of the donor level is under debate. Both interstitial cations [78] or oxygen vacancies [81], possibly merging together into Anderson–Hyde disks of shear [82] at higher concentrations [79] might lead to the donor level. All of them are related, however, to stoichiometric defects. The spatially resolved scanning tunneling spectroscopy on polycrystalline WO_3 revealed the very low conductance in the crystallite center with respect to the boundary suggesting a complete elimination of oxygen vacancies from the bulk [83]. An inversion layer is not considered.

Titanium dioxide, TiO_2, $([Ar]3s^23p^6 \otimes 2[He]2s^22p^6)$

TiO_2 has been thoroughly investigated as rutile single crystal. Conductance dependencies on the oxygen partial pressure are expected with the powers $-1/4$ and $-1/6$, but also $-1/5$. The first two powers have been mainly related to the $V_0^{\bullet}$ and $V_0^{\bullet\bullet}$ ionized states of the oxygen vacancy [84,85], in agreement with Eqs. (3.4) and (3.5). For the power $-1/5$, there are several competing models. Yahia [86] suggests the interstitial Ti_i^x as an electron donor in low oxygen partial pressures

($p_{O_2} < 12$ mbar), while Kevane [84] considers a mixture of different charge states of oxygen vacancies ($V_0^{\bullet}$ and $V_0^{\bullet\bullet}$) could lead to the same effect in the intermediate pressure range. Even the power $-1/6$ has several explanations, being attributed either to $V_0^{\bullet}$ [86], or to the more complex defect $\left[V_0^{\bullet\bullet}{\cdot}2\left(Ti_i^{3+}\right)'\right]^x$ [87]. Different activation energies have been measured and some have been attributed to certain donor defects: 1–18 meV to $\left[V_0^{\bullet\bullet}{\cdot}2\left(Ti_i^{3+}\right)'\right]^x$ and 110–250 meV to $V_0^{\bullet\bullet}$ [87], 325 meV and 515 meV to $V_0^{\bullet}$ and $V_0^{\bullet\bullet}$ [85], and 30–60 meV to "oxygen vacancies" [88]. Hasiguti and Yagi [89] and later Yagi et al. [90] estimated ~1 meV and 30–50 meV over three temperature ranges $T<4$K, 4–110K and 210–250K for interstitial Ti from EPR data. Nowotny provides a diagram of the intrinsic levels in TiO_2 on the basis of previous investigations [91]. $V_0^{\bullet\bullet}$ and $Ti_i^{\bullet\bullet\bullet}$ reportedly lay at 1.18 eV and ~1.5 eV below the CBM while the acceptor level of V''''_{Ti} is at 1.15 eV above the maximum of the VB (VBM). In the case of the anatase phase, the low activation energy of 4 meV has been linked to oxygen deficiencies but has not been assigned to a given defect [92]. Deeper levels of 100–200 meV in the same phase were attributed to oxygen vacancies [93].

Ternary and quaternary oxides

In ternary or quaternary compounds like ITO ($In_2O_3{:}Sn$), GITO ($In_2O_3{:}Sn,Ga$) or ZITO ($In_2O_3{:}Sn,Zn$) the extrinsic cation leads to the formation of complex donor centers, for example, the Frank-Köstlin defect $\left(2Sn_{In}^{\bullet}O_i''\right)^x$ [94–97]. The quasi-chemical reaction of this defect with oxygen $\left(\left(2Sn_{In}^{\bullet}O_i''\right)^x \Leftrightarrow \frac{1}{2}O_2 + 2Sn_{In}^{\bullet} + 2e'\right)$ requires a dependence of the free electron concentration $\left[e'\right]$ on the oxygen partial pressure at the power $-1/8$ and has been found experimentally. Ref. [98] considers the donor level in ternary compounds in the system $SnO_2{-}In_2O_3{-}CdO$ as a result of oxygen vacancies, cation substitutions (with higher valence ones) and cation interstitials.

p-type SMOXs

Although not yet as common in gas sensing as *n*-type, the *p*-type SMOXs have become interesting for this application over the last decades. The generic formula $\left(Me_a^{\frac{2b}{a}}O_b^{2-}\right)$ is the same, but the configuration $a \cdot [NG(n-1)] : (n-1)d^y \otimes b \cdot [He] : 2s^2 2p^6$ includes different occupation degrees y of the d orbitals. This results in a larger variation of their properties. The gap is smaller and optical processes often result in coloration or strong light absorption (black "color").

$$Cr_2O_3, \left(2{\cdot}\,[Ar]\,3d^3 \otimes 3{\cdot}\,[He]\,2s^2 2p^6\right)$$

Cr_2O_3 could present both the n and p native conduction type [99]. In the chromium deficit case, the defect reaction would be: $\frac{3}{2}O_2 \Leftrightarrow 2V'''_{Cr} + 6h^{\bullet} + 3O_O^x$ with a hole concentration proportional to the power $3/16$ of the oxygen partial pressure. In the oxygen-poor case, the reaction $2Cr_{Cr}^x + 3O_O^x \Leftrightarrow 2Cr_i^{\bullet\bullet\bullet} + 6e' + \frac{3}{2}O_2$ could describe the equilibrium with chromium interstitials as a shallow donor and a dependence on the power $-3/16$ of the oxygen partial pressure. Even the possibility of a "standard" oxygen vacancy donor through reaction 3 has been considered. A dependence on the oxygen partial pressure with the power of $-1/6$ is expected. Hagel obtained an oxygen partial pressure power $1.7/16$, which led him to believe that his samples have chromium vacancies as acceptor levels [100]. He found an activation energy of ~100 meV at 300–500°C.

Holt and Kofstad [101] reported an even lower power for the oxygen partial pressure dependency (1/20) of the conductivity which they related to the slow kinetics (i.e., incomplete equilibrium) of the defect reactions in the middle temperature range.

Fe_2O_3, $\left(2[Ar]3d^5 \otimes 3{\cdot}[He]2s^2 2p^6\right)$

For polycrystalline Fe_2O_3 Gardner et al. measured an activation energy of ~100 meV below 1000°C with an independent carrier concentration on the oxygen partial pressure [102]. Below 450°C the activation energy increased to 700 meV upon oxidation of the grain boundary [102]. The same values were recorded on filtered arc deposition layers of high purity Fe_2O_3 by Glasscock et al. [103]. On natural hematite crystals, an intrinsic conduction above 1000°C has been reported [104]. Moreover, the diffusion coefficient of the ^{55}Fe iron tracer increases when the oxygen partial pressure in the ambience decreases. These results have been confirmed by Hoshino and Peterson [105] and Atkinson and Taylor [106]. Hoshino and Peterson [105] additionally observed an extrinsic influence (from unintentionally present *Co*, *Cr* and *Y*, mainly at lower temperatures) and an easier displacement of the Fe^{2+} ions in interstitial positions with respect to Fe^{3+}. It was also suggested that the defects responsible for iron and oxygen diffusion are iron interstitials and oxygen vacancies [106] with a dependence on the power $-1/2$ of the oxygen activity (proportional to the partial pressure) [107]. Ling et al. ascribed the increase in the photocurrent and photo capacitance of the monocrystalline Fe_2O_3 samples prepared in an oxygen-poor atmosphere to the generation of oxygen vacancies [108]. As result of its relatively small gap, the intrinsic conduction becomes relevant at high temperatures.

Co_3O_4, $\left([Ar]3d^7 \otimes 2{\cdot}[Ar]3d^6 \otimes 4{\cdot}[He]2s^2 2p^6\right)$

Cheng et al. assigned an acceptor level at ~300 meV above the VBM which led to a polaronic hopping of holes with a measured activation energy of ~150 meV at 220–400K [109].

CuO, $\left([Ar]3d^9 \otimes 4{\cdot}[He]2s^2 2p^6\right)$

UID *CuO* showed an acceptor level at 120 meV above the VBM whose nature has not been assessed, but seems to be related to the nonstoichiometry of the material (its ability to form Cu_2O) [110].

The above overview shows that a large number of authors found oxygen vacancies and cation interstitials to be the main candidates for the donor role in SMOXs. They assumed $V_O^{\bullet\bullet}$ and $Me_i^{\bullet\bullet\bullet}$ (*Me* being the metal providing the compound cation) as the most probable charge states. Several intrinsic defect complexes have also been considered, but to a lower extent.

3.1.5.2 Extrinsic doping of SMOX

Unintentional/accidental extrinsic doping

During their preparation or characterization, the SMOXs come in contact with gases other than oxygen, often at high pressures or/and temperatures. Many researchers focusing on certain topics (like the behavior of SMOXs under reducing conditions) disregarded the role of foreign gaseous species in the compound as extrinsic doping, most commonly hydrogen (as reducing agent or coming from water vapors) and nitrogen (usually from air). There are, however, several significant reports addressing this topic.

In their significant article dedicated to "Defect structure and electronic donor levels in stannic oxide crystals," Samson and Fonstad elaborated on the risk of hydrogen contamination during the annealing of UID samples in gaseous ambience and performed a controlled exposure to H_2 partial pressures [45]. They observed a new donor with an activation energy of 50 meV. Watanabe et al. analyzed the hydrogenation of SnO_2:*In* as a result of annealing in deuterated water (^2H_2O) vapor dosed in pure argon [111]. They found an increased electron concentration indicating that the inserted deuterium plays a donor role, but only in the specimens strongly doped with indium. Though correlated, the electron concentration was two orders of magnitude less than that of the hydrogen tracer, indicating a complex compensation mechanism.

Hofmann et al. obtained two very clear EPR resonances in UID ZnO. Based on ENDOR data the strongest EPR line was attributed to a shallow hydrogen donor [112] which has been reconfirmed by subsequent research [59]. Anatase TiO_2 reduction in hydrogen changed the depth of the level observed in oxygen deficit from 100–200 to 70 meV [93]. However, the authors considered this shift an error in the evaluation of the defect reaction enthalpy and not a hydrogen induced level. In another experiment a few percent of hydrogen added to the sputtering atmosphere of ITO (In_2O_3:*10at.%Sn*) almost doubled the electron concentration in the specimen. Nevertheless, the possible formation of a hydrogen donor was not addressed [113]. The filling with hydrogen of the surface oxygen vacancies by thermal treatment is ascertained in Ref. [114].

Referred to as protium, the first and most abundant isotope of hydrogen (^1_1H) has been assessed as a shallow donor (CBM—60 meV) in ZnO by μSR [115,116]. Analogous μSR data have been reported for SnO_2 with a donor level at CBM—(18–36 meV), In_2O_3 at CBM—(47–94 meV) [117] and Ga_2O_3 at CBM—(15–30 meV) [118]. The spread in the μSR experimental data is due to the difficulty in attaining equilibrium during the short lifetime of the muons

(2.2 μs). The depth of the donor level is in good agreement with the rough hydrogenic model predictions.

The optical methods also provided evidence on the role played by hydrogen related states in the conductivity of large gap SMOXs. In combined photoconductivity and optical absorption measurements on *ZnO* two photoconductivity lines at 180 and 240 cm^{-1}, respectively, 22 meV and 30 meV have been observed and attributed to hydrogen related shallow donors [119]. Two donor levels at 53 and 47 meV due to bond-centered hydrogen H_{BC} and to anion substituted hydrogen H_O were identified using complementary Raman scattering and photoluminescence investigations of hydrogen impurified *ZnO* [120]. H_{BC} stability above 190°C remains questionable.

UID Cr_2O_3 samples at several hundred degrees Celsius in both hydrogen and water vapor showed increased conductance when compared to a dry, hydrogen free and oxygen containing atmosphere. Moreover, a strong dependence of the conductance on the partial pressure of both H_2 and O_2 occurred, although the samples had a very weak dependence on the O_2 partial pressure when exposed to this gas alone. The authors concluded a conduction dominated by point defects, the kinetics of defect reactions being poor in the absence of hydrogen [101]. McCluskey et al. provide a thorough overview of hydrogen in oxide semiconductors [121].

N^+ implantation led to an increased electron concentration in *ZnO* and to the emergence of a new relatively shallow donor level at 120 meV in DLTS spectra supposed to be connected to V_O [122]. The additional donor level at 16–22 meV (detected by temperature dependent Hall measurements) and the electron trap at 310 meV (DLTS) are presumably related by Brauer et al. to the first and second ionization states of Zn_i. Look et al. consider that in nitrogen implanted *ZnO*, the Zn_i shallow donor completely transforms in a complex defect $Zn_i{-}N_O$ which is also a shallow donor and the accidental contribution of the $H{-}N$ pairs can be disregarded because this defect would be neutral [55]. Excitation photoluminescence spectra on a *ZnO:N* single crystal revealed the deep acceptor character of N_O whose charge transition state lays at 1.45 eV above the VBM [123]. The expected shallow acceptor role of an anion vacancy with less valence electrons appeared as practically unrealizable.

Annealing of ITO (In_2O_3*:10at.%Sn*) in nitrogen lead to increased conductivity (however, similar with the one obtained by annealing in air or oxygen) so that no indication of a nitrogen related state could be inferred [124].

Other residual impurities like *In*, *Al* and *Ga* accidentally included in *ZnO* crystals during manufacturing have been put in evidence by EPR and ENDOR [59,112].

Intentional extrinsic doping

Intentionally impurified SMOXs with foreign elements are currently used to increase the conductivity of transparent electrodes and sometimes that of gas sensor materials (here the catalytic additives are disregarded). Because the origin of the dopant is known in this case, the related investigations mainly address the transport parameters and the depth and nature of the local states (i.e., if they are due to simple point extrinsic defects or complexes, which bind the intrinsic defects with the impurifying elements).

$SnO_2:Sn$ is often addressed in literature with Sb_{Sn} being a preferred substitutional donor. Antimony doping leads to high electron concentrations of $\sim 3 \cdot 10^{20}\,\text{cm}^{-3}$ with roughly a 100 percent doping efficiency because of a low activation energy (27 meV) [125]. The induced local level is $\sim$30 meV below the CBM [47,125]. Indirect information concerning the location of *Sb* in the SnO_2 gap has been obtained from the upwards shift of the Fermi level from 400 meV below the CBM in the nominally pure material to $\sim k_BT$ in the *Sb* doped one [126]. This value is consistent with earlier reported ones (35 meV) [47]. The properties of the intentionally impurified *ZnO* detected by EPR and SIMS are thoroughly addressed by Ref. [127].

The properties of the intentionally impurified Fe_2O_3 and Cr_2O_3 with transitions metals (*Ti*, *Ni*, *Cr*, *Mn*, *V*, *Mg* and *Ni*) through powder mixing and firing have been evaluated from activation energy measurements by de Cogan and Lonergan [128] and by Kim et al. [129] for $Fe_2O_3:Cd$. Hagel reports on $Cr_2O_3:Li$ [100].

Insertion of foreign metals in SMOXs can result not only in additional doping, but also in charge carrier compensation. For example deep *Al* and *Pd* centers in SnO_2 shift the Fermi level from 80 meV below the CBM for the undoped material to respectively 270 meV and 290 meV below the CBM for the impurified ones [130]. The same authors report even larger shifts for *Pt* in SnO_2. WO_3 is sometimes alloyed with alkali metals or *H* towards (*Li*, *Na*, *K*)WO_3 or HWO_3 bronzes. They have evident metallic character and good conductivity [131].

3.1.5.3 Alternative causes of the native conduction of SMOXs

Several investigators found additional reasons for the native conduction of SMOXs, which, according to their experimental evidence, could explain the observed differences between the charge carrier and donor/acceptor concentrations.

The surface/interface accumulation layers

Through XPS, IR reflection spectra and Hall Effect experiments on UID In_2O_3 completed by theoretical evaluations, King et al. detected an (for large gap SMOXs) unusual inversion layer at the material surface [132,133]. This occurrence was explained by the low location on the energy scale of the CBM around the Γ point, below the branching point (situated at a higher energy), as follows from the recipe of Tersoff [134]:

$$E_{BP} \approx \frac{1}{2}\left(\langle E_V \rangle + \left\langle E_C^{\text{indirect}} \right\rangle\right)$$

Here, $\left\langle E_C^{\text{indirect}} \right\rangle$ is the average of the indirect CB minimum. The doping with tin induces semiconductor degeneration and the Fermi level is pushed up above branching point and well above the CBM resulting in known upwards band bending at the surface. The surface accumulation layer can provide a low resistivity channel for electronic conduction, independent of the nature and depth of the bulk local levels circumventing the lack of native suitable doping. Nagata et al. brought PES proves of a surface accumulation layer and Fermi level pinning inside the CB in antimony doped/unintentionally doped SnO_2 thin films [135]. Increased conductivity has been reported also for the interface sheets at the SMOX/substrate junction [136].

3.1.6 Discrepancies and open questions emerging from the experimental investigation of defects in SMOXs

3.1.6.1 Inadvertences in the experimental data

Experimental investigations of the localized states made a better understanding of the role they play in the conduction and other relevant processes (photoconduction, light absorption, gas sensing) of SMOXs possible. For the extrinsic materials, the link between the impurity and induced defect(s) is rather straightforward, even though complex centers could appear. The intrinsic SMOXs, on the contrary, raised great difficulties in the interpretation of the recorded data because of multiple less directly connected technological parameters controlling the compound synthesis/ deposition/ growth and the nature of the detected local levels. The main concerns arise from:

1. The discrepancies (sometimes very large) between the concentration of the defects and that of the conduction electrons.
2. Different assignations given by different authors to the leading defect in similar samples of the same material.
3. Large differences between the energy levels and other parameters allocated to the same defect across the literature. For example oxygen vacancies accounted for as shallow donors in activation energy measurements appeared to be deep centers in EPR and ENDOR [54,57,59].
4. The defects' degree of stability in the lattice under temperature gradients and applied electrical fields (very critical for the gas sensors, which, in principle, have to operate under such conditions).

3.1.6.2 Principle scientific queries

Nearby the above underlined inadvertencies, several general scientific questions still remain open and have for a long time [137]:

1. Why should V_O result in deep levels in insulators (alkali halides, SiO_2) and "standard" semiconductors (Si, $GaAs$), but would form shallow levels in SMOXs (actually SnO_2 in the referred article)?
2. The oxides of main group elements (CaO, MgO, SiO_2) are stoichiometric. Why should the SMOXs not be stoichiometric? Especially SnO_2, In_2O_3 and Ga_2O_3, which do not contain transition metal cations. Is this related to the multivalence of metals?
3. The location of the Fermi level inside the CB would allow for optical transitions from the bottom to upper CB states, which will render the material opaque. Surprisingly, electron-rich SMOXs are still highly transparent in the visible range (*ITO* and SnO_2 being the most

employed transparent electrodes) even though they are degenerated and the Fermi level lies inside the CB.

3.1.6.3 The origin of the discrepancies among experimental parameters relevant for the conduction of SMOXs

The disagreements among the experimental data overviewed in Section 3.1.4 (and listed in Section 3.1.5.1) deserve careful consideration. Going beyond the inherent statistical spread in material and sample preparation or the accuracy of the investigation tools, some principle reasons can be identified.

The thermodynamic equilibrium/nonequilibrium conditions under which the experimental method operates

There is a huge amount of data acquired in thermodynamic equilibrium of the semiconductor with one of its components in pure phase (gas for oxygen and solid state usually for metals), as described in Section 3.1.4. Because this type of equilibrium seems only to occur for temperatures higher than half of the compound melting point [50] the values inferred in thermodynamic equilibrium measurements cannot be extrapolated at lower temperatures without risking large errors. In many cases, such an extrapolation could be completely erroneous. For instance, a Zn_i–N_O complex defect has a weight of $[Zn_i]/[Zn_i - N_O] \sim 10^2$ at 950°C, but only $[Zn_i]/[Zn_i - N_O] \sim 1$ at 500°C and $[Zn_i]/[Zn_i - N_O] \sim 0$ at <250°C [55].

Therefore, evaluating the donor/acceptor depth from the enthalpy of the quasichemical reactions in dependence on the oxygen partial pressure is inadequate for almost all current applications of SMOXs operating at room temperature or up to one third of the melting temperature for most of the gas sensors.

The energy levels extracted from spectroscopic methods might show two types of disagreements when compared with those acquired under equilibrium conditions. The first one, shortly addressed in Section 3.1.3 and resumed in more detail in Sections 3.1.7.2 and "Comparing the experimental data from nonequilibrium measurements with the theoretical simulations", is related to the nonequilibrium character of such estimations. One encounters a second issue for the techniques requiring low temperatures (like EPR) because, again, the extrapolation to room temperature might be wrong. This topic will be included as well in the discussions on the theoretical approaches.

The erroneous identification of the evaluated defect in indirect experimental methods

Both equilibrium and nonequilibrium experimental methods often use indirect identification of the defects. For instance, "activation energy" measurements relay on the simplistic assumption that the defect identity is unequivocally provided by the quasichemical reaction (the dependence of the molar fraction/concentration of defects/free charge carriers on the partial pressure of oxygen, for example—as inferred in Section 3.1.4). There is, however, no direct evidence for this assumption. A decrease in the oxygen content of the compound can either result from an increased concentration of oxygen vacancies or through an increase in the interstitial cation concentration. Additionally, if the oxygen partial pressure is controlled through dosing with hydrogen or nitrogen, unintentional doping can occur leading to similar consequences. The situation is similar for spectroscopic approaches. The methods based on electronic state transition, like OA and PL, only require a simple resonance (a fit between energy exchanged by the employed radiation with the energy of the electronic transition of the defect) to evaluate the dopant depths in the gap. In certain cases, recognition can be improved if the peculiarities of the spectrum related to the defect configuration or to its environment (vibrational bands, magnetic vicinity and additional interactions) are considered.

The accidental coincidence of the native energy level with the energy of an extrinsic center

This type of error is less frequent, but cannot be excluded. It is the case, for example, for SnO_2:Sb, where the extrinsic donor level of Sb, located at 35 meV below the CBM [47], almost overlaps with the native donor level at 30 meV below the CBM, assigned by Samson and Fonstad to $V_O^{\bullet}$ [45]. The contradictory experimental data will be considered later in the context of results from theoretical simulations. A good review, which also includes the theoretical approaches, is available in Ref. [138].

3.1.7 Theoretical approaches to the nature and level position of localized states in SMOXs

3.1.7.1 General features of the theoretical approaches employed in the simulation of the SMOXs' defects

Theoretical methods can provide the thermodynamic, electronic, and optical parameters of artificially built defects, making them indispensable in understanding complex situations. Simplistically, theoretical approaches simulate an ideal neutral crystal in which a certain defect is "introduced." To validate the method and to identify necessary corrections, the equilibrium crystal lattice constants and the gap are then evaluated and compared to experimental values. Usually, the band structure and the density of states are provided. The defect formation energy/free enthalpy is computed as the difference between the values for the ideal crystal and for the defective one. The equilibrium with two reservoirs of particles at constant[5] chemical potentials (typically for gaseous oxygen and for the solid metal) is assumed. Through the chemical potential values, the equilibrium can be shifted towards oxygen-rich or to metal-rich "preparation conditions." In addition, the electron exchange with an electron reservoir having the electrochemical potential (Fermi level) $E_{F,R}$ is ensured. $E_{F,R}$ is swept from the VBM to the CBM and the defect formation energy/free enthalpy is displayed across the gap. The defect formation energy/free enthalpy is linear in $E_{F,R}$ and its slope is proportional to the charge state of the defect. The charge transition points give the limits of the defect

[5] The compound formation reaction and other "chemical" conditions may introduce additional restrictions on the available values of the chemical potential for one of the compound constituents.

stability in a given charge state. The computed values of the defect diffusion barrier and the associated vibration modes can be compared to experimentally obtained values. Compact descriptions of the computation paths can be found in [65,139] as well as in many of the theoretical contributions cited. From [65], a typical presentation of the results has been reproduced in Fig. 3.3. An exhaustive and pertinent overview concerning the "first-principles calculations for point defects in solids" can be found in Ref. [140]. Because of an optimal ratio between the computing volume (cost and time) and numerical accuracy, the most common theoretical approaches are based on density functional theory (DFT) [141–144]. It is implemented in the standard local density approximation (LDA) and other approximation variants, such as generalized gradient approximation (GGA) [145–147], Meta-GGA (including also the second derivative of the density) [147], and hybrid functional [148–151]. Many electrons methods like Hartree-Fock (HF) with configuration interaction (CI) or many body perturbation theory (MBPT), as well as random algorithms based on Monte-Carlo simulations [152] are, in principle, better suited for solid-state problems, but their use is prohibited by the large demand of computing resources. Nevertheless, owning its relative simplicity, DFT is extensively employed to predict or explain the behavior of defects in SMOXs, appealing to already consecrated functionals (B3PW91 [153] and B3LYP [153,154]) and codes ([155,156] www.vasp.at, http://dft.sandia.gov/Quest/DFT_codes.html). The next section addresses the most significant theoretical results in the field of SMOXs defects.

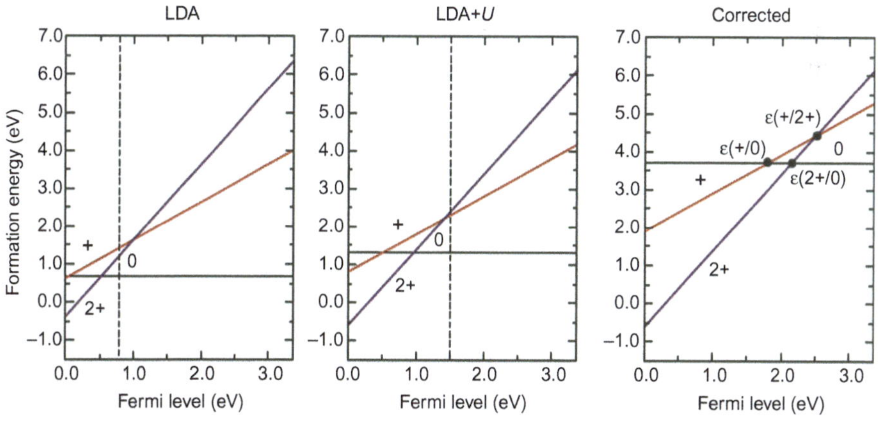

Figure 3.3 Calculated formation energies as a function of the Fermi level for the oxygen vacancy in ZnO: LDA results, LDA + U results, and extrapolated formation energies. The zero value of the electron reservoir Fermi level corresponds to the valence-band maximum and the dashed lines in LDA and LDA + U panels indicate the band gap in the respective approximation.

Source: Reprinted with permission from A. Janotti, C.G. Van de Walle, Native point defects in ZnO, Phys. Rev. B. 76 (2007) 165202, doi:10.1103/PhysRevB.76.165202.

3.1.7.2 Theoretical approaches to intrinsic defect levels in SMOXs

n-type, large gap SMOXs

The common/comparable configuration (with $3d^{10}/4d^{10}/4f^{14}$ metal last occupied shell) and similar properties of this class of compounds enabled a relatively unitary theoretical approach for large gap SMOXs. Most of the authors used "ab initio" methods based on DFT and its implementations (LDA, LDA + U, GGA). The basic goal was to estimate the character, the formation energy and the location in the gap of the states induced by the main point defects. The following is an overview of the literature outcomes.

V_O: In most cases, V_O was found to result in a deep level, which does not provide free electrons in the CB. However, there is no consensus regarding its other properties. Using DFT-LDA/GGA-EC, Lany et al. report the formation of a deep acceptor in SnO_2 [157].

Based on DFT-GGA + U-VASP, Singh et al. found that V_O is a deep donor (1.8 eV below the CBM) in SnO_2 [158]. The defect location has been confirmed by Janotti and Van de Walle (DFT-LDA/LDA + U/EC-VASP) [65,159,160], and by Oba et al. [161]. The V_O is evaluated to be a deep center in SnO_2 at 1 eV above the VBM in the bulk [162]. With DFT-GGA, Tanaka et al. calculated the formation energies of the oxygen vacancies in all three charge states for SnO_2, ZnO and In_2O_3 [163]. Although the gap errors have not been corrected, the trend of oxygen vacancies to form deep centers in the gap, with the exception of SnO_2 in fluorite phase where they are shallow acceptors or VB states, becomes visible. These numerical estimations were made under O−rich conditions. In TiO_2, the oxygen vacancy results in a level at 1−2 eV from the VBM (simulation made on O−poor Ti_8O_{15} with LCAO-HF) [164]. The V_O states in TiO_2 are both in the middle of the gap for the rutile phase, but in anatase they are far away from each other, one is just under the CBM (DFTLSDA-GGA + U /PBW functional) [165]. Reunchan et al. identified a donor level at 200 meV below the CBM, but could not determine if it is in fact shallow (will provide electrons in the CB) or deep (remaining unionized) [166].

Some earlier theoretical studies based on less performant models and algorithms predict a shallow donor role for V_O. Kilic and Zunger assessed a relatively shallow level ($E_{V_O^{2+/0}} = CBM - 114 meV$) in SnO_2 with a low formation energy as a result of loosely bound electrons [137]. However, their DFT-LDA data lacked the gap correction (theoretical E_g ~1.2 eV) meaning that the actual level depth could be up to 300 meV below the CBM as Fig. 3.3 suggests. Using a similar formalism, which provides a theoretical gap of only E_g ~0.95 eV, Oba et al. reported a shallow donor character for V_O in ZnO [167]. Based on tight-binding Green's-function calculations, Robertson found that the $V_O^{\bullet}$ is a shallow level in SnO_2, in agreement with the activation energy based measurements [168]. Rantala et al. showed, using DFT-LDA-LCAO, that the oxygen vacancies in SnO_2 are neutral defects at and near the surface and only have donor character in the bulk [169].

Me_i: Cation interstitials have low concentrations in In_2O_3, SnO_2 and ZnO [157]. This stands in contradiction with the Refs. [170] and [137], which report the metal interstitial as main donor. Based on DFT + LDA, Kilic et al. predicted a $Sn_i^{4\bullet}$ level inside the CB ($E_{Sn_i^{4+/3+}} = CBM + 205meV$) [137]. According to Singh et al., $Sn_i^{4\bullet}$ is a shallow mobile donor in SnO_2, but because the ionization enthalpy is very high ($\Delta_f H_{Sn_i^{4\bullet}} = 12$ eV), its concentration is very low (cannot exceed the equilibrium one) [158]. Comparable results were obtained for Zn_i in ZnO which should be a shallow donor with an increased formation energy [65,160,161] that can diffuse easily [65,160]. Sn_O in SnO_2 is similar to $Sn_i^{4\bullet}$ [158].

V_{Me} and O_i: According to Janotti and Van de Valle, the metal vacancies and oxygen interstitials are deep acceptors in ZnO which can compensate shallow donors [65,160]. Oba et al. concluded the same for V_{Zn} but did not evaluate O_i [161]. In In_2O_3, V_{In} was found to be an acceptor up to the charge state 3 + with a large formation energy [166].

Me_O and O_{Me}: The antisite defect O_{Zn} is deep within the gap and does not provide charge carriers [65,160]. Zn_O lies in the CB of ZnO but its formation enthalpy is so high that no significant electron concentration will result [65].

The attractive interaction of constitutive defects results in a low formation energy for the Intrinsic Complex $Sn_i + V_O$ [137].

It has been shown, using DFT-LDSA/VASP, that the coexistence of V_O and In_i leads to shallow donors [171]. The defects were also found to influence the conduction, which stands in contrast to other studies [157].

If they act as intrinsic electron producers, V_O and cation interstitials (for example, In_i and Zn_i) do not lead to more than 10^{14} cm^{-3} free electrons at room temperature. Therefore, the previous assumption fails to explain the experimentally observed electron concentrations of $\sim 10^{18}$ cm^{-3} [157].

p-Type, middle and small gap SMOXs

Theoretical approaches resulted in a larger variation for *p*-type than for *n*-type SMOXs, making a common feature extraction difficult. Therefore, each compound will be addressed separately.

Fe_2O_3: Lee and Han analyzed (DFT,GGA + U) the point defects in Fe_2O_3 [172]. They found Fe_i and V_{Fe} to be shallow level dopants (donor, respectively acceptor) and were able to reproduce the high temperature behavior of the material in oxygen ambience established experimentally [107]. V_O was seen as neutral and not involved in the conduction.

Cr_2O_3: Lebreau et al. considered the characteristic point defects for Cr_2O_3, namely the chrome vacancy V_{Cr}, chrome Frenkel defect V_{Cr} and Cr_i, and the oxygen vacancy V_O, using DFT + U formalism [173]. The formation energy of V_{Cr} and Cr_i is the lowest ($\sim$2.5 eV), followed in order by V_{Cr} and V_O (both about 5 eV). The defect stability has the reversed order. V_{Cr} results in a level at 0.3 eV above the Fermi level, while the chrome vacancy from the Frenkel defect (V_{Cr} and Cr_i) induces a level at 0.5 eV below the Fermi level. V_O does not yield a localized state within the gap.

CuO: Among the native defects of CuO the most stable charge states in a *p*-type material are, according to Wu et al., O''_{Cu} and V''_{Cu} [174]. V''_{Cu} has the lowest formation enthalpy because of its maximal negative charge even under *n*-type material conditions. The authors attribute the native *p*-character of *CuO* to V''_{Cu}. Roughly the same conclusion emerged from a reference dedicated to extrinsic doping [175].

The efforts to confirm the experimentally assessed origin of the *n*-type conductivity in large gap SMOXs via theoretical methods was largely unsuccessful. For instance, Lany and Zunger examined (DFT-GGA-EC) the cause of *n*-tape conductivity presented by such materials (*ZnO* and In_2O_3) but found no gap levels associated with native defects able to provide the electron concentrations delivered by measurements [157]. Theoretically, the oxygen vacancies, indicated as shallow donors by the activation energy experiments in reducing ambience at high temperature, appeared as deep centers in the gap. An alternative explanation is required for the experimental results.

3.1.7.3 The persistent photoconductivity as source of native conduction in SMOXs

One possible, rather exotic, reason for native conduction, suggested in [176], could be "persistent photoconductivity" (PPC) after illumination. Using DFT-LDA, Zhang et al. show that PPC appears in *ZnO* due to the localized states of the oxygen vacancy [170]. Lany and Zunger confirmed the work of Zhang et al. They dedicated a whole investigation using DFT-LDA to examine oxygen vacancies in II-VI (e.g., *ZnO*) and chalcopyrite semiconductors [176]. They additionally found that V_O behaves like a DX-center. DX-centers are complexes of an extrinsic donor D and an unknown intrinsic defect X, normally encountered in III-V semiconductors [177]. According to Lany and Zunger [176] (see Fig. 3.4), in II-VI semiconductors under thermal equilibrium, the V_O has a deep stable "defect localized state" (DLS) in the gap (type α). If the material is illuminated the charge state of the oxygen vacancy changes from V_O^0 to $V_O^{\bullet\bullet}$ and theoretically lies within the CB as DLS type

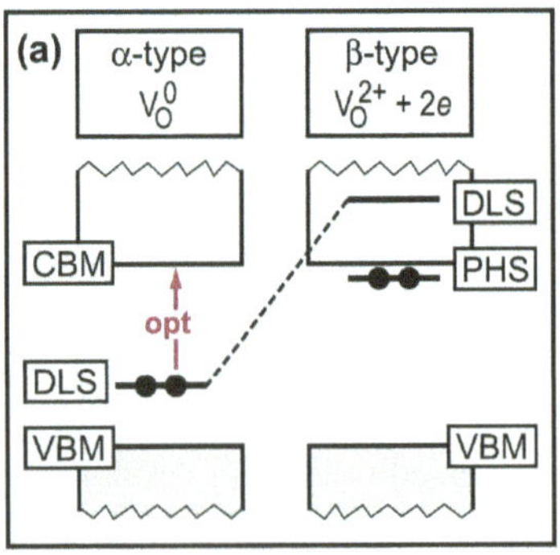

Figure 3.4 The localized states of the oxygen vacancy in ZnO without (left panel side) and during/after illumination (right panel side).
Source: Reprinted with permission from S. Lany, A. Zunger, Anion vacancies as a source of persistent photoconductivity in II–VI and chalcopyrite semiconductors, Phys. Rev. B. 72 (2005) 35215, doi:10.1103/PhysRevB.72.035215.

β, in agreement with the defect reaction $V_O \Leftrightarrow V_O^{\bullet\bullet} + 2e^{'}$. The electron associated to the resonant state DLS-β can relax to the bottom of the CB into a "perturbed host state" (PHS). PHS is a delocalized, hydrogenic like, state with CB origin which enables the electron to participate in conduction. EPR data on ^{67}Zn containing ZnO, revealed the presence of a F^+ center in the crystal, that is, a singly ionized oxygen vacancy $V_O^{\bullet}$ located deep within the gap [178]. The oxygen vacancy in In_2O_3 crystals has the same features [176]. To undergo the reversed transition from the PHS to the DLS-α ground state the electron has to overcome an energy barrier which bestows to the DLS-β resonant state a metastable character. Therefore, the nonequilibrium electrons which are optically "pumped" in the CB (on the PHS actually) have long lifetimes (even days) and can contribute to the conduction. Janotti and Van de Walle disagree with the conclusions of Lany and Zunger for two reasons [65]. Firstly, the energies of the charge transition states of V_O, $V_O^{\bullet}$ and $V_O^{\bullet\bullet}$ obtained by Lany and Zunger using the DFT-LDA + U approach were not corrected for the gap error (only the gap value was rigidly shifted upwards). This procedure implicitly implies VB character for the defect states which, thus, have to be handled as VBM states. In addition, the energy barrier impeding the electrons located on the charge state $V_O^{\bullet}$ of the F^+ center (the highest in energy) from escaping is small (300 meV) [159]. Therefore, the meta-stability of PPC would occur for low temperatures, but not at 300K or over. Consequently, Janotti and Van de Walle find the PPL origin of the native *n*-type conductivity unrealistic. The combined optical and magnetic resonance experiments [62] provided two different but equivalent models for $V_O^{\bullet}$, one closer to the estimation of Lany and Zunger ($V_O \rightarrow V_O^{\bullet}$ at VBM + 0.9 eV) the other ($V_O \rightarrow V_O^{\bullet}$ at VBM + 2.48 eV) to the one of Janotti and Van de Walle. Under these circumstances, the different theoretical approaches, even supported by the existing experimental evidence, remain inconclusive and contradictory. However, in $SrTiO_3$ at room temperature PPL has been already demonstrated [179].

3.1.7.4 Theoretical approaches for extrinsic doping in SMOXs: unintentional doping

The unintentional doping with hydrogen has been experimentally identified as a source of electrons in large gap SMOXs. Consequently, it has generated theoretical interest. Using DFT-GGA-PAW-VASP, Singh et al. considered H_i and H_O as possible donors in SnO_2 [158]. Since H_i becomes very mobile after the ionization into $H_i^{\bullet}$, it is not appropriate for real doping. The only "viable" donor is H_O three times coordinated with Sn_{Sn} because it preserves the bond length ($Sn_{Sn}-O_O$: 207 *pm*; $Sn_{Sn}-H_O$: 214 *pm*), has a high energy barrier of dissociation ($H_O^{\bullet} \rightarrow H_i^{\bullet} + V_O$: 2.2 eV) in a +1 charge state and its concentration follows the oxygen partial pressure at high temperatures, as revealed by almost all experiments [47]. In Ref. [180], the hydrogen coming from hydration is assumed to appear as $OH_O^{\bullet}$, $H_O^{\bullet}$ and $(H_2)_i^{'}$, the most energetically favorable being $OH_O^{\bullet}$, which, for the Fermi level near the CBM of TiO_2, will have a formation enthalpy even lower than $V_O^{\bullet\bullet}$. In addition, hydrogen acceptor complexes $\left(V_{Ti}^{''''} \cdot OH_O^{\bullet}\right)^{'''}$ that could trap a free electronic charge are possible (they could explain the differences between calculated and measured concentrations in intrinsic oxides). The computation was made in the frame of DFT/VASP with GGA-PBE and HSE06 for interstitial hydrogen.

Varley et al. analyzed the hydrogen behavior in SnO_2 impurified with acceptor impurities (Al, Ga and In). The results indicate $H_i^\bullet$ as possible donor in pure semiconductors, which will compensate the acceptors in the p-doped material [181]. In Ref. [182] the hydrogen is allowed to (theoretically) interact with the cation vacancies in SnO_2, In_2O_3 and Ga_2O_3 using the DFT with screened hybrid functional HSE06-VASP code [149]. The simulations show again that $H_i^\bullet$ act as shallow donor in all three materials. The cation vacancies have large formation energies in SnO_2 and In_2O_3 being improbable as single point defect. However, the complexes of the type $(V_{Sn}{-}H)'''$ and $(V_{In}{-}H)''$ are much more favored with regard to their formation energy and their stability up to 700−1000°C. In the case of Ga_2O_3 the situation is slightly different. The intrinsic V'''_{Ga} has a low formation energy and under oxygen-rich conditions appears spontaneously. The hydrogen complexes $(V_{Ga}{-}H)''$ have even lower formation energies, which would make them very probable, also under Ga-rich conditions. Multiple symmetrical bonding of hydrogen in metal oxides was theoretically (LDA + U) put in evidence by Janotti and Van de Walle [183]. In ZnO, the four dangling bonds of the Zn ions surrounding an oxygen vacancy combine towards symmetrical state a_1, which strongly interacts with the 1 s orbital of hydrogen. Thus, H_O becomes as favorable in energy as H_i but more stable at high temperatures and more probable under O-rich conditions. Like H_i, the complex defect $H_O{-}4Zn$ has a donor character. Additionally, its concentration is related to the oxygen partial pressure as required by experimental data.

If the SMOXs already contain oxygen sites substituted with nitrogen, adding hydrogen will result in strong $N{-}H$ complexes which will passivate the N_O deep acceptor, as shown for SnO_2 in Ref. [184].

Look et al. estimate, using DFT-LDA, that the $Zn_i{-}N_O$ complex, which can accidentally form in ZnO, is a shallower donor than the isolated Zn_i [55]. Lany and Zunger found that $(Zn_i{-}N_O)^\bullet$ is not able to provide electrons for the CB but could act as a compensating donor [157].

3.1.7.5 Theoretical approaches for extrinsic doping in SMOXs: intentional doping

Zunger discusses the "practical doping principles" of SMOXs theoretically [185]. Three groups of doping rules are identified. They emerge from (α) the necessity to hinder the spontaneous compensation effects induced by the position of the Fermi level (β), the adjustment of chemical potentials of different elements in order to maintain the approximate host material stoichiometry and avoid the formation of auxiliary compounds, and (γ) the effects of the local dopant bond within the lattice. Consequently, for example, for n-doping, the donor insertion into the host material will raise the Fermi level leading to the spontaneous formation of compensating acceptors, which in turn will limit the level of impurification. Despite this principle expectation, SMOXs are natively good conductors or are easily doped undergoing mechanisms, which limit the compensation. Lany and Zuner show that the "electron killers" (actually compensating states) like O_i in In_2O_3 and V_{Zn} in ZnO are very unstable allowing for an electron concentration of $\sim 10^{21}\ cm^{-3}$ to be produced via Sn or Al doping, respectively [157].

The energy levels and character of numerous transition metal impurities in In_2O_3 and ZnO have been estimated by DFT LDA + U with the CB correction in [186]. For ZnO:In, a shallow donor level at 36 meV from the CB leading to degeneracy is theoretically possible [187].

The metal-site acceptor states (Li, Na, Cu, and Ag) in wide gap oxides (SnO_2, In_2O_3, and ZnO) are found to be deep levels in the frame of DFT + U and nonlocal external potential [188] while appearing as shallow delocalized levels in standard DFT approaches.

The elements of the 15th group (N, P, As, and Sb) have amphoteric character in SnO_2. However, P, As, and Sb prefer to incorporate as substituents of Sn inducing shallow donor levels (lower formation energy). Interestingly, N rather replaces the oxygen creating a deep acceptor. These results were obtained with a DFT, HSE hybrid function [184].

The magnetic properties of TiO_2 impurified with several transition metals (V, Cr, Mn, Fe, Co, and Ni) are addressed in Ref. [189]. The calculated energy spectrum of the dopants indicates shallow donor features for V and Co and shallow acceptor features for Mn and Co. The degree of compensation in TiO_2:Co, however, is not discussed nor are its electrical properties considered. Resonant states in the VB and/or CB have been obtained for all the listed transition metals above.

DFT-LDA simulations for (Li, Na, K) WO_3 or HWO_3 bronzes predict metallic character for the resulting compounds (even though the gap error is significant, that is, 0.8 eV instead of experimentally 2.6 eV) [190]. Unlike the alkali atoms, the hydrogen impurity does not occupy the central position of the unit cell of cubic WO_3, but prefers a location closer to one oxygen (at 103 pm) forming the hydroxide ($O-H$) group (free group bond length of 96 pm). This peculiarity could explain why the density of states and the position of the Fermi level in the CB of HWO_3 is different from that of other bronzes (DFT-GGA computation) [191].

The features of CuO containing different classes of elements is considered by Peng et al. in the DFT-LDSA + U frame [175]. In general, the alkaline metals create shallow acceptors in CuO (30 meV above the VBM for Li and Na) resulting in a p-type semiconductor. The elements of group 13 (Al, Ga and In) and 4 (Ti, Zr, and Hf) lead to deep (Al_{Cu}, Ga_{Cu}, In_{Cu}, and Ti_{Cu}) or relatively shallow (Zr_{Cu} and Hf_{Cu}) donors. The formation energies of Zr_{Cu} and Hf_{Cu} are, however, too large prohibiting an effective n-doping.

There are two major outcomes from the presented theoretical achievements:

- The numerical simulations generally do not confirm the experimental results, mainly concerning the native defects, or if they do, it is only in part.
- Different theoretical approaches can produce similar but also incompatible pictures of the defects in SMOXs.

The next section will be devoted to the analysis of the most significant and intricate issues emerging from the literature data on defects in SMOXs.

3.1.8 The contradictions of different theoretical approaches among themselves and in the context of experimental results; reasons and possible solutions

The numerical simulations of model defects in SMOXs provided a heterogeneous set of results including similar trends, comparable numerical values and compatible interpretations as well as significant differences and almost irreconcilable disagreements. Identifying the reasons for these discrepancies within the general, self-consistent framework of the theory would be possible, but an approach, which also considers experimental evidence, would be more practical and efficient. Finally, any theoretical outcome has to be addressed in the light of the experimental data, even if it is difficult. Understanding the correlation of the computed parameters with the measured ones is not always straightforward and requires a proper procedure, which is not always error-free.

3.1.8.1 Disagreements originating in the limited performance of the theoretical approaches

As mentioned in Section 3.1.6.1, DFT based approaches are generally used to simulate defects in SMOXs. The method is typically applied to a model crystal built up from periodically repeated supercells consisting of several dozen/hundreds of host atoms with the ideal lattice periodicity. The defect is "immersed" in the middle of each supercell. Both DFT and cell methodologies bring specific errors.

DFT errors: the underestimation of the band gap

According to Mott [192], the fundamental gap and gap center of a semiconductor are:

$$E_g = I - A \quad (VBM - CBM) \tag{3.10}$$

$$E_g^{1/2} = \frac{1}{2}(I + A) \quad \left(\frac{1}{2}(VBM + CBM)\right) \tag{3.11}$$

whereby the ionization energy I and the electron affinity A in the ground state are given by:

$$I(N) = E(N-1) - E(N) \quad \text{corresponds to the VBM} \tag{3.12}$$

$$A(N) = E(N) - E(N+1) \quad \text{corresponds to the CBM} \tag{3.13}$$

The fundamental gap in the frame of the DFT calculations is, however, not the true one (Eq. (3.10)) because DFT has been devised to correctly output the ground state of a quantum system with only a fixed number of particles [141,142]. Calculating the gap involves changes in the total number of system electrons and implicitly a functional derivative of the ground state energy upon density, resulting in errors [193,194]. A thorough approach to this problem is given by Perdew in [195]. The self-interaction resulting in the exchange-correlation potential is another reason for the gap underestimation [195,196].

A discontinuity in the functional derivative of the exchange correlation energy produces a discontinuity in the exchange correlation potential and, consequently, in the effective potential (see [194], reconsidered in Ref. [197]). Some newer approaches, for example, L(S)DA + U, include an additional potential for the *d*/*f* states of heavier elements or (in the case of hybrid functions) use an empirical mixture of exact (HF) exchange with density functional [148–151] and have improved the computational accuracy. Fig. 3.3 displays, as vertical dot lines, the "theoretical" position of the CBM with respect to the VBM, that is, the calculated gap value, for *ZnO* in LDA (left panel) and LDA + U (middle panel) DFT versions. It is reasonable to think that a procedure containing errors in the gap value will also produce errors in the formation energy of the defects, their electronic states and transition levels. To overcome the issue, an ad hoc empirical solution has been proposed and largely implemented: the shift of the calculated gap value towards experimental values together with a shift of the calculated local levels. While there is no trouble with the gap itself, the procedure employed for the defect levels and similar parameters was, and still is, a matter of debate. The issues surround the nature/origin of the localized state. If they behave like VB states, then there is no reason to shift them and one has to consider the theoretical values as true. Conversely, should the localized states be similar to the CB ones, then the shift is justified and appropriate adjustment schemas are required [65,139]. The authors rarely agree on the type and magnitude of the corrections, making a straightforward comparison of their results difficult [65,159]. Sometimes only the trends of the defect states/levels are discussed, and any correction or doubts are disregarded [163]. A pertinent overview on the DFs is given in [143].

DFT errors: supercell and computational issues

In addition to the intrinsic method errors, the errors introduced by the supercell size, the defect concentration and the numerical algorithms must be considered. The supercell approach has the advantage to produce a periodic structure which is helpful in calculating delocalized states, but also has some drawbacks [140,198]. Because a supercell larger than 1000 atoms is prohibited by the computing effort, the realizable defect concentration in the model is at least 10^{-3} of the regular atom number. At such large concentrations, the defect–defect interaction, which actually is an artefact, must be considered resulting in additional complications and errors. It is even more difficult to evaluate the parameters of the charged defects because, on one hand, the Coulomb potential has a long range and, on the other hand, the neutrality has to be preserved in the supercell (usually by setting the macroscopic electrical potential to zero). Furthermore, the weak localization of the hydrogenic states of the ionized shallow levels makes the associated wave function exceeding the supercell size. A larger supercell diminishes these effects but requires increased computing resources. Other types of restrictions arise from the choice of the basis function and the used numerical algorithms. In order to integrate into the reciprocal space, the standard approach is the Monkhorst-Pack mesh which circumvents the high symmetry points [199]. More details on these subjects are available in [140].

3.1.8.2 Disagreements originating in the mismatch between the theoretical and the experimental data

Comparing the experimental data from equilibrium measurements with the theoretical simulations

At the beginning of the section, the difficulties of the direct comparison of theoretical and experimental data were highlighted. In principle, the quantum mechanics formalisms, disregarding the particular approach, deliver the eigenfunctions and the energy spectrum of closed and isolated systems. This is the case for the DFT values for different defect parameters. The formation energies $\Delta_f E_d$ are computed as indicated in Section 3.1.6. To obtain the corresponding enthalpies of formation the defect formation work under constant pressure has to be evaluated as $W_{f,d} = p\Delta_f V_d$ (whereby $\Delta_f V_d$ is the volume of formation) and added to the formation energy. Experimental evaluations occur normally under constant temperature and pressure, so that the free enthalpies of the defects are required. They are attained by adding the entropic contribution $-T\Delta_f S_d$ (whereby $\Delta_f S_d$ is the entropy of formation) to $\Delta_f E_d$. The free enthalpy of the defect formation $\Delta_f G_d = \Delta_f E_d + p\Delta_f V_d - T\Delta_f S_d$ contains the temperature dependency and can be validated by experimental data. The main contribution to the entropy of formation is brought by the configurational term, which is, in most cases, several orders of magnitude higher than other entropic terms (vibrational, electronic) [140]. Because $\Delta_f G_d$ is not experimentally available, a direct confirmation of the theoretical data is not possible. The defect concentration under thermodynamic equilibrium is, however, given by a dependence similar to Eq. (3.7):

$$c_d = g \cdot e^{-\frac{\Delta_f G_{md}}{k_B T}} = g \cdot e^{\frac{\Delta_f S_d}{k_B}} \cdot e^{-\frac{p\Delta_f V_d}{k_B T}} \cdot e^{-\frac{\Delta_f E_d}{k_B T}} \tag{3.14}$$

whereby g is the degeneration of the defect level (provided several equivalent positions for the defect in the elementary cell exist). For a small volume of formation, the equation above has an Arrhenius plot with $\Delta_f E_d$ as activation energy and the defect formation entropy divided by Boltzmann constant as the vertical intercept. Therefore, if the measured concentration follows an Arrhenius dependence on temperature, the experiment–simulation comparison can be made. This implicitly means that: (1) the experimental data is acquired under equilibrium at temperatures above half of the compound's melting point, (2) the theoretical values are calculated for 0K and (3) the temperature dependence of the theoretical concentration results from the entropy of the defect formation (the work of the defect formation being neglected).

Comparing the experimental data from nonequilibrium measurements with the theoretical simulations

For this purpose, the optically induced electronic transition is the representative example because all other equivalent experiments are using the same methodology. As shown in Fig. 3.5, the electron from the equilibrium ground electronic state of the defect absorbs a photon with appropriate energy E_{abs} and reaches the first excited state (possible with another charge state, like when ionizing a donor). The defect configuration with the electron in the new state will differ from the initial one. After a certain time, roughly the relaxation time, the final defect configuration will relax towards a new equilibrium configuration. However, the transition occurs so fast that the electron reaches its final electronic state with the initial defect configuration. Because the final state is not a relaxed one, it has a higher energy than under equilibrium conditions. After relaxation, the electron state adjusts to the final equilibrium configuration of the defect, minimizing its energy (this is a nonradiative process). A deexcitation of the electron will follow the reversed path, but this time the ground state will be the final state (however, out of equilibrium during the electron transition). Consequently, the electron energy in the nonequilibrium final state (nonequilibrium ground state) will be higher than in the equilibrium ground state and the emitted photon will have a smaller energy $E_{em} < E_{abs}$ than the absorbed one. The transition energy estimated in an equilibrium measurement E_{therm} will lay between the above limits: $E_{em} < E_{therm} < E_{abs}$. The process resembles the classic Frank-Condon effect. It also involves the defect vibrations, which are coupled with electronic transitions. The defect configuration is often described by the configuration coordinate Q, a unique parameter similar to the interatomic distance (bond length) in diatomic molecules. A clear presentation of the optical transitions between the electronic states of the defects in the frame of the configuration coordinate can be found in [200]. The theoretical approaches to this topic are also discussed and examples, like nitrogen in ZnO, are considered.

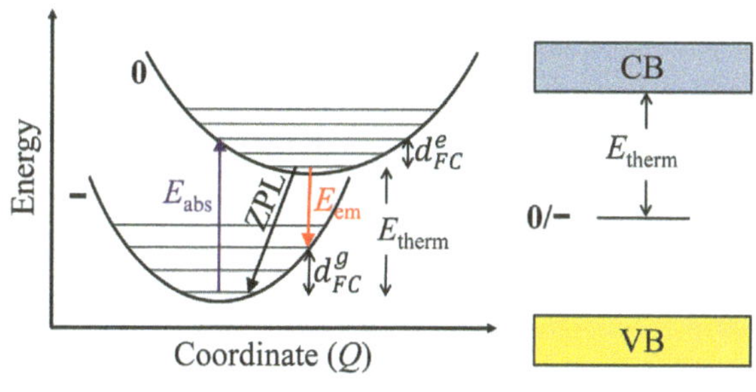

Figure 3.5 The optical electronic transitions between defect's electronic states in the frame of the configuration coordinate Q.

Source: Reprinted with permission from H.J. Monkhorst, J.D. Pack, Special points for Brillouin-zone integrations, Phys. Rev. B. 13 (1976) 5188–5192, doi:10.1103/PhysRevB.13.5188.

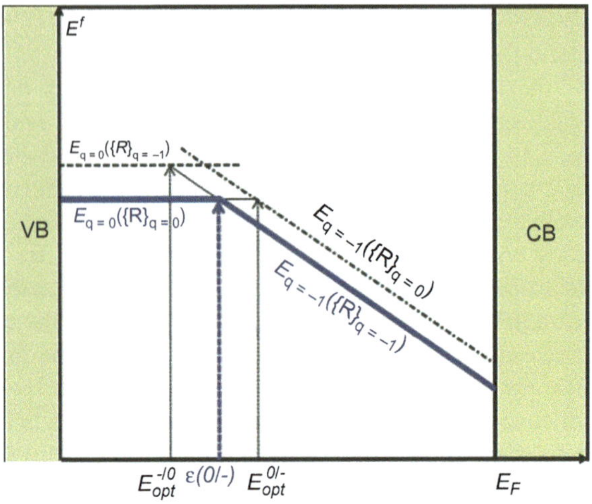

Figure 3.6 Charge state transition levels for equilibrium processes (full line, blue) and for optical processes (nonequilibrium ones) (dash-dot line, black).
Source: Reprinted with permission from C. Freysoldt, B. Grabowski, T. Hickel, J. Neugebauer, G. Kresse, A. Janotti, et al., First-principles calculations for point defects in solids, Rev. Mod. Phys. 86 (2014) 253–305, doi:10.1103/RevModPhys.86.253.

Fig. 3.6 shows the differences between the charge transition levels for equilibrium configurations, which are unique, and nonequilibrium (optical) ones, which depend on the transition direction (due to the Frank-Condon effect for optical transitions). The theory must simulate the right experimental situation in order to produce reliable data.

3.1.8.3 Improving the experimental and theoretical accuracy

The experimental precision depends on the performance of the instrumentation, but also on the ability of the operator to design suited experiments under optimal experimental conditions. On the side of the theoretical approaches, one legitimately expects upgraded methods and codes. The investigations referred to in Section 3.1.6, already showed positive trends in the evaluation of the gap. The improvement over the time is visible when comparing simulations from the 80 and 90 s with modern work. Controversially, predictions on the character of the same defect in identical materials decreased in the latest years.

An overview of the large statistical basis on the theoretical gap evaluation [201], is very encouraging: only 0.28 eV absolute mean deviation in the gap values with the B3PW91 function (over 64 semiconductors/insulators with the gap between 0.014 and 7 eV). It is a clear indication that the increase of the computing power/cost ratio well compensates the drawbacks of the density functionals. Such favorable circumstances are expected to allow larger supercells with more realistic defect concentrations to be employed. For instance, the doping concentration of SMOXs rarely exceeds 10^{19} cm^{-3}, which means that it is typically smaller than 10^{-5} from the regular atom number.

New theoretical approaches/viewpoints were also proposed. The idea has been advanced that many problems in DFT formalism, including gap underestimation, arise because all previous DFT simulations used a single basis to self-consistently calculate the system ground state without requiring an absolute energy minimum [202]. Therefore, if from the possible basis function sets, the one, which satisfies the absolute minimum condition, is chosen then the results should be thorough. Such optimistic solutions must pass the time test, in order to prove the generality of the method and reproducibility of the results.

3.1.9 Conclusion and perspectives

The main literature outcomes about the defects in SMOXs appear to be:

1. Traditional experimental methods, when correctly performed, provide mainly thermodynamic equilibrium data, which must be converted in order to be comparable to the theoretical ones.
2. The identity of the defects inferred from equilibrium measurements is under question (oxygen vacancies and metal interstitials).
3. There are experimental results obtained in inappropriately devised experiments (under not well-specified conditions, neither equilibrium nor adiabatic) which have a reduced relevance and are difficult to interpret.
4. The theoretical approaches, mainly based on DFT, are affected by severe gap underestimation and use empirical correction schemas, whose legitimacy and accuracy are questionable.
5. The computed energy levels of the defects do not reproduce the experimental values.
6. The theory assigns other roles to the defects than the supposed ones. The oxygen vacancies have been mainly thought as shallow donors, with contribution to the free charge in semiconductors and finally appeared to be deep centers in the gap, significant for photoluminescence or optical absorption.
7. The differences between the theoretical and experimental results are far more than accidental, but seem to follow a trend.
8. Under the above circumstances, alternatives for the sources of conductivity in SMOXs are urgently required, but until now, the theoretical approaches do not provide viable solutions. The persistent photoluminescence and the surface/interface accumulation layers do not fit most materials, structures and morphologies.
9. The most reasonable remaining possibilities are:
   a. The unintentional doping of SMOXs (hydrogen, nitrogen, metals and transition metals).
   b. The formation of defect complexes, with or without extrinsic contribution.

Improvements in the scientific instrumentation, the increase of the computational power and the interdisciplinary investigations might help to improve the quality of the scientific results concerning the native conduction in SMOXs and its origin, because, in spite of high interest, allocated resources and practical needs, the actual understanding level in the field is far from what is needed.

3.2 Experimental investigations

Since the first reports on the gas-sensing effect of SMOXs [203,204], numerous efforts were made to describe their working principle [205–212]. Due to the sheer complexity of the chemical and physical processes ranging from the atomic level to the macro scale, the huge quantity of different sensing materials and the many influencing factors, for example, temperature, pressure or atmospheric composition, the understanding of SMOX-based gas sensors is a highly challenging task. The state of the art quantitative model of the working principle is achieved by combining fundamentals of semiconductor physics with elementary chemical reactions, giving a satisfactory explanation of the gas-sensing effect for well-defined cases, for example, CO sensing with pristine SnO_2 in dry air. The further understanding of gas sensing under more realistic, and therefore more complex, conditions or the development of models for less uniform materials, for example, doped SMOXs or mixed oxides, requires the verification and refinement of the assumed material properties and the involved chemical and physical processes, through experimental techniques. Many works are based on experimental findings obtained by ultrahigh vacuum techniques applied to model specimens like pressed pellets or single crystals [206,209]. It is evident that the experimental conditions barely match the operation conditions (like pressure and temperature) and real sensing materials, for example, layers of nano-sized grains. Consequently, many surface science techniques suffer from a pressure, temperature and material gap, which limits the informative value for the understanding of SMOX-based gas sensors. Such uncertainties are avoided by applying experimental techniques to model systems or, preferably, to real gas sensors under application relevant conditions [213,214]. As in catalysis research [215–217], the experimental conditions are categorized as ex situ, in situ and operando. A clear classification is certainly controversial, however, a simplified one could be the following [213,214]:

- Ex situ experiments include all sample forms and conditions that are not relevant to their operation.
- In situ experiments are performed on any type of specimen under conditions relevant for the gas sensor's operation (pressure, temperature, atmospheric composition), but do not involve a real-time evaluation of the sensor performance of the probed specimen.
- Operando experiments are performed under application-relevant conditions on real sensor devices and involve the real-time evaluation of the sensor performance of the probed specimen.

The operando methodology for gas sensor research includes spectroscopic, as well as electrical techniques [213,214]. Over the last decades, the utilization of in situ and operando methods increased by the steady improvement of numerous spectroscopic and electrical techniques. This process and selected examples will be discussed in Chapter 4.

3.2.1 Spectroscopic techniques

Regarding the suitability of spectroscopic techniques for operando research on gas sensors, a first selection can be made by the atmospheric conditions, that is, many surface science techniques involving (charge) particles, for example, electrons, require vacuum conditions and do not fulfill the needs of in situ or operando research on gas sensors, but still give valuable information on fundamental material properties. A comprehensive overview of these techniques applied to various metal oxides relevant for gas sensing is given in the book of Heinrich and Cox [218] or the review articles of Diebold and Batzill [219–221]. However, spectroscopic techniques involving only photons are suitable for operando research on gas sensors and still have some restrictions, like temperature ranges or specimen types (sensor or powder). The development of common spectroscopic techniques applied to gas sensors or gas-sensing materials made a huge progress as it is shown in Fig. 3.7. In 1995, in situ infrared spectroscopy was applied to sensing materials [222] and, in following years, emerged as an important tool for operando studies on gas sensors [223,224]. A similar progress is found for XAS, which was developed from an in situ [225] to a powerful operando technique for gas sensor research within a few years [226–228]. More recently, further spectroscopic techniques, namely UV/vis [229], Raman [230,231] and X-ray emission spectroscopy [232], were successfully applied to gas sensors bringing new insights in the gas-sensing process of various materials. However, other spectroscopies like EPR, solid state Nuclear Magnetic Resonance (NMR) or Mössbauer spectroscopy were not part of this development.

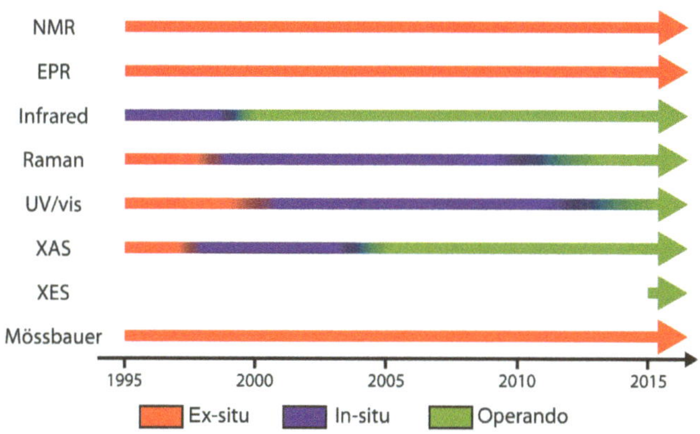

Figure 3.7 Progress of operando spectroscopic techniques used to study gas sensors. A detailed discussion of the different techniques is given in the corresponding sections of this chapter.

3.2.1.1 Radio and microwave spectroscopies

Spectroscopies in the radio and microwave frequency range are NMR and EPR, respectively [233]. To date NMR spectroscopy is no common technique to study the working principle of SMOX-based gas sensors. A few works use NMR spectroscopy to characterize SMOX-based sensing materials, for example, recording ^{119}Sn solid-state NMR [234,235]. However, as many common analyte gases contain protons (e.g., H_2, H_2O and organic compounds) or carbon (e.g., CO, CO_2 and organic compounds), ^{1}H or ^{13}C NMR spectroscopy presents a suitable technique to identify adsorbed species and the related surface chemistry of various volatile organic compounds on gas-sensing materials under in situ conditions. Gurlo and Riedel described the potential of solid-state NMR in their review article, taking ethanol as an example [214]. The progress of in situ NMR spectroscopic techniques for the study of heterogeneous catalysis demonstrates the applicability of solid state NMR spectroscopy in continuous flow conditions and at elevated temperatures in the typical ranges for SMOX gas sensor operations [236,237]. Thus, in situ solid-state NMR spectroscopy may play a role in future works investigating the gas reception of gas-sensing materials. Unlike NMR spectroscopy, the use of EPR spectroscopy has a long history in the study of gas-sensing materials. EPR spectroscopy is used by various authors to detect paramagnetic oxygen species (e.g., $OH^{\cdot}$, O^{-}, or O_2^{-}) or to follow the formation and ionization of oxygen vacancies [238–242]. However, these works have to be considered as ex situ studies, since the measurements were done via pretreatment of the powder samples at elevated temperatures and in different atmospheres, followed by quenching and measuring at low temperatures (4–298K). On the one hand, EPR spectroscopy is a technique suitable to probe various species, which are expected to be involved in the gas-sensing process [212]. On the other hand, the huge temperature gap of several hundred °C between typical sensor operation (100–400°C) and the actual measurement conditions (at best room temperature, in many cases −269 to −163°C) requires at least the critical review of the experimental conditions and drawn conclusions.

3.2.1.2 Infrared spectroscopy

Since the gas reception is strongly linked to the surface chemistry of SMOXs, the reactive species and reaction intermediates play a key role in order to understand the gas-sensing effect. The most common methods to identify and study molecular surface species are vibrational spectroscopies, like FTIR or Raman spectroscopy. Investigations on the surface chemistry of tin dioxide (SnO_2)—the benchmark material for gas sensors—using IR spectroscopy have a long history and various works were published in the past few decades. Most of these early works have to be considered as ex situ studies, since the gases were dosed after the evacuation and most of the samples were self-supporting disks measured at room temperature [243–250]. An advanced IR study was presented by Harrison and Willett. By combining transmission IR measurements of a self-supporting disk with conductance measurements, they could correlate the observed species with the changes in the conductance. However, the spectroscopic measurements were still done far below the typical operation temperature, that is, ex situ [251]. The first work that fulfills in situ conditions in terms of temperature and atmospheric conditions was published in 1995 by Lenaerts et al. [222]. The work was performed on self-supporting SnO_2 disks measured in transmission mode, but the atmospheric compositions were far from realistic concentration levels for reducing gases. The essential role of the temperature for the interaction of SnO_2 with different gases, however, was demonstrated. The role of the temperature and the presence of oxygen on the gas reception of SnO_2 was further investigated by Baraton and Merhari [252,253], as well as Lenaerts et al. An important finding of these works is the role of the free charge carrier absorption in the IR region, which correlates with the changes in the conductance. The first step of studying real gas sensors was done in the same year by Benítez et al. By measuring a CdGeON-based gas sensor in different oxygen backgrounds using Diffuse Reflectance Infrared Fourier-Transform Spectroscopy (DRIFTS) [223], they demonstrated the great potential of DRIFTS. However, the material and atmospheres (N_2 or dry air) studied did only little to improve the general understanding of SMOX-based gas sensors. In 2000, a quite successful attempt to study SMOX gas sensors under working conditions was made by Pohle et al. They used Infrared Emission Spectroscopy (IRES) to investigate the interaction of water vapor and hydrogen containing gases with Ga_2O_3 [254] or the interaction of oxygen with WO_3, $AlVO_4$, and Co_3O_4 [255]. However, this technique is mainly limited to sensors operated at high temperatures (300–600°C) and is therefore not suitable for many SMOX-based sensors operated at or below 300°C. Further works on SnO_2 were done by Emiroglu et al. using DRIFTS, first on SnO_2 powders [256] and later, in 2003, on a Pd-doped SnO_2 gas sensor [224]. The work of Harbeck (née Emiroglu) et al. has to be considered as the first real operando DRIFTS study on gas sensors, regarding the sample form (gas sensor), temperature (300°C), atmospheric composition (ppm-range of CO dosed in dry and humid air) and the real-time read out of the sensor performance. In situ and operando DRIFTS was successfully applied to various sensing materials [257–263]. Further understanding of the surface chemistry of SMOX was achieved by the use of isotopic labeling [257,258,260,263], for example, H_2/D_2, H_2O/D_2O, or $^{12}CO/^{13}CO$, the use of on-line residual gas analysis [257,259,261–263], and time-resolved measurements [254,263,264]. Good examples for the use of operando DRIFT spectroscopy are studies investigating the gas reception mechanism of CO on WO_3 and SnO_2. M. Hübner et al. exposed a WO_3 based gas sensor to 250 ppm CO in air and in the absence of oxygen. In both conditions the recorded IR spectra (see Fig. 3.8) show decreasing bands at 2063 and 1845 cm^{-1}, which correspond to overtone/combination modes of lattice W = O vibrations [265]. This finding indicates that WO_3 is being reduced during the detection of CO forming CO_2, which was found in the exhaust even in the absence of O_2. Thus, the reception of CO on WO_3 is based on the formation of oxygen vacancies, a process which is much more pronounced in the absence of O_2, since without atmospheric oxygen the healing of oxygen vacancies is hindered. The formation of possible reaction intermediates, such as carbonates or carboxylates, is not observed on WO_3.

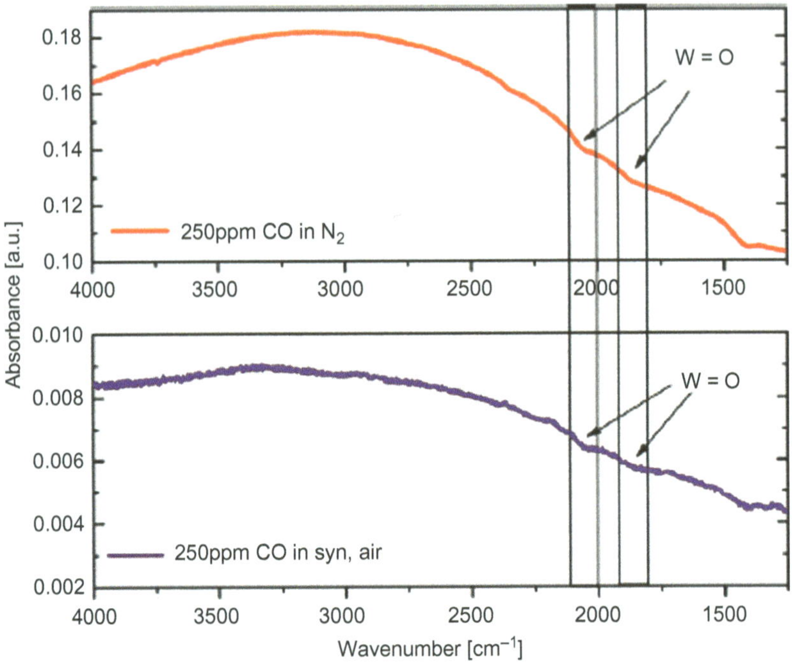

Figure 3.8 DRIFT spectra of a WO_3 sensor exposed to 250 ppm CO in the absence of O_2 (top) and in 150 ppm O_2/N_2. The decrease of the W = O overtone/combination modes at 2063 and 1854 cm^{-1} indicates the reduction of the material.
Source: Reprinted with permission from M. Hübner, C.E. Simion, A. Haensch, N. Barsan, U. Weimar, CO sensing mechanism with WO_3 based gas sensors, Sens. Actuators B Chem. 151 (2010) 103–106, doi:10.1016/j.snb.2010.09.040.

This is different for SnO_2, as already early ex situ studies report the formation of carbonates and carboxylates on SnO_2 during CO exposure [244,246]. The formation was confirmed by later operando DRIFTS studies on SnO_2 gas sensors [260,261]. Hence, these species were assumed to be an intermediate product of the CO oxidation on SnO_2, for example, caused by the reaction of CO and a superoxide anion (O_2^-) [260]. However, other in situ [222] and operando [257] studies did not observe the formation of carbonates or carboxylates on some SnO_2 materials. A comparative operando DRIFTS study of two SnO_2 sensing materials (one forming carbonate species during CO exposure, one showing no carbonates or carboxylates at all) revealed that the carbonates may be formed during a reaction with CO_2, which was formed during the exposure to CO and not by a direct reaction of CO with SnO_2 [257]. Using isotopically labeled ^{13}CO, the absence of carbonates was confirmed since for one material only the bands of gaseous CO and CO_2 were shifted during the isotopic exchange experiment (Fig. 3.9, top). The decreasing bands in this region were identified as Sn–O lattice overtones, indicating the reduction of SnO_2 due to the exposure CO. The other material shows a series of bands

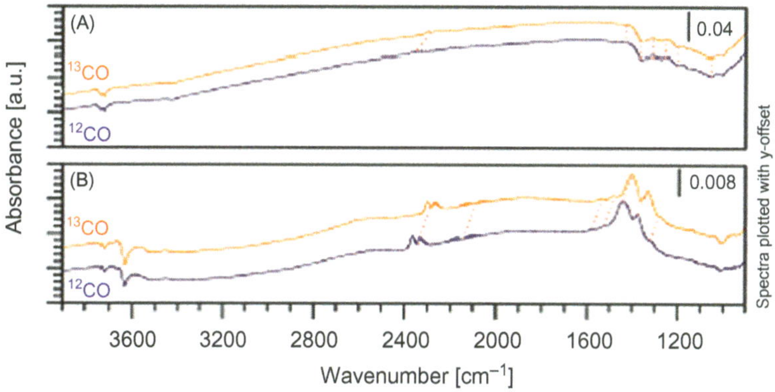

Figure 3.9 DRIFT spectra of two SnO_2 materials exposed to 500 ppm ^{12}CO (blue line) and ^{13}CO (orange line). The top spectra were taken from the SnO_2 material calcined at 1000°C and bottom spectra from the one calcined at 450°C.
Source: Reprinted with permission from D. Degler, S. Wicker, U. Weimar, N. Barsan, Identifying the active oxygen species in SnO_2 based gas sensing materials: an operando IR spectrsocopy study, J. Phys. Chem. C. 119 (2015) 11792–11799, doi:10.1021/acs. jpcc.5b04082.

being shifted by a factor of 0.968–0.978 (theoretical shift 0.978), confirming the formation of carbonates (Fig. 3.9, top) [257]. This finding implies that carbonates are not an essential surface species for the gas-sensing process as both materials show significant resistance changes upon CO exposure. The role of the carbonates was further evaluated by the exposure to CO_2, revealing that the same carbonate species are formed, however, with a negligible effect on the sensor resistance. In conclusion, these findings show that carbonates are not an active species in terms of CO reception on SnO_2, but rather a spectator species. The reduction of both materials was shown using H_2 as a reducing gas. However, the identified Sn–O overtone bands are located in the same region as the carbonate bands. In case of CO, the decreasing Sn–O bands are presumably overlaid by formed carbonates. This example demonstrates two important facts:

- Isotopically labeled gases are a suitable tool to identify surface species, especially in the spectral region where bands of various bands appear. Besides $^{12}CO/^{13}CO$, the use of the isotope pairs H_2/D_2 or H_2O/D_2O is proven to be convenient for surface studies on gas-sensing materials [257,258,263].
- Not all of the observed surface species are necessarily involved in the gas-sensing process or show only a small contribution. Indeed, this complicates the analysis of spectroscopic data and one should critically review all findings. Elaborate experiment planning (different materials, gases and temperatures), the comparison with theoretical works or time-resolved spectroscopy are suitable approaches to identify active and spectator species.

The feasibility and benefits of time-resolved operando DRIFTS, that is, the simultaneous time-resolved recording of spectra and AC conductivity, were demonstrated by Simons et al. studying zeolite-based NH_3 sensors [264]. An excerpt of this study is shown in Fig. 3.10. In the thermal desorption (Fig. 3.10, left) and the SCR experiment (Fig. 3.10, right) two different NH_3 species are found on the zeolite surface, corresponding to weakly bound NH_3 (2520 cm^{-1}) and NH_3 species strongly bound to Brønsted sites (1450 cm^{-1}). The proton conduction on unloaded zeolites is attributed to a thermal activated proton hopping between Brønsted sites, while on NH_3-loaded zeolites two mechanisms occur, namely a Grothuss-like proton conduction along chains of secondary adsorbed NH_3 in between Brønsted sites or NH_3 bound to Brønsted sites forming vehicles for proton motion. The contribution of the different adsorbed NH_3 species can be examined by correlating the changes in the AC conductivity of the sensor with the spectroscopic intensities of the different NH_3 species. This is done by extracting the time constants of the decay during the thermal desorption or SCR reaction, respectively. The calculated time constants during the thermal desorption are 4.5 min for the decay of the AC conductivity and the weakly bound NH_3 species and 5.5 min for the NH_3 bound to Brønsted sites. The comparison of the time constants indicates a stronger contribution of the weakly bound species to the proton conductivity. During the SCR reaction of NO, the AC conductivity first drops similar to the thermal desorption, then follows a sigmoidal curve and finally approaches the value of the unloaded zeolite. After an initial decay, which is found for both NH_3 species, the normalized intensities of the weakly and strongly bound NH_3 show different decay characteristics. While the intensity of the strongly bound NH_3 follows a linear decay, the weakly bound NH_3 species adopts the sigmoidal curve found for the AC conductivity. The work of Simons et al. clearly demonstrates the benefits of time-resolved

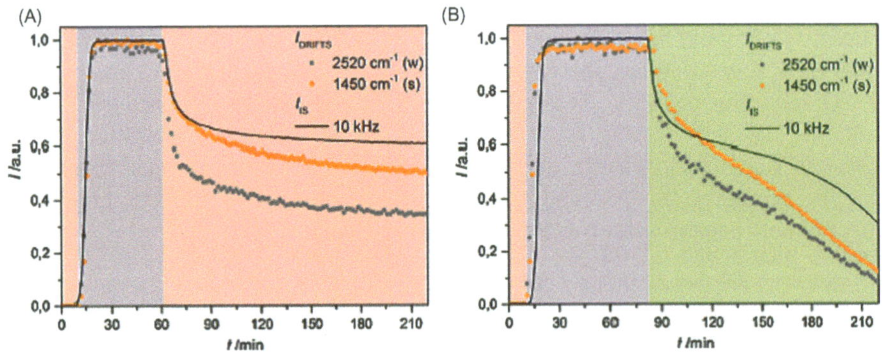

Figure 3.10 Time-resolved evolution of the normalized AC resistance at 10 kHz (I_{IS}) and IR bands (I_{DRIFTS}) of a ZSM5-based NH_3 sensor device. Left: the thermal desorption; right: the SCR conversion; both at 175°C. The IR bands at 2520 and 1450 cm^{-1} are attributed to the NH_3 species that are weakly and strongly bound to Brønsted sites, respectively. The different atmospheric conditions are indicated by the background colors of the graphics: red—N_2, blue—NH_3 in N_2 and green—SCR conditions (NO and O_2 in N_2).
Source: Reprinted with permission from T. Simons, P. Chen, D. Rauch, R. Moos, U. Simon, Sensing catalytic conversion: Simultaneous DRIFT and impedance spectroscopy for in situ monitoring of NH_3^- SCR on zeolites, Sens. Actuators B Chem. 224 (2016) 492–499.

spectroscopy as it allows to discriminate the influence of different species on the conductivity and thus on the sensing process itself.

3.2.1.3 Raman spectroscopy

Compared to the large number of IR studies on SnO_2 and other sensing materials, the number of Raman spectroscopy studies on sensing materials exceeding a simple ex situ material characterization is rather small [230,231,266,267]. Possible reasons are the little spread and the late development of practicable Raman accessories [268]. The first in situ Raman study on a gas-sensing material was reported in 1999 by Pagnier et al., investigating the H_2S sensing mechanism of SnO_2 [267]. Simultaneous spectroscopy and electrical measurements were performed on pellets or powders at operation relevant temperatures. The H_2S exposure was done in the absence of oxygen, limiting the informative value for H_2S sensing in real live, that is, the presence of oxygen. Nevertheless, the occurrence of sulfur-related species correlates with the electrical changes, indicating that the surface is reduced by H_2S forming oxidized sulfur species. A comparable study on the gas-sensing mechanism of tungsten oxide was published in 2001 [266]. But still, pellets, instead of real sensors, were exposed to test gases in the absence of oxygen [266]. Although the method had shown its potential to study changes in the sensing material and the formation of surface species, it took more than 10 years until the first operando study was reported by Sänze et al. [230,231]. These works represent a good example for an operando study of a SMOX gas-sensing material, in this case In_2O_3. Simultaneous spectroscopic and electric measurements were done on a thick film sensor operated at different temperatures, dosing realistic test gas concentrations in the presence and absence of oxygen. Further information was obtained by analyzing the residual gases. Gathering the resistance and spectra of the sensing layer as well as analyzing the exhaust at different temperatures and gas atmospheres, results in a huge quantity of information (see Fig. 3.11). The Raman spectra reveal a series of bands assigned to bixbyite-type In_2O_3 (304, 361, 493, and 625 cm^{-1}) and vibrations of other vibration modes of In_2O_3 (δ(InO_6, octahedron) at 304 cm^{-1}, ν(In-O-In) at 361 cm^{-1} and ν(InO_6) at 493 and 624 cm^{-1}). As shown by this example, Raman spectroscopy allows the identification of the studied oxide under various operation conditions. This is a difference to IR spectroscopy, since for pure compounds the absorption of the material's fundamental modes is too strong and leading to a total absorption of the IR radiation. Furthermore, some spectra (N_2 and air at 190°C, and air and ethanol (EtOH) in air at 325°C) show two ν(OH) vibrations at 3643 and 3659 cm^{-1}, which are assigned to bridging hydroxyl groups. Upon exposure to 250 ppm EtOH in N_2 at 190°C the surface of In_2O_3 is reduced as indicated by the formation of a new band and a shoulder at 325 and 407 cm^{-1}, respectively. Furthermore, the ν(OH) vibrations disappear, while a new band at 2937 cm^{-1} is formed. This band is assigned to the ν(CH) stretch mode of acetate formed at the surface. EtOH exposure in air at 190°C has similar effects, however, the reduction of the surface is less strong, as the reoxidation reaction with O_2 takes place at the same time and an additional surface species is formed, which is indicated by a new

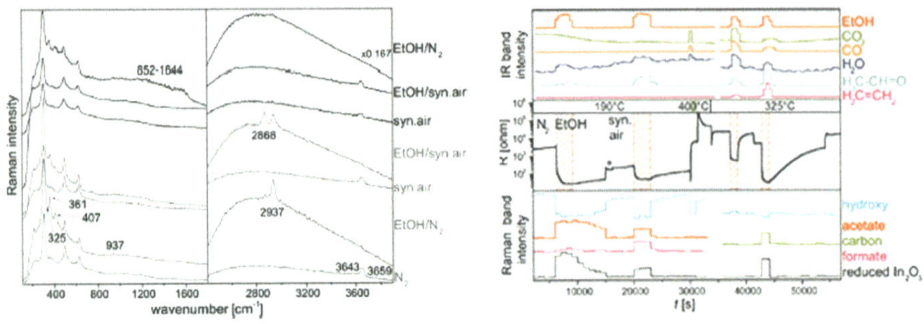

Figure 3.11 Left: Operando Raman spectra of an In_2O_3-based gas sensor recorded at 190°C and 325°C in different atmospheres (N_2, air, EtOH/N_2, and EtOH/air). The spectra are plotted with an *y*-offset. Right: Correlated spectroscopic and resistance measurements of the performed experiments: Exhaust analysis using FTIR (top), resistance measurement (middle) and Raman spectra of the sensing layer (bottom).
Source: Reprinted with permission from S. Sänze, A. Gurlo, C. Hess, Monitoring gas sensors at work: operando Raman-FTIR study of ethanol detection by indium oxide., Angew. Chem. Int. Ed. Engl. 52 (2013) 3607–3610, doi:10.1002/anie.201207258.

band at 2868 cm^{-1} corresponding to a formate-like surface species. Additional differences of the EtOH exposure in N_2 and in air are found when looking at the exhaust analysis (Fig. 3.11, top right). In the presence of oxygen, a larger fraction of acetaldehyde is formed compared to one upon exposure to EtOH in N_2. At 325°C, a different picture is found. In air, no reduction of In_2O_3 or surface adsorbates are detected by Raman spectroscopy upon EtOH exposure, while the exhaust analysis shows the formation of various oxidation products (CO, CO_2, H_2O, acetaldehyde, and ethylene). The exposure to EtOH in N_2 shows some remarkable differences. Firstly, the Raman spectra indicate the deposition of carbon (852–1644 cm^{-2}) and the reduction of In_2O_3. Secondly, the exhaust analysis shows a decreased formation of total oxidation products (CO, CO_2, and H_2O) and an increased formation of acetaldehyde and ethylene. In summary, the study of Sänze et al. identifies the reduction of In_2O_3, that is, the formation of vacancies, as one reason for the decreasing resistance of the material. However, the absence of a clear sign of the formation of reduced In_2O_3 at 235°C in air, while a clear decrease in the resistance is observed, does not fully exclude additional reactions or surface species influencing the resistance. Furthermore, these works emphasize that the gas reception strongly depends on the temperature and the atmospheric composition, in addition to having a rather complex surface chemistry with analytes more complex than CO or H_2, also involving the only partial oxidation of the analyte, as it was shown by the simultaneous analysis of the exhaust.

3.2.1.4 Ultraviolet, visible and near infrared spectroscopy

Spectra recorded in the UV/vis/NIR region probe electronic transitions providing valuable information on the optical band gap of semiconductors (valence to CB transitions), oxidation states of metal centers (d-d or charge-transfer transitions) or the presence and size of metallic clusters (localized surface plasmone resonance) [269]. UV/vis/NIR spectra of SMOXs are mainly recorded in diffuse reflectance mode (Diffuse Reflectance Spectroscopy, DRS) using integrating spheres or specialized mirror optics [269]. Early examples for the use of UV/vis spectroscopy to study gas-sensing materials are usually based on the ex situ characterization of pretreated powders using integrating spheres, for example, identifying the nature of transition metal dopants and the effect of different pretreatments on the SMOX materials or the transition metals [249]. In 2001, Amalric-Popescu et al. presented a study on changes in the conductivity and the UV/vis/NIR spectra of SnO_2 undergoing evacuation and oxygen exposure at different temperatures. The electric and spectroscopic experiments were done on two different systems. The spectra were recorded using a mirror optic and an in situ cell for powders, which allows the control of the temperature and the atmospheric conditions. The results reveal a correlation of increasing conductance and vacancy formation. The latter was verified by ex situ EPR measurements after different pretreatments [239]. In 2005 and 2006, a series of UV/vis/NIR studies on SMOX sensing materials was published using a homemade quartz cell which allows the pretreatment of powders or thin films. However, the spectroscopic experiments were still carried out at room temperature [270–273]. The first successful attempt to use UV/vis spectroscopy to study gas sensors under realistic conditions was made in 2015, using a homemade sensor cell with a quartz window which allows operando measurements. The work demonstrates that a material reduction, as proposed by the results of Amalric-Popescu et al., is found under application relevant atmospheres at 300°C, but is most likely limited to the surface or subsurface of SnO_2 [229].

3.2.1.5 X-ray absorption and emission spectroscopies

XAS is a unique tool to study gas-sensing materials. The information obtained by XAS allows the identification of oxidation states, local chemical surroundings and the electronic structure. The information on oxidation states and the chemical structure are extracted from different regions of the X-ray absorption spectrum. The region close to the edge step is called X-ray Absorption Near Edge Structure (XANES) and contains information on the oxidation state of the absorbing element. The most prominent feature, which is related to transitions from core levels to occupied states and thus sensitive to the oxidation state, is the so called whiteline [274]. Above the absorption edge an oscillatory structure is found which can be extracted. The analysis of the Extended X-ray Absorption Fine Structure (EXAFS) gives detailed information on the nature, number and distance of the surrounding atoms and the disorder of the material [274–276]. Therefore, XAS is recognized as the method of choice to study the structure of gas-sensing materials. Unlike other common techniques used to characterize materials (like X-ray diffraction), electron microscopy or photoelectron spectroscopies, XAS has several advantages. Since only photons are used, XAS does not require a vacuum, which is mandatory for photoelectron spectroscopies and electron microscopies. As no vacuum is needed, it is possible to study gas-sensing materials under in situ and operando conditions, making it its biggest advantage [276]. In contrast to X-ray diffraction methods, which require long-range ordered structures, XAS is sensitive to the local chemical surrounding, that is, it can be used to study small or amorphous phases. In addition, XAS is sensitive, even to low concentrated elements, for example, 0.2 wt% Pt in SnO_2 (~0.15 at%) [227]. Thus, XAS is a more suitable technique to study the structure of noble metal loadings in gas-sensing materials, which are typically low concentrated and do not form ordered or bulk-like phases [276]. Furthermore, XAS is an element-specific technique making it an interesting method for studying one element selectively in mixed or doped SMOXs [277]. Hence, XAS is used to study base materials [278–281], noble metal loadings [227,228,241,279,282–289] or filter layers [290]. However, only a small number of these works was done under operando conditions. Nevertheless, ex situ studies of gas-sensing materials provide valuable information. Serrini et al. used XAS to study the influence of the preparation conditions on the sensing material itself [280]. They showed that EXAFS analysis is suitable to study very small grains (<2 nm), even detecting phases not found by XRD. The advantage of XAS for the determination of the oxidation state and structure of noble metal dopants was demonstrated for SnO_2 doped with Pd or Ru, respectively [286]. Using HR-TEM, Marikutsa et al. found that Ru and Pd form a segregated phase on rather large SnO_2 particles. However, no such phase was found on smaller SnO_2 nanocrystals doped with Pd or Ru, even though the successful loading was confirmed by EDX. In case of nanocrystalline SnO_2, the existence of a RuO_2 and a disordered PdO phase was found with XAS, respectively [286]. One of the first in situ measurements on a gas-sensing material was done by Gaidi et al. studying Pt-doped SnO_2, correlating results from XAS obtained from thick films with separately measured resistance data [225]. The first operando XAS cell for gas-sensing devices was introduced by Safonova et al. [226] and used to study Pd-doped SnO_2. In their work, the sensor resistance and X-ray absorption spectra were simultaneously measured in different atmospheres and at different temperatures. This setup allows the correlation of changes in the resistance with changes of the oxidation states of the noble metal loading and the base material. However, the experimental conditions, namely the exposure to rather high H_2 concentrations in the absence of O_2, do not fulfill realistic sensor conditions, as O_2 is essential for the gas-sensing process. The huge influence of the atmospheric composition was shown by Koziej et al. [228], demonstrating that the Pd-loading is reducible in the absence of O_2 (1000 ppm H_2 in He), but remains oxidized upon exposure to 30 ppm H_2 or 50 ppm CO in air. Pd-doped SnO_2 is a good example for how strong the influence of the chosen temperature, and rather more important, the atmospheric composition on the obtained spectroscopic results is, and thus the experiments should be performed under conditions as close as possible to real conditions. The use of the High Energy Resolved Fluorescence Detection (HERFD) mode further improved the information which can be obtained from XAS [291], for example, by revealing that Pt in Pt-doped SnO_2 is even more oxidized than Pt in bulk PtO_2 [227,289]. Another potential application of hard X-ray spectroscopy is the study of the electronic structure of gas-sensing materials and the changes during gas exposure, as demonstrated by Hirsch et al. [232]. While (HERFD-)XANES provides information on unoccupied states of the material, X-ray emission spectroscopy (XES) gives insights into the occupied states of the material [274,291].

One of the sensing materials most widely studied by XAS is Pt-doped SnO_2, which is a good example to discuss the aspects of gas sensor research with XAS. The Pt-doped materials studied by XAS were prepared by various methods including aerosol pyrolysis [225,292], flame spray pyrolysis (FSP) [285], wet impregnation [227,289], and magnetron sputtering [287]. The earliest samples studied by XAS are the ones prepared by aerosol pyrolysis. In the materials with 3 at% and 6 at%, Pt is found in an oxidized state, while for 12 at%, Pt is found to be metallic. The comparison of the EXAFS oscillations of the 6 at% Pt-doped SnO_2 sample with PtO_2 show similar features, however, a detailed analysis of the EXAFS data is missing. The appearance of metallic Pt for high loadings is in agreement with TEM measurements for these materials [293], but one should keep in mind that such high loadings are not typical for gas-sensing materials. The in situ XANES spectra recorded at 350°C during the exposure to air, 300 ppm CO in air and 900 ppm CO in N_2, showed that in air the oxidation state of Pt undergoes only minor changes, while during the exposure to CO in N_2 the decrease of the whiteline intensity indicates a reduction of Pt, even though the spectrum does not resemble the one of metallic Pt [225]. The Pt-doped SnO_2 sensing materials obtained by FSP had a nominal loading of 0.2 and 2.0 wt%. In both cases, Pt was found to be highly oxidized and, as indicated by only a few backscatterers in the Fourier-transformed EXAFS spectrum, is highly dispersed in the supporting SnO_2 [285]. The tendency of Pt to be incorporated in SnO_2 was also found for Pt being introduced in SnO_2 as metal chloride prior to the grain size determining calcination step [227]. Hübner et al. found that Pt is found in chemical surroundings resembling SnO_2 rather than PtO_2 and concluded that Pt replaces Sn in the SnO_2 lattice. The EXAFS analysis is supported by HERFD-XANES recorded during reduction experiments, which indicate that the Pt is still highly oxidized even in 2 vol% H_2 in He at 600°C (see Fig. 3.12, left). As reported for the materials prepared by aerosol pyrolysis during the in situ exposure to CO in air [225], the operando XAS experiments reported by Hübner et al. did not show strong changes in the oxidation state of Pt during the sensor operation (see Fig. 3.12, right) [227]. However, if Pt is introduced in SnO_2 after the grain size determining calcination step, the resulting material shows distinct differences to the one studied by Hübner et al. [227,289]. The HERFD-XANES recorded during the temperature programmed reduction showed, that Pt is easily reduced in 0.5 vol% H_2 in He already at room temperature (see Fig. 3.12, middle). Furthermore, the EXAFS analysis showed Pt in a chemical surrounding containing O in the first and Pt, as well as Sn, backscatterers in the second coordination shell. The authors concluded, that Pt forms a separated oxide phase at the surface [289]. Despite Pt being easily reduced during the reduction experiment, the operando XAS are similar to the ones reported by Hübner et al., that is, Pt is dominantly oxidized under operation conditions (see Fig. 3.12, right) [289]. The sum of in situ and operando XAS rule out a

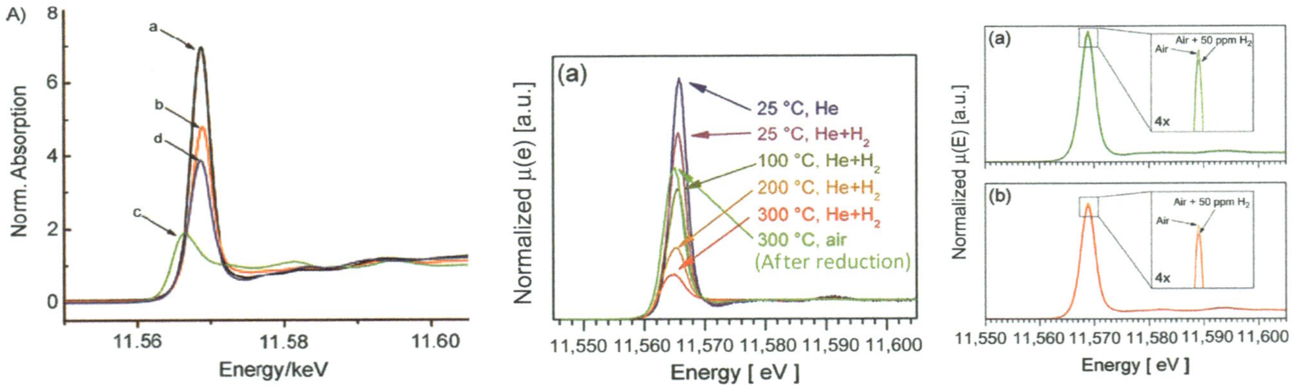

Figure 3.12 HERFD-XANES spectra of 0.2 wt% Pt-doped SnO_2 gas sensors recorded at the Pt L_3 edge. The materials were obtained by different wet impregnation routes. The left figure shows the reduction experiment of the materials, prepared by introducing Pt before the calcination step, (A) in air at 300°C, (B) in 2 vol% H_2 in He at 600°C, (C) metallic Pt reference, and (D) PtO_2 reference. The figure in the middle shows the reduction experiment of the material obtained by introducing Pt after the calcination step. The right figure shows the operando experiment (50 ppm H_2 in air at 300°C) of the materials—top: Pt introduced after the calcination; bottom: Pt introduced before the calcination.
Source: Reprinted with permission from M. Hübner, D. Koziej, M. Bauer, N. Barsan, K.O. Kvashnina, M.D. Rossell, et al., The structure and behavior of platinum in SnO_2-based sensors under working conditions., Angew. Chem. Int. Ed. Engl. 50 (2011) 2841–2844, doi:10.1002/anie.201004499; D. Degler, H.W. Pereira de Carvalho, K.O. Kvashnina, J.-D. Grunwaldt, U. Weimar, N. Barsan, Structure and chemistry of surface-doped Pt:SnO_2 gas sensing materials, RSC Adv. 6 (2016) 28149–28155, doi:10.1039/C5RA26302F.

metallic Pt phase being responsible for the enhanced gas-sensing performance, for example, by a spill-over of H_2 as proposed in earlier studies [294]. A rather interesting observation of these works is that Pt was not found forming separated PtO_2 phases, but rather is found incorporated in SnO_2 or forming an oxide phase with a strong contribution of Sn. Murata et al. corroborated that trend studying a series of Pt-doped SnO_2 by magnetron sputtering. Pt-doped SnO_2 was found to form solid solutions with a rutile structure up to 10 at% Pt in SnO_2 [287]. The discussed examples show that noble metal loadings not only separate off metal or metal oxide clusters, but may form atomic sites or mixed phases with the host oxide. In addition, even small changes in the preparation of sensing materials may result in different structures and material properties, as demonstrated for the materials obtained by wet impregnation of SnO_2.

The work of Hirsch et al. demonstrates the potential of combining HERFD-XANES and resonantly exited valence to core XES (vtc-XES) to study the electronic structure of $La_2O_2CO_3$ and its changes upon gas exposure [232]. Operando HERFD-XANES and vtc-XES were recorded at the La L_3 edge in different atmospheres at 250°C. The exposure to water did not influence the HERFD-XANES (Fig. 3.13A), while the exposure to CO_2 causes considerable changes in the HERFD XANES (Fig. 3.13B) and vtc-XES (Fig. 3.13C). This finding reveals that La sites are selective to CO_2 and do not interact with water. However, since water causes a change in the resistance, it is likely that water interacts with another adsorption site not related to La and, therefore, is not detectable by an element-specific technique like XAS. A different situation is found when examining the HERFD-XANES and vtc-XES recorded during CO_2 exposure. The sharpening of the whiteline (Fig. 3.13B) and the changes of the peaks at −10 and −50 eV in the vtc-XES indicate that CO_2 has an oxidative character by increasing the oxygen coordination of La at the surface [232].

3.2.1.6 Mössbauer spectroscopy

Mössbauer spectroscopy can provide structural information of SMOX materials, if the metal compound exhibits a Mössbauer effect. Examples relevant for sensing are ^{57}Fe, ^{119}Sn, ^{121}Sb, ^{182}W, or ^{197}Au [214,295]. However, the method is not only limited by the low number of suitable elements, it is also limited by the experimental conditions, for example, the low temperature (−173°C) which requires a quenching of the specimen before the spectroscopic experiment [296]. Conversion electron Mössbauer spectroscopy (CEMS) [297] is considered as a possible operando technique for gas sensor research, for example, in the review article of Gurlo and Riedel [214], but since then, no further reports on in situ or operando Mössbauer spectroscopy of gas sensors are found. As stated in the frequently cited work of Niemantsverdriet and Delgass [298], Mössbauer spectroscopy can be used to access the oxidation states and the phase (chemical surrounding) of a material. Similar results can be obtained from XAS, which is element-specific as well, but applicable to a large number of elements and an established operando technique for gas sensor research.

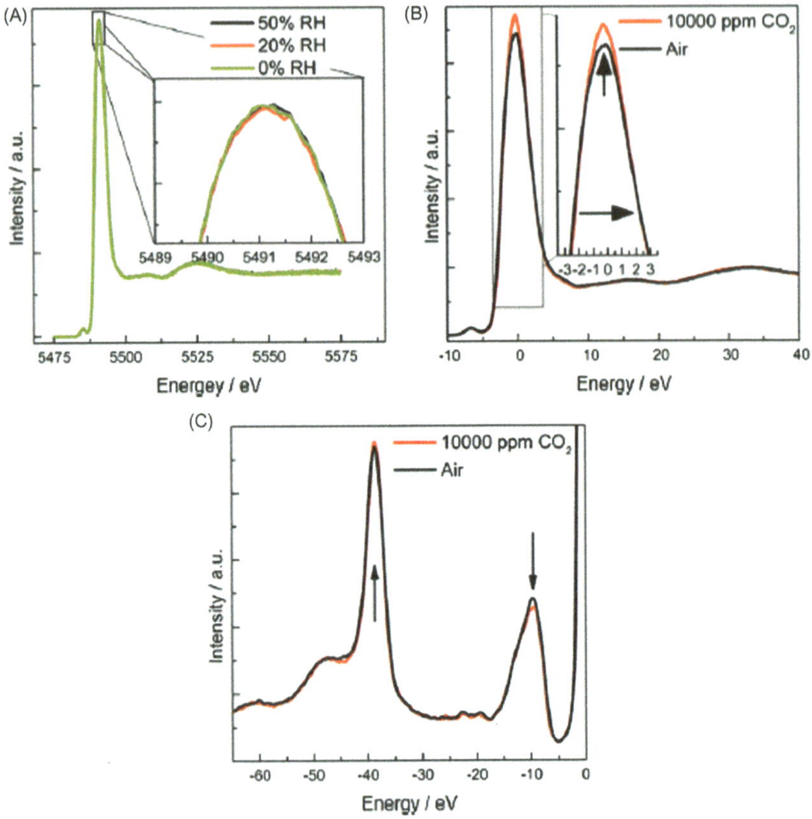

Figure 3.13 In situ X-ray absorption (A, B) and X-ray emission spectra (C) of a $La_2O_3CO_3$-based gas sensor at 250°C, exposed to different humidity backgrounds (A) and the exposure to 10,000 ppm CO_2 in humid air (50%r.h.).
Source: Reprinted with permission from O. Hirsch, K.O. Kvashnina, L. Luo, M.J. Süess, P. Glatzel, D. Koziej, High-energy resolution X-ray absorption and emission spectroscopy reveals insight into unique selectivity of La-based nanoparticles for CO_2, Proc. Natl. Acad. Sci. 112 (2015) 15803–15808, doi:10.1073/pnas.1516192113.

3.2.2 Electrical techniques

Doubtless, electrical measurement techniques are the most vital techniques for gas sensor research, ranging from basic DC and AC resistance measurements to rather sophisticated measurements, like simultaneous work function and resistance measurements or combined conductance and Hall effect measurements. Unlike for the operando spectroscopic techniques, the electrical measurements do not require the complex development of operando cells, which embed a gas sensor in the beam path of a spectrometer. However, the measurement setups, for example, in case of the Kelvin probe, or data evaluation was steadily improved. Furthermore, as discussed in the following paragraphs various parameters, for example, like the homogeneity and morphology of the sensing layer, have an influence on the obtained results and have to be considered when performing these measurements.

3.2.2.1 DC-resistance measurements

Most certainly, DC-resistance (DCR) measurements are the most widely spread technique for gas sensors based on SMOX, which are used to evaluate the gas-sensing performance of a material. Although, DCR measurements are a common technique, it is important to consider the different factors influencing the measurement. The homogeneity and morphology of the sensing layer itself, the electrode geometry, electrode material, the number of sensors measured at the same time and the electrical measurement parameters itself influence the outcome of DCR measurements [205,212,213]. Already the way in which the resistance is measured, that is, the applied voltage or current itself will affect the obtained results [299]. The resistance of a typical porous sensing layer is commonly described as a series of potential barriers. In that case, the potential drop used to measure the resistance is distributed over the different series elements and concentrated over the high resistance elements, namely the potential barriers at the grain-grain and grain-electrode contact. It is shown, that the measured resistance depends on the applied voltage, and more interestingly that this dependency is different for different gas atmospheres. Consequently, one will obtain different relative resistance changes, that is, sensor signals, for different potentials [299]. To avoid such effects, a defined polarization and constant measurement potential should be guaranteed during a measurement. How the layer morphology and the SMOX electrode contact influence the sensing process is nicely demonstrated when comparing compact and porous layers [212] with different electrode spacing [299,300]. Hoefer et al. studied the effect of the SnO_2 electrode contact for thin film layers and found a distinct influence of the SnO_2 electrode contact region on the sensor signals to CO and NO_2 [300]. The work of Bauer et al. on thick, porous layers found an approximately linear dependence of the resistance on the electrode spacing in air and under exposure to CO and NO_2 [299]. These findings are explained by the differences in the layer morphologies: In case of a thick porous layer, the overall conduction is controlled by back-to-back Schottky barriers at the grain-grain boundaries, which outnumber the SMOX-electrode contacts by far, that is, the influence of the latter is minimized as the SMOX-electrode contact resistances are in series with a number of grain-grain contact resistances several orders of magnitude higher. A different situation is found for compact thin film layers, for which the resistance is approximately the sum of the bulk resistance and the SMOX-electrode contact [212], that is, the contribution of the SMOX-electrode contact is dominant. The effect of the electrodes can be both electrical and chemical [212]. When comparing Pt and Au electrodes, strong differences in the chemistry are observed. While metallic Pt leads to a parallel reaction not contributing to the sensing effect, metallic Au improves the oxygen adsorption on SnO_2 enhancing the sensing effect [301]. In conclusion, the DCR measurements taken as "simple" are rather complex as various factors—besides the interaction of the sensing material with the target gases—determine the outcome of such measurements. Consequently, one should carefully consider these factors when performing gas-sensing measurements.

3.2.2.2 Impedance spectroscopy

Impedance spectroscopy, also electrochemical impedance spectroscopy (EIS), can be used to identify single elements of a complex system or device, or, depending on the AC frequency, determine the nature and contribution of the charge carrying species (ions or electrons/holes, e.g., proton conduction) [264]. In case of a thick film SMOX gas sensor, different contributions of the surface and bulk of the sensing material, as well as the influence of grain-grain and grain-electrode contacts can be studied under realistic operation conditions. The interpretation of the obtained equivalent circuit allows the analysis of the changes in the bulk resistance, in the depletion/accumulation layer, as well as the influence of the SMOX-electrode contact [207,212,213,302]. A good example is the impedance spectroscopy study of a Pd-doped SnO_2 thick film gas sensor [302]. As shown in Fig. 3.14 (on the left side) the resistance of a porous sensing layer is described as a series of different elements, namely the resistance of the bulk of the sensing material represented by a resistor (R_b), the grain-grain contact described by a parallel resistor (R_{gb}) and capacitor (C_{gb}), and the SMOX-electrode contact also described by a resistor (R_c) and capacitor (C_c). As indicated in Fig. 3.14, the bulk resistance is proportional to the concentration of free charge carriers (n_b) and constant. The resistance of both contact regions (R_{gb} and R_c) depend on the surface concentrations of the free charge carriers, which are limited by the respective surface band bending. The surface band bending of the SMOX-electrode contact depends on the work function difference of the sensing and electrode materials and is not dependent on the gas atmosphere [212]. However, the surface-band bending of the SMOX material, and thus at the grain–grain contacts, depends on the concentration of ionosorbed oxygen and is strongly affected by the atmosphere composition [212]. This model for the charge transport in a thick film SMOX layer is resembled by the equivalent circuit of the experimental data (Fig. 3.14, top right). The impedance data is fitted by a resistor R_0 corresponding to R_b, a parallel resistor and constant phase element (CPE) representing the grain-grain contact and a parallel resistor and capacitor for the SMOX-electrode contact. An additional parallel capacitance (C_{geom}) owing the capacitance of the sensing substrate and electrodes themselves is used for the fit. The fitted value for C_{geom} is independent of the gas atmosphere and corresponds to the one of a blank sensor substrate. The fitted resistance R_1 (grain-grain contact) decreases with increasing CO concentration (Fig. 3.14, bottom right), while the fitted capacitance of the CPE increases [302]. The observed trends for R_1 and CPE indicate the decrease of the band bending at the SMOX surface, that is, the back-to-back Schottky barriers at the grain-grain boundaries. It should be mentioned that analysis of the data in more complex atmospheres, for example, the presence of water vapor, is rather complex and additional effects have to be considered [302]. However, the discussed example clearly demonstrates the insights in the charge

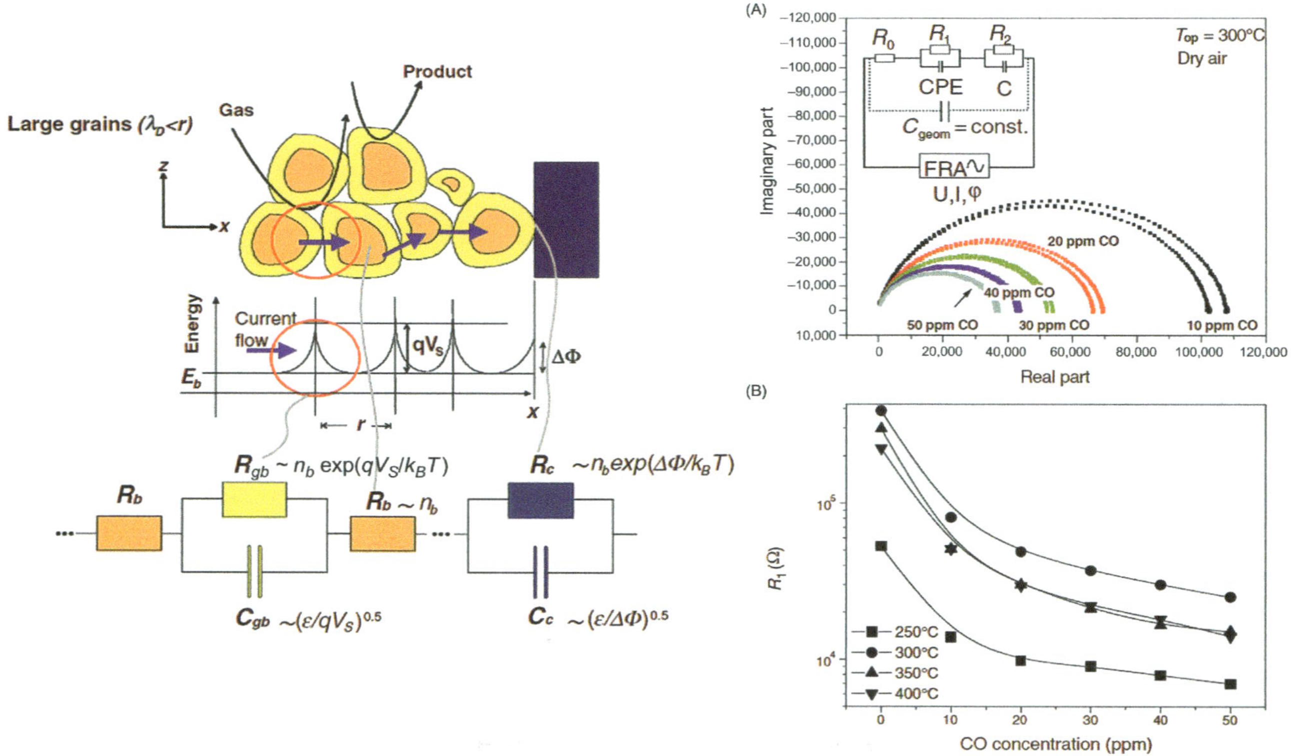

Figure 3.14 Insights of impedance spectroscopy measurements in the charge transport in a porous thick film sensing layer. Left: Equivalent circuit for the different contributions found bulk resistance (R_b), grain–grain contact (R_{gb} and C_{gb}), and SMOX-electrode contact (R_c, C_c). Right: Impedance data of a Pd-doped SnO_2 sensor exposed to dry air, including the equivalent circuit used for fitting the data (top); and the decrease of R_1 (corresponding to R_{gb}) as a function of the CO concentration.

Source: Reprinted with permission from N. Barsan, U. Weimar, Understanding the fundamental principles of metal oxide based gas sensors; the example of CO sensing with SnO_2 sensors in the presence of humidity, J. Phys. Condens. Matter. 15 (2003) R813–R839.

transport in sensing layers obtained by impedance spectroscopy, which are not accessible by DCR measurements.

3.2.2.3 Operando work function measurements

The work function of a gas sensor provides information on the occupation of the surface states, for example, caused by ionosorption of gases forming charged adsorbates, as well as on uncharged adsorbates exhibiting a dipole momentum. The work function ϕ of a semiconductor is defined by

$$\phi = (E_C - E_F) + qV_S + \chi \tag{3.15}$$

with $(E_C - E_F)$ as the distance of the CB edge E_C in the bulk and the Fermi-level E_F, the surface band banding qV_S and the electron affinity χ [212,303]. Accordingly, the occupation of the electronic surface states is derived from qV_S, while the presence of dipoles is associated with χ [303]. The absolute work function can be determined by photoelectron spectroscopies, but due to the pressure gap these techniques are not suitable to study the adsorption of gas species on SMOX surfaces or even gas sensors under operation conditions. A more suitable approach is a Kelvin probe (KP) setup, which is used to measure the contact potential difference (CPD) between the sensing layer and the reference with a work function not affected by changes in the atmosphere. However, this approach does not provide the absolute work function, but the changes of the work function [303]:

$$\Delta\phi = q\Delta V_S + \Delta\chi \tag{3.16}$$

The combination of simultaneous CPD and DCR measurements has proven to be an adequate and efficient tool to study gas-sensing materials [126,130,262,284,302,304–308]. Although this approach was first applied in the 1980s by Mizsei and Harsányi on sputtered SnO_2 thin films [309], the development of simultaneous CPD and DCR measurements with the KP setup took several steps within the last decades. In 1991, the first extended and systematic study of a Taguchi-type gas sensor in operation conditions was reported by Schierbaum et al. [310]. An essential aspect of the KP setup is the use of a reference material inert to changes in the gas atmosphere. Doll et al. compensated the work function changes of an Au-KP-tip by referencing their measurements to a silica nitride layer [311]. A more convenient solution is the use of passivated tips [262,302], which are in use to date. Besides the role of the reference material there are two different approaches to measure the CPD: zeroing/lock KPs, which are rather stable, but prone to systematic errors, and nonlocking KPs, which are less prone to systematic errors. A detailed explanation of the KP setup and the different measuring principles is given by Oprea et al. [303]. The simultaneous measurement of the work function and the DCR offer the possibility to probe changes of the electron affinity under operation conditions, which is not possible using other electrical techniques. Changes of the electron affinity are obtained by subtracting the change in the surface band bending from the measured change of the work function. The change of the surface band bending is calculated from the DCR measurement [212]:

$$q\Delta V_S = kT \cdot ln\left(\frac{R_{gas}}{R_0}\right) \tag{3.17}$$

R_0 is the resistance of the sensing layer in reference conditions (e.g., pure carrier gases) and R_{gas} is the resistance of the sensing layer under test gas exposure. The obtained equation

$$\Delta\chi = \Delta\phi - kT \cdot ln\left(\frac{R_{gas}}{R_0}\right) \tag{3.18}$$

allows the calculation of changes of the electron affinity. This is demonstrated by Koziej et al. who investigated the mechanism of propane sensing with undoped SnO_2 in dry air [262]. However, the major changes of the work function were found to be related to changes in the band bending, while only small changes of the electron affinity were found. A different situation was found by Hübner et al. investigating the CO detection of undoped CuO in dry and humid conditions [306]. While upon CO exposure in dry air the electron affinity remained almost unaffected (Fig. 3.15, left), the exposure to water vapor caused a strong increase of the electron affinity (Fig. 3.15, right), assigned to the formation of hydroxyl groups on the surface. Furthermore, a strong decrease of the electron affinity was found upon CO exposure in humid conditions (Fig. 3.15, middle), indicating a removal of the hydroxyl groups. This example demonstrates the possibility to access the formation of adsorbate species by electrical, that is, nonspectroscopic, techniques.

Another important application of simultaneous CPD and DCR measurements is the investigation of the transduction mechanism in SMOX-based gas-sensing materials and studying the relation between the changes in the surface band bending and the resistance of the sensing layer. However, as the work function changes also include the electron affinity, this approach is only valid as long as the electron affinity is not changed, that is, only in dry conditions. Thus, in cases where $\Delta\chi \approx 0$, changes of the surface band bending are equal to changes of the work function. A good example for this approach is the switch from depletion layer to accumulation layer controlled transduction, reported by M. Hübner et al. [304]. In this work, undoped SnO_2 was exposed to O_2 and reducing gases (CO, H_2) in different O_2 backgrounds. The left part in Fig. 3.16 shows the relation between the changes in the surface band bending ($q\Delta V_s$) and the resistance. The reference point, or flat-band situation ($q\Delta V_s = 0$), is taken in pure N_2, that is, the absence of any adsorbate species, which could cause a surface band bending. The curve progression shows three regions with a different relation between the surface band bending and the resistance of the material. One region is above the flat-band situation ($q\Delta V_s > 0$) and two below. The region above the flat-band situation is caused by the exposure to O_2 and the resistance R is proportional to qV_s by $exp\left(\frac{qV_s}{kT}\right)$. This

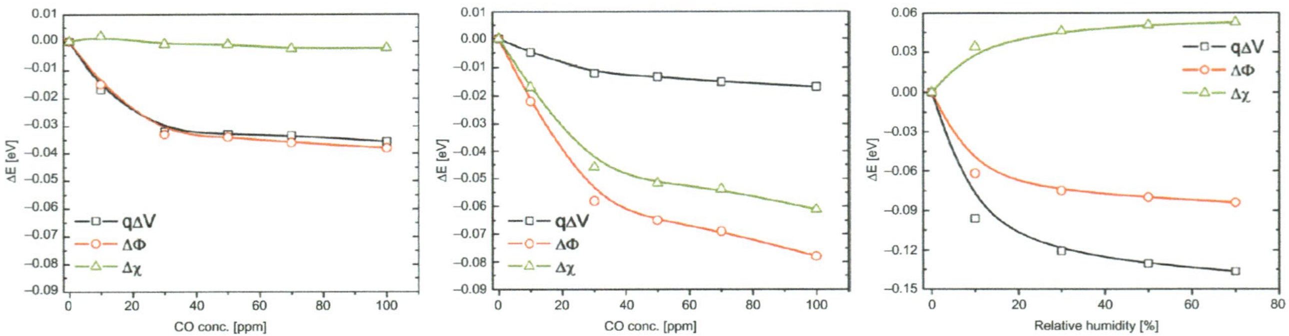

Figure 3.15 Results of simultaneous CPD and DCR measurements of CuO exposed to CO in dry air (left), CO in 50 %r.h. (middle), and different concentrations of water vapor (right).

Source: Reprinted with permission from M. Hübner, C. Simion, A. Tomescu-Stănoiu, S. Pokhrel, N. Barsan, U. Weimar, Influence of humidity on CO sensing with p-type CuO thick film gas sensors, Sens. Actuators B Chem. 153 (2011) 347−353, doi:10.1016/j.snb.2010.10.046.

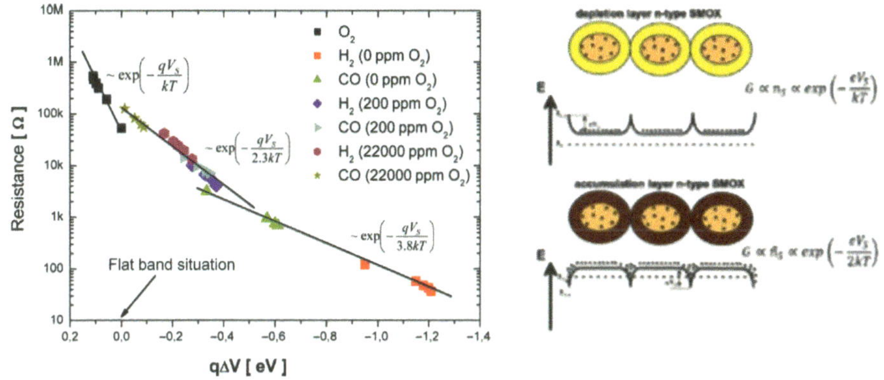

Figure 3.16 Experimental relation between the sensor resistance and the change of the surface band bending found by simultaneous CPD and resistance measurements (left) and the corresponding conduction models (right).
Source: Reprinted with permission from N. Barsan, M. Hübner, U. Weimar, Conduction mechanisms in SnO_2 based polycrystalline thick film gas sensors exposed to CO and H_2 in different oxygen backgrounds, Sens. Actuators B Chem. 157 (2011) 510–517, doi:10.1016/j.snb.2011.05.011; N. Barsan, J. Rebholz, U. Weimar, Conduction mechanism switch for SnO_2 based sensors during operation in application relevant conditions; implications for modeling of sensing, Sens.Actuators B Chem. 207 (2015) 455–459, doi:10.1016/j.snb.2014.10.016 [312].

relation is in agreement with the formation of a depletion layer at the surface of SnO_2 and an upwards band bending controlling the conduction in the sensing layer (Fig. 3.16, top right). The exposure to reducing gases in low oxygen backgrounds results in a relation of $R\sim exp\left(\frac{qV_s}{2.3kT}\right)$, which is close to the theoretical value of $R\sim exp\left(\frac{qV_s}{2kT}\right)$ for a conduction controlled by a downwards band bending, that is, formation of an accumulation layer (Fig. 3.16, bottom right). The changing slope in the accumulation layer-controlled regime (at -380 meV) corresponds to the point when the CB edge pins the Fermi level. In case of undoped SnO_2, the change from depletion layer to accumulation layer-controlled conduction occurs close to N_2 conditions indicating intrinsic surface states causing an additional band bending, that is, the assumption of the flat-band situation is valid. A shift of the change of transduction models indicates the presence of an initial band bending, which was experimentally proven for various noble metal loadings by combined CPD and DCR measurements [126,130,305].

3.2.2.4 Hall-effect measurements

In order to understand the charge transport within the sensing layer, it is important to evaluate whether the changes in the resistance are solely caused by changes in the charge carrier concentration or also by changes of their mobility. Hall-effect measurements are the most widely used technique to determine the components of the conductivity, that is, carrier concentrations and their mobility, in conducting materials. Hence, this technique has been successfully applied to various gas-sensing materials

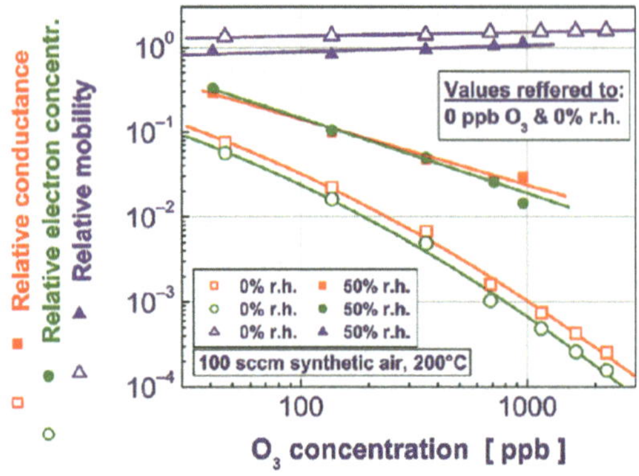

Figure 3.17 Relative values of the effect electrical parameters of corundum-type In_2O_3 exposed to O_3 at 200°C in dry and humid conditions.
Source: Reprinted with permission from A. Oprea, A. Gurlo, N. Barsan, U. Weimar, Transport and gas sensing properties of In_2O_3 nanocrystalline thick films: a Hall effect based approach, Sens. Actuators, B Chem. 139 (2009) 322–328, doi:10.1016/j.snb.2009.03.002.

like Ga_2O_3, In_2O_3, SnO_2, ZnO, and WO_3, as summarized by Oprea et al. [313]. A description of the experimental setup and the different approaches to extract and treat data is also given by Oprea et al. [313]. While the analysis of Hall-effect measurements on single crystals allows a straightforward extraction of the charge carrier concentration, the analysis of data obtained from sensing layers (in general porous thick films) is rather complex encountering problems related to the structure and morphology of sensing layers and the operation of SMOX-based gas sensors. Hence, additional considerations have to be made for the data analysis of gas sensors utilizing solid-state or the more suitable chemo-physical modeling procedures [313,314]. Even though there are various works and approaches to analyze Hall-effect measurements, no general solution is available due to the complexity of the gas-sensing process and the varying factors related to the sensing layers themselves. Still, there are some good examples using Hall-effect measurements to study gas-sensing materials. One of these examples is the study of In_2O_3 used as an ozone sensor [315]. A key finding for corundum-type In_2O_3 is the correlation of the changes in the conductance (Fig. 3.17, red squares) and the electron concentration (Fig. 3.17, green circles), while the electron mobility (Fig. 3.17, blue triangles) shows only minor changes (factor 1.5 or less). This trend is in line with the common model for SMOX-based gas sensors which ascribes the resistance changes/sensor signals to changes in the surface concentration of charge carriers [212].

3.2.3 Residual gas analysis

In various works, the exhaust gas was analyzed [230,231,262,265]. Since the available techniques involve not only spectroscopies, the residual gas analysis will be discussed separately.

The fundamental step, which is the basis for gas sensors, is the interaction of the analyte gas with the sensitive layer, often called reception. The reception can either be simply the adsorption/desorption of gases, that is, the analyte gas is not transformed during the sensing process, or a reaction of the gases on or with the surface of the sensing material, that is, the analyte gas is decomposed and transformed into another gas.

Adsorption/Desorption mechanism:

$$NO_2 + e^- \leftrightharpoons NO_2^- \quad (3.19)$$

Decomposition of the analyte gas:

$$NO_2 + V_O + 2e^- \rightarrow O^{2-} + NO \quad (3.20)$$

$$2NO_2 + 2V_O + 4e^- \rightarrow 2O^{2-} + O_2 + N_2 \quad (3.21)$$

To differentiate and further understand both reception mechanisms, the qualitative and quantitative composition of the gas phase provides fundamental information, whether and to which extent the analyte gas is changed by the interaction with the gas sensor. Since a gas sensor consists of different materials (substrate, sensitive layer, electrodes, and heaters), the measurement should take into account that not only the sensing layer interacts with the gases. In practice, this can be done by measuring powders of the sensing material, using passivated sensor substrates or by measuring blank samples for comparison [316]. The range of available techniques to analyze the gas phase is large and the preferred technique depends on the target and the interfering gases as well as the concentration range. The analyzing techniques either can be selective/specific (e.g., chemiluminescence based devices and liquid or solid-state electrochemical cells) or nonselective (e.g., infrared [130,231,262] or mass spectrometers [259,263]). The combination of residual gas analysis with electronic and/or operando-spectroscopic experiments is considered a powerful approach to understand materials, as changes of the sensing material—electronic properties, material composition or surface species—can be directly correlated with changes in the atmosphere.

3.2.4 Examples for operando studies of doped SnO_2 sensing materials

Operando studies of gas-sensing materials provide insights in the relation between the structure, material properties and the function of the gas-sensing materials. The understanding of structure-function relationships enables the knowledge-based design of gas-sensing materials as an alternative to the trial-and-error-based development of sensing materials. The potential of this approach is demonstrated by investigations on the gas detection mechanism of undoped and doped SnO_2 gas-sensing materials, which will be discussed in the following section.

Understanding the working principle of undoped, that is, pristine, SnO_2 is the basis for investigations on doped, that is, noble metal loaded, SnO_2 sensing materials. The current model for undoped SnO_2 was successively improved based on theoretical calculations/modeling and experimental results obtained with operando methods. The detection of reducing gases on undoped SnO_2 is based on the reaction of the analyte gas with preadsorbed oxygen (Fig. 3.18A). This reaction is based on the ionosorption of atmospheric oxygen, which causes an electron depleted surface layer. The extent of this depletion layer depends on the concentration of ionosorbed surface oxygen, which is determined by the adsorption and desorption of atmospheric oxygen and the reaction of ionosorbed oxygen with the reducing gas [212]. However, undoped SnO_2 also responds to reducing gases in the absence of oxygen. The presence of reducing gases causes an electron accumulation layer, which initially was ascribed to the formation of analyte gas related donor species, for example, CO^+ or H^+, in the absence of atmospheric oxygen [304,317]. The transition from depletion to accumulation controlled conduction is demonstrated by simultaneous work function and resistance measurements (see Fig. 3.16). The formation of such an accumulation layer in ambient conditions, that is, in the presence of oxygen and water vapor, casted doubts on the existence of an analyte gas related donor species. In order to determine the nature of the electronic donor species, which causes the accumulation layer, the surface chemistry of undoped SnO_2 was studied by

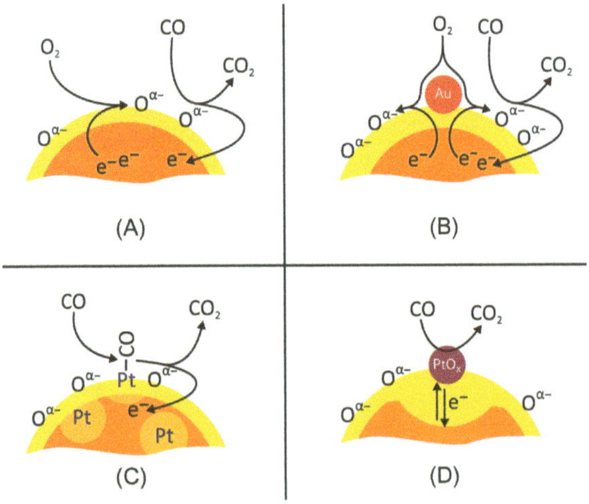

Figure 3.18 Current models for the gas reception on undoped and doped SnO_2-based gas sensors: Gas detection on undoped SnO_2 (A), sensitization by oxygen spill-over on Au-loaded SnO_2 (B), activation of lattice oxygen by atomically distributed Pt, and (C) Fermi-level controlled gas detection on Pt oxide loaded SnO_2. The models are based on the operando studies of the corresponding materials. The electron depleted surface layer is shown in yellow and the unaffected bulk in orange. A detailed discussion is given in the text. *Source*: Adapted from D. Degler, Spectroscopic Insights in the Gas Detection Mechanism of Tin Dioxide Based Gas Sensors, Eberhard Karls University of Tübingen, 2017 [318].

operando IR spectroscopy. The spectra revealed that in dry and humid air, as well as in the absence of oxygen, the gas reception is based on the same surface reactions, namely the reaction of the analyte gases (CO, H_2, and H_2O) with surface oxygen forming oxygen vacancies, which act as electron donor. The surface concentration of oxygen vacancies depends on the reduction of the surface by the analyte gases and the (re-)oxidation by atmospheric oxygen [257]. If the surface concentration of oxygen vacancies is lower than in the bulk of the SnO_2 grains, a depletion layer is formed, while when the concentration of oxygen vacancies at the surface exceed the one in the bulk an accumulation layer is formed. The gas reception via the reduction of the SMOX sensing material is also shown for WO_3 [265] and In_2O_3 [230,231] by using operando IR or Raman spectroscopy, respectively. This reception mechanism can be altered or completely changed by the presence of noble metal loadings (Fig. 3.18).

The influence of noble metal loadings on SMOX sensing materials is rather complex, but the general sensitization mechanisms are distinguished: A solely chemical sensitization, for example, by the improved adsorption and spill-over of the reactive species, or a dominantly electrical effect based on the electronic coupling of noble metals with the SMOX; the latter case may also involve changes of the chemical reactions [294,319,320].

Loading SnO_2 with Au particles is a well understood example for a spill-over sensitization mechanism, which solely changes the chemical processes without affecting the electronic properties of the SMOX [284,321]. It was found, using operando XAS, that Au is always present in the metallic state and that the oxidation state is not affected by the composition of the atmosphere [284,321]. The presence of these metallic Au clusters improves the CO oxidation and CO sensing [284]. The sensitization by an electronic interaction, that is, charge transfer, between SnO_2 and the Au clusters was excluded by operando work function measurements and, like for undoped SnO_2 (Fig. 3.16), the flat-band situation is found in nitrogen. Thus, there are no intrinsic electronic states and the space charge layer at the grain surface is determined by the ionosorption and the reactions of surface oxygen species [284]. The comparative study of undoped and different Au-doped SnO_2 materials shows, that there is a strong enhancement of the oxygen ionosorption rather than an improved CO adsorption [321]. This is in line with operando XAS and DRIFTS measurements, which indicate an interaction of O_2 with Au and an increased oxidation of the SnO_2 surface, but do not show adsorbed CO species [321]. The sensitization mechanism of Au on SnO_2 is summarized in Fig. 3.18B: O_2 is adsorbed on metallic Au clusters and subsequently dissociated, transferred to and ionized by SnO_2; the increased concentration of surface oxygen increases the reactivity of the surface and consequently the sensor signals, while the basic reception and transduction mechanism of undoped SnO_2 is not changed.

Significant changes in the chemical and electronic properties are observed in the case of Pt-doped SnO_2 materials. The changes, and consequently the gas-sensing performance, depend on the Pt structure, which is determined by the introduction method for the noble metal. Taking sol-gel-made SnO_2 as an example, the Pt precursor can be introduced before [322] or after [323] the calcination of the as prepared powder. If Pt is introduced before the calcination, the sensor signals to CO are only slightly affected by humidity and are decreased compared to the ones of undoped SnO_2 in dry conditions, but in humid air higher than the ones of undoped SnO_2 [289]. However, if Pt is introduced after the calcination, the sensor signals to CO in dry air are significantly lower than the ones of undoped SnO_2, but in humid air the materials show the highest sensor signals to CO [289]. The fundamental difference of both Pt-doped SnO_2 materials is found in the Pt structure: The evaluation of the EXAFS revealed, that if Pt is introduced before the calcination, Pt is atomically distributed in the SnO_2 lattice [227], while if Pt is introduced after the calcination it forms a separate Pt oxide phase, which is in close contact with the supporting SnO_2 [289]. XANES spectra recorded during the temperature programmed reduction (Fig. 3.12, left and middle) show, that in case of the atomically distributed Pt, no reduction is observed, since Pt is incorporated in the bulk and the surface of SnO_2 [227], while the Pt oxide clusters are easily reduced and reoxidized [289], since they are in close contact with the atmosphere and thus at the surface of the SnO_2 grains. In both cases, the operando XANES spectra recorded during gas sensing show no reduction of Pt (Fig. 3.12, right) [227,289].

The electronic and chemical effect of both Pt structures is different. In the case of atomic Pt sites, Pt, on the one hand, acts as an electron acceptor decreasing the concentration of free charge carriers and thus increasing the resistance of the material [130]. On the other hand, the Pt incorporated in the SnO_2 surface acts as a reactive site increasing the surface reactivity [259]. Operando DRIFTS combined with analysis of the exhaust gases showed, that the atomically distributed Pt (Fig. 3.18C) increases the adsorption of CO and activates lattice oxygen bound to Pt [259]. The selective enhancement of CO adsorption on Pt and the increased reactivity explain the low impact of water vapor on the CO sensor signals. In the case of Pt oxide clusters on the SnO_2 surface, operando DRIFTS and further resistance measurements show that the reaction takes dominantly place on the Pt oxide clusters, which are electronically coupled with SnO_2, so that changes of the Pt oxide clusters' stoichiometry change the electronic interface of Pt oxide and SnO_2 and, thus, the resistance of SnO_2 (Fermi-level control mechanism) [324]. In dry air, the oxidation of CO on the Pt oxide clusters causes only small changes in the stoichiometry due to the feasible reoxidation of Pt oxide. Consequently, there are only small changes in the resistance of SnO_2. However, in humid air, the reoxidation is inhibited and the stoichiometry of the Pt oxide clusters is changed, releasing electrons back to SnO_2, which were trapped in the Pt oxide clusters, and thus lowering the resistance of SnO_2 (Fig. 3.18D). The Fermi-level control mechanism and the lack of changes in the Pt oxide phase in dry air explain the observed gas-sensing behavior of the material.

In summary, these examples demonstrate the advantage of combining suitable ex situ, in situ and operando methods to investigate the structure-function-relationships of gas-sensing materials and, thus, provide the basis to prepare tailor-made gas-sensing materials. And yet, these examples represent only a small fraction of all the different SMOXs, dopants and structures. A general understanding of the structure-function-relationships of SMOX-based gas sensors will still require extensive and systematic attempts to study gas-sensing materials by the above described methods.

3.2.5 Summary and outlook

Without any doubt, the techniques used to study SMOX-based resistive gas sensors developed over the last decades. Moreover, this development is beyond the sole refinement of available techniques. Much more essential, the approach to study the gas-sensing effects of SMOXs changed. At the end of the 1980s, gas-sensing materials and essential mechanisms of gas sensing were studied by vacuum techniques mainly on idealized materials, that is, single crystal surfaces [206]. The current approach aims to study real sensing materials, that is, porous thin and thick films, under application relevant conditions, while simultaneously recording the sensor response [213,214]. These in situ and operando methods have proven their high potential in providing essential insights in the gas-sensing process, as discussed for selected examples in the previous sections. Today, many common spectroscopic techniques are available as operando methods to study gas sensors, but still, the refinement of these methods and the application of sophisticated experimental methods, for example, time-resolved measurements, isotopic labeling or modulation excitation spectroscopies [257,258,263,264,325,326], offer ways to enhance the information that can be extracted from operando measurements on gas sensors and thus allow the identification of the active species. Unlike for a catalyst, in which the active species defines the ones actively involved in the catalytic reaction, the active species involved in charge transfers, that is, gas sensing, may differ from the most chemically active one. Besides operando studies of gas sensors, an extensive and detailed knowledge of the material and device structure, that is, crystal structure, electronic states as well as layer morphology, is essential to understand the sensing process. The structural information should be gathered as close to operation conditions as possible. The feasibility of this was, for example, demonstrated by determining the (temperature dependent) crystal structure of a WO_3 sensing layer at various operation temperatures [327]. Neither the progress of many known vacuum technologies, like photoelectron spectroscopies or electron microscopies, should be disregarded, as many approaches for (near)ambient pressure photoelectron spectroscopies or environmental electron microscopies are getting available at synchrotron light sources or even at laboratories [328,329]. However, it remains questionable whether these new vacuum methods will bridge the pressure, temperature and material gaps, but at least the gaps will be narrowed.

And now, how should one investigate gas sensors? Apparently, a general and all-embracing answer is hardly possible, but there are essential requirements and considerations for gas sensor research:

- Gas sensors have to be understood as a whole, with all the complex influencing factors, ranging from the selected sensing material itself, the sensor design and the layer preparation to the selected methods and parameters to extract the information.
- Investigations on gas sensors should address all relevant chemical, physical and electrical aspects as well as the different length scales ranging from atomistic processes to macroscopic.
- The conditions of the measurements should be as close as possible to the actual application conditions, in terms of operation temperature, analyte gas concentration ranges,

oxygen level and major interfering gases like water vapor. Although, with increasing complexity, for example, of the atmospheric composition, additionally involving simpler conditions might be useful to understand the complex ones.
- Characterizing the sensing material and sensitive layer is essential for analyzing the results of operando measurements. This characterization should involve suitable material characterization techniques (ex situ, in situ or operando) such as XRD, spectroscopies (e.g., Raman, IR or XAS) and electron microscopy; as well as the sensor performance (DCR measurements) in application-relevant conditions.
- Gathering information on the same object of research with different (operando) methods, for example, spectroscopic and electrical, provides complementary information and is essential to obtain the full picture of the gas-sensing process.
- Investigating trends within a series of rational selected samples (pristine and doped oxides) or conditions (operation temperatures as well as atmospheric compositions) provides deeper insights in the gas-sensing process. Generally, the selection of a reference or benchmarking increases the comparability of a study.
- A better interpretation or understanding of the results from operando measurements can be achieved by involving results obtained from idealized systems, either from experimental or theoretical works, for example, surface science or computational chemistry studies, respectively. However, with increasing complexity of the real sample the applicability of these results should be critically reviewed.

Matching all criteria remains a rather challenging task, although it is considered to be essential to improve the understanding of gas sensors based on reliable and consistent experimental proofs. The future developments of improved or new tools to investigate gas sensors is hard to predict, but clearly the operando approach has proven to be successful and, thus, operando methods will be essential for future gas sensor research. The discussed techniques and examples may give inspiration for the investigation of sensing materials and the steady improvement of research methods for gas sensors.

3.3 Computational atomic-scale calculations based on first-principles density functional theory (DFT)

Being able to understand the fundamental principles of reactions and processes occurring at the surfaces of the sensing layers of metal-oxide based gas sensors would help to enhance their macroscopic properties toward better practical applications, notably when talking about the sensitivity and the selectivity. Providing atomic scale insights and design guidance will benefit the optimization of the sensor operation for dedicated applications. Some advanced characterization experiments at the atomic scale can be conducted, but cannot reach a detailed description of the microscopic multistep process of a reaction as it provides only data on the initial reactants and the final products.

Computational atomic-scale calculations based on first-principles density functional theory (DFT) can help to unravel the basic mechanisms involved in chemical sensing. With the development both in computers becoming faster and in the electronic structure theory and computational approaches becoming accurate and reliable, DFT calculations are valuable and appear as versatile tools in the research and development process. In the case of gas-sensor applications, DFT is a great approach to understand the fundamental principles involved in the interaction between the gas molecules and the surface, especially to describe the changes in both the geometric and electronic structures in the sensing layers during the detection. Nevertheless, unraveling the atomic scale chemical reactions that can occur at the surface of the sensing layer is not an easy task as a convoluted link exists between each type of active site on the surface of the sensing layer, the gas species and the sensing response. It requires the consideration of the structure of the sensing layer in terms of crystallography, defects, and also the gas atmosphere with the presence of various chemical gaseous substances such as humidity, for instance.

The remainder of the chapter is focused on how DFT is used to identify and depict atomic scale chemical reactions in the context of gas sensing. Also, with the objective to achieve a rationale design of the sensor operation, the description of how the transient/dynamical operation of a metal oxide gas sensor can be modeled from the knowledge of the atomic scale chemical surface reactions is drawn.

3.3.1 Multilevel-modeling of metal oxide gas sensors

If theoretical calculations and simulations are requested to help the development of metal-oxide gas sensors, they have to address the main challenges faced by the technological developments, that is, the sensitivity, the selectivity, the long-term stability and the response time. Each of these features can be examined under DFT investigations. The selectivity is correlated to the ability of the gas molecule to interact and/or react with an active site of the sensing layer forming metastable or stable adsorbed states. The sensitivity is correlated to the charge transfers occurring between the gas molecules and the sensing layer. Charge transfers are strongly dependent on the studied gas. The long-term stability and the response time are time dependent and necessitate the access to the *kinetics* of the chemical reaction and to the careful characterization of the multistep reactive processes.

In other words, DFT calculations are used to get a detailed description of the atomic scale chemical reactions occurring at the surface of the sensing layer and to evaluate the resulting changes in the electronic structure of the metal-oxide layer. The following data is investigated:

- the identification of the atomic scale reaction involved during the detection as a function of the surface topology.
- looking at the bond making and breaking.
- the determination of the stability of the adsorbed species even through a multistep reaction.
- the possible reaction between several adsorbed species.
- the calculation of the interaction energies between the gas molecule and the surface.
- the calculation of the adsorption energies.
- the effect of the surface crystallography and topology.
- the effects of the preexisting surface defects on the adsorption energies.

- the calculations of the activation barriers involved in surface adsorption and recombination.
- the charge transfers.

More specific issues can be addressed such as the effect of doping or humidity on the sensing properties. Other properties to compare with experimental data can also be obtained using DFT such as spectroscopic data (IR, XPS).

Nevertheless, DFT calculations cannot depict a realistic detection of the gas molecule since neither the temperature effect nor the competitive dynamic behavior between gaseous molecules are considered in such calculations. Developing the means of predicting the sensor response and controlling the chemical reaction at the nanometer scale can, thus, help to push forward the detection process of a metal-oxide based gas sensor. The predictive simulation of the detection process is strategic since it could notably help to master the selectivity of the metal-oxide layer under thermal cycles. A macroscopic model based on the chemical rate theory can be developed to simulate the operation of the sensor by climbing the scales where DFT results are used as input data to parameterize the model. Such a multiscale modeling methodology is detailed in [330].

In summary, theoretical calculations and simulations on sensing nanomaterials meet between chemistry and condensed matter physics and require a multilevel modeling strategy.

3.3.2 Atomic scale investigations using DFT calculations

The changes in the resistance of the sensing layer are related to the chemical reactions that occur at its surface. Therefore, a deep knowledge of these reactions is crucial and a thorough investigation of the reactivity of the gaseous molecules is required to obtain a detailed balance of the chemisorbed species on the surface.

The reactivity between the gas molecules and the metal oxide surface is dependent on several features of the surface itself (the crystallography and the electronic properties of the surface, the presence of defects, the presence of dopants, the surface pretreatment, the preadsorbed molecules or atoms, the diffusing adsorbed species, etc.) and on other physical parameters (the surface temperature, the incidence angle of the molecule, etc.).

The potential energy surface (PES) is the central quantity in any theoretical description of the reactivity between a gas and a surface because all chemical processes are governed by the underlying PES. This PES can be described by the knowledge of the stability of the adsorbed species and the path between those stable or metastable adsorbed states. This path called minimum energy path (MEP) defines the activation barriers of a local mechanism likely to occur on this surface.

Thus, to describe a chemical reaction, we need to know the potential energy of the system as a function of the coordinates of the reactants. But considering all the above features and parameters, mapping out the PESs is a challenge.

3.3.3 Reactivity

The identification of the active sites is the first evidence of a sensing process. The adsorption is performed by approaching gas molecules toward the investigated reactive sites by taking into account the topology of the surface. Then, a minimization of the system is carried out to determine the adsorption mode and to characterize the energetics. Several adsorption modes can be observed. A physisorption is an adsorption process in which the electronic structure of the atom or molecule is barely perturbed. The main interacting force involved in such a process is caused by the van der Waals forces. In contrast, a chemisorption is an adsorption process that involves a chemical reaction and that implies the formation of a covalent bond between the molecule and one or more atoms on the surface. Strong electronic changes can be observed in the case of a chemisorbed species.

The adsorption energies are then calculated using the following formula:

$$E_{ads} = E(\text{adsorbed state}) - E_{surface} - E_{molecule} \tag{3.22}$$

With E(adsorbed state) as the total energy of the system, where the molecule is relaxed on the surface, $E_{surface}$ as the reference energy of the system with the surface only (no molecule) and $E_{molecule}$ as the reference energy for the gas molecule calculated in the same simulation box as the one containing the surface.

Conventionally, the larger the adsorption energy, the stronger is the reaction.

Typically, the adsorption energy of a physisorption is less than 0.20 eV, while the adsorption energy of a chemisorption mode is higher than 0.20 eV. These values are given as a guide and must be read in association with the electronic transfers and the bond making or breaking associated to the adsorption process.

In order to discriminate such adsorption modes and to understand in detail the formation of a bond, an analysis of the density of states (DOS) for the different adsorbed configurations is also very helpful in understanding why and where the bond breaking or making processes occur. The DOS is a useful output of DFT calculations in many domains of material sciences, notably in terms of band gap characterization and electronic properties of defects.

From physiosorbed or chemisorbed states, the adsorbed molecules are susceptible to leave the surface under thermal/vibrational effects. In that case, we talk about desorption processes. This reaction can be associated to further complex processes such as the etching of the surface or the reduction of the surface. In some cases, an atomic reorganization of the surface such as a reconstruction is observed.

In Fig. 3.19, some typical adsorption states of CO(g), CO_2(g), and H_2O(g) on a SnO_2 (101) surface are shown. CO(g) adsorbs on a Sn atom of the surface with an adsorption energy of −0.45 eV (Fig. 3.19A). When the CO(g) adsorption occurs into an existing vacancy (Fig. 3.19B), shown as a structural defect, or in the vicinity of an existing vacancy (Fig. 3.19C), the adsorption energies are −0.24 eV and −0.30 eV, respectively. From these three adsorption energies, the adsorption of CO(g) occurs preferentially on a clean part of the SnO_2 surface compared to defective parts. The adsorption of CO_2(g) above a Sn atom is less favorable with a typical adsorption

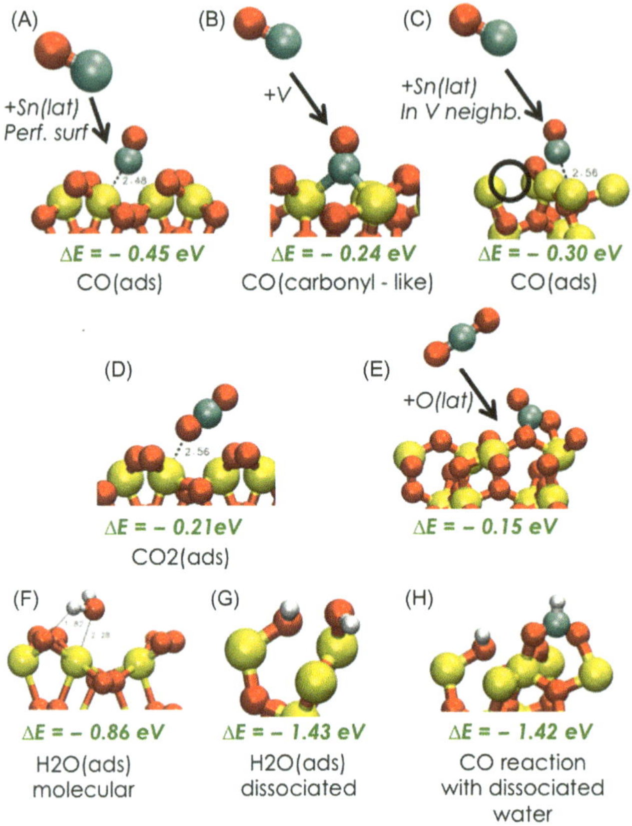

Figure 3.19 Side views of different adsorbed states of CO(g), CO_2(g) and H_2O(g) on the SnO_2 (101) surface and associated adsorption energies: (A) chemisorption of CO(g) on a Sn atom, (B) chemisorption of CO(g) in an oxygen vacancy V, (C) chemisorption of CO(g) in an oxygen vacancy in close vicinity, (D) chemisorption of CO_2(g) on a Sn atom, (E) CO_2(g) reaction with a bridging oxygen atom forming carbonate species, (F) molecular adsorption of water (g) dissociative adsorption of water resulting in two –OH fragments, and (G) reaction of CO(g) with dissociated water. Yellow, red, blue, and white spheres represent tin, oxygen, carbon and hydrogen atoms, respectively.

energy of -0.21 eV (Fig. 3.19D), but the adsorbed CO_2(g) can also react with an oxygen atom from the surface to form carbonate-like species, with a typical adsorption energy of -0.15 eV (Fig. 3.19E). The adsorption of a water molecule on the (101) SnO_2 surface is a multistep process where first a molecular adsorption mode is observed (Fig. 3.19F) before the dissociation of the water molecule occurs (Fig. 3.19G). That results in one OH adsorbed on a tin atom and one H adsorbed on a bridging oxygen atom of the surface. The dissociation is a favored reaction since it stabilizes the system by -0.57 eV compared to the molecular adsorption. In Fig. 3.19H, a surface reaction is shown between an adsorbed CO molecule with OH coming from dissociated water. The resulting structure is highly stable with an adsorption energy of -1.42 eV.

A strong correlation exists between the reactivity of gas molecules and the nature and the structure of the sensing layer. Changing the nature of the metal oxide is thus an efficient way to improve the sensitivity and/or the selectivity toward a targeted gas since its reactivity with the surface will be modified due to different chemical interactions [256,265,331–334].

Controlling the structures of the surface can also help to design the sensitivity of the sensor: the different structures of the surface of the same oxide exhibit different reactivities because of the different atomic arrangements on the surface, the relaxation of the surface and associated reconstructions. Thus, in a polycrystalline system, each surface structure will then contribute to the overall rate of the response and the most active one will typically dominate. Other technological developments to improve the nanostructuring of the oxide also propose new material architectures to improve the gas sensing [335–338].

It is proven experimentally that defective surfaces of metal oxides are the most reactive ones. The surface defects, notably the oxygen vacancies, play a major role in the detection. Here we point out that the surface oxygen atoms are of crucial importance in the operation of the sensor, as they are the more reactive sites on the surface. Using DFT calculations performed on the adsorption of CO(g) on the SnO_2 (101) surface [330], we demonstrate that the oxygen atoms that are present at the surface of the metal oxide, maintain a catalytic activity on the surface through successive reduction and oxidation of the surface, revealing a key role played by the appearing/disappearing vacancies under CO(g) exposure (Fig. 3.20) [257].

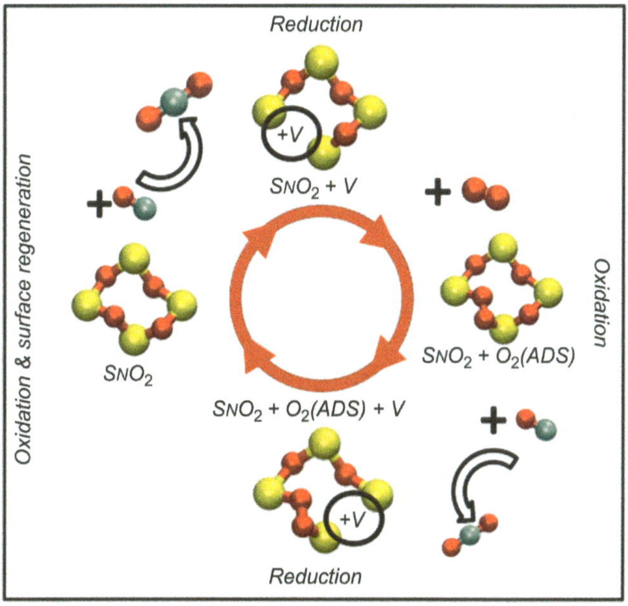

Figure 3.20 Catalytic activity observed on the SnO_2 (101) surface, where the successive reduction and oxidation of the surface occurs under CO(g) and O_2(g) exposure. Reduced-size pictures are top view of the SnO_2 (101) surface.

In Fig. 3.20, the catalytic cycle on the SnO_2 (101) surface is schematized: the CO(g) adsorption on the SnO_2 (101) surface reduces the surface by reacting with a bridging oxygen atom of the surface, leaving a vacancy on the surface and creating a CO_2(g) molecule. This so-formed vacancy then reacts with molecular oxygen species, present in the gas phase, that finally help with the surface regeneration [257,330].

Finally, hydroxyl groups are usually present on the surface of metal oxides because of the humidity in the ambient air. The hydroxyl groups play a major role in the detection since they can form highly stable adsorbed states on the surface that will be trapped on the surface and finally pollute the surface. The hydroxyl groups disturb also the detection of targeted gases by preventing them to react with the reactive sites on the surface [258,302,312,339–346].

3.3.4 Getting into kinetics

As previously mentioned, a full description of the PES requires the knowledge of the MEP combining the region of reactants with the region of products, and the one between the several adsorbed states of the products. We talk about reaction pathways. This concept is of prime importance in the discussion of the chemical reactions since it will define the rate of the reactions. We will also see that this concept is the main ingredient in the understanding and the modeling of the dynamical response of the sensor using the transition state theory.

The determination of the reaction pathway requires the identification of the adsorption sites of the reactants, and the stability of the products, and sometimes the identification of the intermediary states, shown as metastable states in a multisteps reaction (Fig. 3.21).

As products can adopt many adsorbed states, it is necessary to carry out the minimizations of both local and global energy minima, and also the search of the saddle point between those states, shown as transition states in the chemical reaction. Once the transition state is located, it determines the activation barriers and draws the MEP. There is no unique way to calculate the activation barriers. Among the mainly used methodologies, we can name the drag method or the Nudged Elastic Band method [347]. Transition rates are then approximated using the transition state theory.

Fig. 3.21 shows the details of the reduction of the SnO_2 (101) surface under CO(g) exposure. A multistep reaction is characterized: from the adsorbed state of the CO molecule, an activation barrier of 0.40 eV is required to allow the reaction of CO (ads) with one bridging oxygen atom from the surface. An intermediate is identified here as a formate-like species, that results easily in a CO_2(ads) species in a vacancy. Because of the activation barriers surrounding the formate like state, it is more favorable for the system to form CO_2(ads) species (0.20 eV) compared to come back to a CO(ads) state where the back-activation barrier is large with 0.73 eV (0.33 eV from the energy of the formation of the formate like species plus the 0.40 eV of the calculated activation barrier of the CO reaction with lattice O). Finally, CO_2(ads)

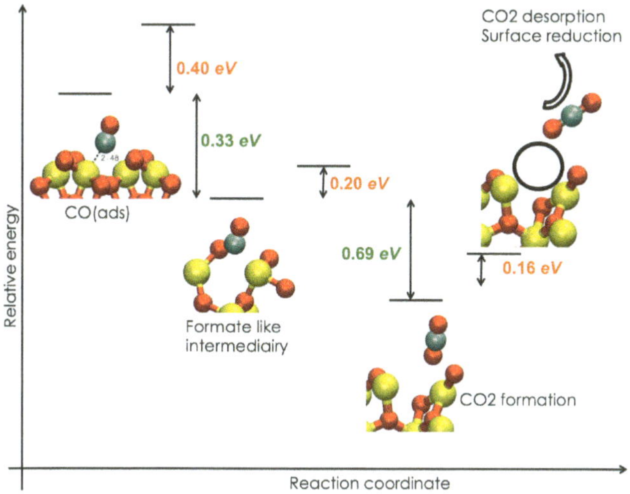

Figure 3.21 MEP of the multistep reaction path of the reduction of the SnO_2 surface by CO(g) giving rise to the formation of a CO_2(g) molecule and leaving a vacancy on the tin oxide surface. Activation barriers are given in red, energy gains in green. Pictures are reduced side views of the tin oxide surface.
Source: Adapted from J.-M. Ducéré, A. Hemeryck, A. Estève, M.D. Rouhani, G. Landa, P. Ménini, et al., A computational chemist approach to gas sensors: modeling the response of SnO_2 to CO, O_2, and H_2O gases, J. Comput. Chem. 33 (2011) 247–258, doi:10.1002/jcc.21959 with permission.

desorbs from the vacancy with a small activation barrier of 0.16 eV and gives rise to a vacancy on the surface.

The activation barriers are thus a key parameter for the response time of the gas sensor. From a dynamic point of view, experimentalists can play with the nanostructures of the metal oxide layer to enhance the sensing properties, notably by acting on the height of the activation barriers to favor targeted reactions by controlling their kinetics. Such effects can be achieved thanks to the doping process during the elaboration of the material [224,348].

3.3.5 Doping

A target gas might not be detected during the detection process due to many reasons: too large activation barriers exist for the reaction on the surface, its reaction with the surface might also be hindered on the surface by another adsorbed species or because of too low associated electronic changes between the gas and the surface. In the presence of a dopant, the activation barrier can be strongly reduced or the charge transfers can be deeply enhanced, so that the reaction proceeds much more efficiently.

One must be aware that more often than not the use of a dopant will change the adsorption mode, the reaction path, and sometimes can involve unknown intermediate reaction steps notably involving the adsorption of gases on the dopant. Moreover, this kind of calculations must be conducted carefully because of the size of the used simulation cell mimicking a modeled system, which can be too small to avoid artefacts due to the interaction of the dopant with its periodic image. Note also, that for such simulation, the modeled system containing the dopant is featured by a high concentration of dopant per nm^2.

3.3.6 Charge transfer in DFT calculations

An analysis of the charge density and an estimation of the repartition of the charges per atom are great tools to analyze the interaction and hybridization of the chemical bonds between the reactants. The Bader approximation is one of the existing methods to estimate the repartition of the charges per atom in a given system [349]. The effects of reducing or oxidizing gases are thus easily calculated.

Coupling a static approach with relaxation calculations, this tool helps the understanding of the low or high sensitivity. A low sensitivity will be associated with low charge transfers, whereas large charge transfers will be related to a high sensitivity.

From a practical point of view, the main goal of a metal oxide-based sensor is to be able to discriminate gaseous species even if these gaseous species react with the same specific reactive site, even if pollutants are in the gas phase. Also, a metal oxide-based sensor must be able to quantify concentrations of gases and to handle temperature variations, and so on.

All of these considerations are dependent on the rate of a given reaction to occur. That must be the fundament for the setup of an efficient modeling of gas-sensing operations.

The DFT calculations are conducted at 0K and, consequently, the thermal effects are not considered. The actual operation of the sensor cannot be accurately and reliably described only by considering the sole DFT results.

Some modeling approaches tried to predict efficient ranges of temperature for the operation of gas sensors using the phase diagram or thermodynamic models. Those approaches are useless since the operation of a gas sensor is dynamic, depending on the mixture of gases and the varying temperature.

For a true comparison with experiments, the modeling of the operation of the gas sensor has to consider first the reaction rates and the probabilities of forming adsorbed species that can be derived from the calculated PESs. In such a strategy, the meaningful atomic scale data, harvested using the DFT calculations, appear as a great source for the development of a macroscopic model, only based on theoretical considerations.

The model can be designed using the chemical kinetics formalism that allows the evaluation the effects of the different external or experimental conditions on the velocity of the chemical reaction occurring on the surface. Especially, using this formalism, the thermal effects can be introduced. This formalism takes into account the rate of a reaction to occur through the transition state theory where the activation barriers have been accurately determined by DFT. For details, referred to [330].

Chemical kinetics considers the formation of a product at the expense of the disappearance of the reactants and define how quick a reaction takes place using Arrhenius' Law.

For a given reaction:

$$A + B \underset{k_{-i}}{\overset{k_i}{\leftrightarrow}} C \tag{3.23}$$

the chemical kinetics follows:

$$\frac{-d[A]}{dt} = \frac{-d[B]}{dt} = \frac{d[C]}{dt} = k_i \cdot [A] \cdot [B] - k_{-i} \cdot [C] \tag{3.24}$$

with k_i being the reaction rate of the reaction known using the following Arrhenius' Law

$$k_i = \frac{k_B T}{h} \cdot e^{\left(\frac{-\Delta E_i}{k_B T}\right)} \tag{3.25}$$

where ΔE is the activation barrier determined by DFT calculations.

By considering the reaction path provided in Fig. 3.21, the variation of the concentration of CO(ads) on the surface can be written as:

$$\frac{d[CO(ads)]}{dt} = \underbrace{\phi_{CO}[Sn(1)]}_{R.I} \underbrace{- k_{-1}[CO(ads)] - k_2[CO(ads)][O(1)] + k_{-2}[I_1]}_{R.I} \tag{3.26}$$

where *R.I* is the reaction of the adsorption of CO(g) on a tin atom as described in Fig. 3.21. This adsorption process is defined by ϕ_{CO} the flux of CO(g) calculated considering Boltzmann Statistics. k_{-1} and k_2, k_{-2} are the rate constants of the desorption of CO(ads) from the tin atom to return to the gas phase, k_2 is the reaction rate from CO(ads) to form the formate like species and k_{-2} is the rate of the back reaction from the formate like species to CO(ads).

A transient selectivity of the gas detection can be identified using the model by playing with the applied temperature, allowing a selective control of the diffusion/reaction process at the surface of the sensing layer.

3.4 Modeling gas sensing with SMOXs

The limitations faced by the traditional try and see development strategy applied in the sixties and seventies of the last century determined the quest for the development of practical sensor models. It was recognized that it is important to explain the experimentally found relationship between sensor resistance/signal and the concentration of the target gases and the way in which the sensing layer characteristics are influencing it. Since it is important to be aware of the main ideas and to understand the origin of the main concepts, this section starts with a historical overview on the efforts made in sensor modeling. It continues with the examination of the very important role of the conduction mechanism in the transduction function and goes into the details of how the state of the art experimental insights are changing the current state of the art modeling.

3.4.1 History

The first attempt to understand the sensing characteristics of porous thick SnO_2 layers was presented by Clifford and Tuma in 1982 [350], after metal oxide based gas sensors were already established on the market. The now-a-days generally known description of the sensing effect—combustible gases consume preadsorbed oxygen ions to form oxidation products—was built in simple mathematical expressions. They provided relations to describe the experimentally found power law response of the electrical resistance as a function of the target gas concentration including H_2, CH_4, O_2, CO, and H_2O. This way, it was possible to understand important drawbacks of the sensors as, for example, the role of oxygen, which explains the lack of selectivity. In a companion paper [351], the influence of the temperature, the kinetics of the oxygen adsorption and the charge exchange with the semiconductor was discussed assuming that the conductance is entirely controlled by the presence of Schottky barriers at the grain boundaries (described in [352]). Later on, Williams et al. compared thick and thin layers of metal oxides and took the morphology of the sensing layer into account [353]. The effects of the temperature, the humidity, the oxygen partial pressure and the combustible gases were interpreted in terms of changes in the surface charge determining the height of the Schottky barriers. However, at this time, it was not yet possible to correlate morphological and electrical properties of the sensing material with the shape of the calibration curve as the link between semiconductor physics and surface chemistry was still missing.

The first comprehensive model, which provides relations linking the sensor response to the concentration of a combustible gas was presented by Bârsan and Weimar in 2001 [212]. Taking into account the grain size, layer morphologies and different surface reactions, the model combines chemical reactions taking place at the surface (described by the quasichemical reaction formalism) to their electrical effects (expressed by using a combination of the Poisson and electroneutrality equations). In accordance to experimental investigations, it predicts a power law dependence and allows for the identification of the reacting oxygen species. Additionally, it suggests that the value of the exponent contains information about the morphology, the grain size and the defect concentration.

Rothschild and Komem investigated the influence of the grain size by concentrating mainly on the effects in the semiconducting material as a result of a change in the surface charge [354]. Because their focus is very much on the semiconductor physics part and there is no link to the surface chemical reactions, the effect of target gas exposure on the sensor signal is not taken into consideration directly.

Later on, Yamazoe and Shimanoe extended the models to the role of shape and size of grains in SMOX-based gas sensors [355,356]. The investigation of the response to reducing gases is limited to cases in which the grains are fully depleted and in flat-band condition (the difference in the energy between the surface and the center of the grains, ΔE, is smaller than the thermal energy). In this work, the link between reception and transduction was considered by simplifying the quasichemical relations dramatically to derive analytical solutions. In contrast to Rothschild and Komem, Yamazoe and Shimanoe found that the sensor signal depends on the square root of the reciprocal of the grain size.

A next step was achieved in 2010, when on the basis of [354] a model for *p*-type SMOXs was presented [211]. The most important result in this work is that the sensor signal in the case of *p*-type SMOXs is just the square root of the signal for *n*-type SMOX and the identification of the impact of the conduction mechanism on the transduction function and sensor performance.

Even if imperfect, the models developed in the past 30 years proved themselves useful by allowing the transition from the purely empiric development of the beginnings to some kind of knowledge based one. Significant advancements were made possible by the development of concepts such as reception and transduction functions revealing the role of the sensitive layer morphology in gas sensing. Additionally, they helped expose the very large gaps in the understanding of sensing, identifying the targets of additional experimental and theoretical research.

Currently, many of the initial assumptions need to be reconsidered because of the breakthroughs obtained by applying modern operando techniques. It is now possible to gain unprecedented insight into the reception and the transduction and into the impact of fabrication technology onto sensor performance.

To model the sensing effect of SMOX based gas sensors, it is crucial to consider both basic functions that describe the sensing effect—reception and transduction. The reception function describes the reactions taking place at the surface of the metal oxide and the associated charge transfer processes. The transduction function deals with the translation of those into measurable changes of the sensing layers electrical resistance. Both contributions are important for the magnitude of the sensor response and its variation with the target gas concentration. The importance of the transduction function was underestimated for a long time as it was thought that the conduction mechanism does not change during sensing. In further studies, this was proven false and a breakthrough was achieved by a combination of modeling and operando DC electrical resistance measurements and measurements of the changes in the work function [304,312,357,358]. This is the reason why we will concentrate mostly on it.

3.4.2 The conduction mechanism

In sensing, the sensor response is based on the magnitude of the change in the conductance in the sensing layer. It is due to the charge transfers occurring between the gas molecule and the surface and also due to the reactions between products on the surface. Existing models and modeling options will be reviewed in the following section.

3.4.2.1 n-type SMOXs

The depletion layer-controlled conduction mechanism: The charge transfer into acceptor states situated below the CB edge appearing at the surface as a result of surface reactions, for example, because of oxygen adsorption, leads to the formation of charged species and a buildup of negative surface charge. As a consequence, the electrons from the CB will be repulsed so that their concentration at the surface will be decreased—only the electrons with sufficient energy will be able to reach the surface. This will determine the appearance of a depletion layer. In the representation of the energy bands the increase of energy needed to reach the surface is symbolized by an upwards bending (Fig. 3.22). Only the electrons possessing enough energy to overcome the potential barriers at the grain boundaries can contribute to the conduction process, which means that the magnitude of the so-formed back-to-back Schottky barriers controls the conduction process. Assuming the grains to be large (1D approach and Schottky approximation), the conductance G of the sensing layer is proportional to the electron concentration in the CB at the surface of the grains (n_S) (for more details see [304]):

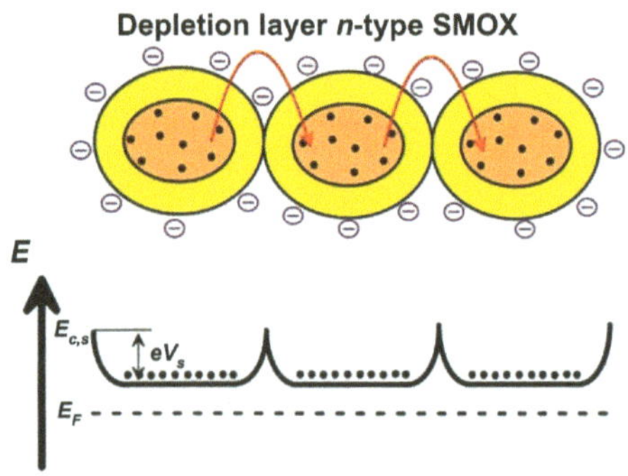

Figure 3.22 The upper panel shows the grains in the sensing layer with the depletion region (yellow) the bulk area and the negative charge trapped at the surface. The lower panel shows the corresponding energy band diagram.
Source: Reprinted with permission from N. Barsan, M. Hübner, U. Weimar, Conduction mechanisms in SnO_2 based polycrystalline thick film gas sensors exposed to CO and H_2 in different oxygen backgrounds, Sens. Actuators B Chem. 157 (2011) 510–517, doi:10.1016/j.snb.2011.05.011.

$$G \sim n_S \tag{3.27}$$

n_S can be correlated to the magnitude of the band bending with the help of the Boltzmann distribution at $V = V_S$:

$$n_S = n_b \cdot exp\left(\frac{-eV_S}{kT}\right) \tag{3.28}$$

Upon gas exposure, the concentration of negative charges and the depth of the depletion layer changes due to surface reactions and subsequent charge transfer processes. Consequently, the resistance changes from R_0 to R_{gas} and the relation between R_{gas} and the band bending change $e\Delta V$ can be calculated by using [304,352]:

$$R_{gas} = R_0 \cdot exp\left(\frac{e\Delta V}{kT}\right) \tag{3.29}$$

Decreasing the negative charge concentration at the surface by increasing the reducing gas concentration in the atmosphere, the upward band bending and also the depth of the depletion layer decreases until flat-band situation is reached. A further increase in reducing gas concentration leads to the formation donor levels above the CB. By the electron transfer from these states into the CB, a positive charge at the surface is formed and an accumulation layer of electrons inside the semiconductor exists. Consequently, the energy bands are bended downwards at the surface. This situation can be described by the accumulation layer model.

The accumulation layer-controlled conduction mechanism: In the case of an accumulation layer controlled conduction mechanism, the space charge layer (brown in Fig. 3.23) is more conductive than the inner part of the grains. This situation is completely different when compared to the depletion layer model as the current flows through the accumulation layer. Consequently, one can simply assume the conductance is proportional to the average electron concentration in the accumulation layer [304]:

$$G \sim \tilde{n}_S \tag{3.30}$$

In the end, this leads to (the full derivation of can be found elsewhere [304]):

$$\tilde{n}_S = n_b \cdot exp\left(\frac{-eV_S}{2kT}\right) \tag{3.31}$$

Like in the case of depletion layers, the relation between R_{gas} and the band bending change $e\Delta V$ can be calculated:

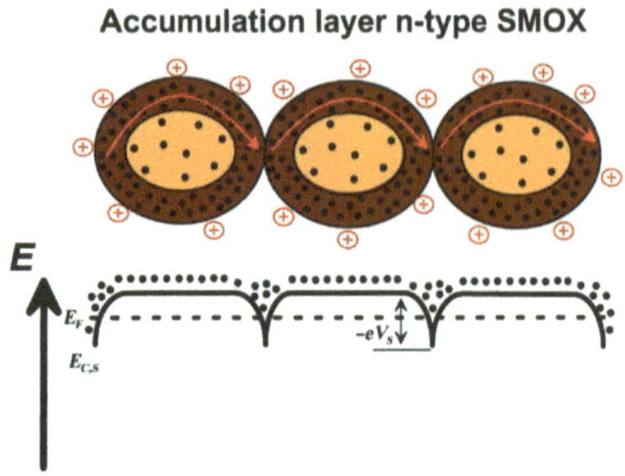

Figure 3.23 The upper panel shows the grains in the sensing layer with the accumulation region (dark brown) the bulk area and the positive charge trapped at the surface. The lower panel shows the corresponding energy band diagram.
Source: After N. Barsan, M. Hübner, U. Weimar, Conduction mechanisms in SnO_2 based polycrystalline thick film gas sensors exposed to CO and H_2 in different oxygen backgrounds, Sens. Actuators B Chem. 157 (2011) 510–517, doi:10.1016/j.snb.2011.05.011.

$$ln\left(\frac{R_0}{R_{gas}}\right) = \frac{e\Delta V}{2kT} \tag{3.32}$$

It is important to note that the slope of the curve describing the relationship between $ln(R_0/R_{gas})$ and the change in the band bending $e\Delta V$ decreases when switching from the depletion to the accumulation layer-controlled conduction mechanism. This means, in comparison to the depletion layer-controlled situation, the same change in surface charge/band bending leads to a lower change in the resistance in the case of accumulation layer-controlled conduction mechanism. In other words, the same change in the surface charge will lead to a lower sensor signal if accumulation layers are present.

Thus, in a first approximation the sensor signal for the same relative change in the band bending in the case of the accumulation layer-controlled conduction mechanism S_{acc} will be just the square root of the corresponding sensor signal obtained in the case of the depletion layer-controlled conduction mechanism S_{dep}:

$$S_{acc} = \sqrt{S_{dep}} \tag{3.33}$$

3.4.2.2 p-type SMOXs

For *p*-type SMOXs, the electron transfer into acceptor states below the VB leads to the build-up of negative charges at the surface and the formation of an accumulation layer of holes inside the semiconductor. In this case, an upward band bending exists. Consequently, the conductivity of the grains' surface is increased in comparison to the bulk and the current will flow through the accumulation layer.

Assuming that the bulk contribution is not too large, the conduction process is dominated by the current flowing through the space charge layer. This allows the application of the same kind of approach as for *n*-type SMOX in the case of downward band bending. Calculating the average hole concentration in the sensing layer yields the dependence of the resistance on changes in the band bending [211]:

$$ln\left(\frac{R_{\text{air}}}{R_{\text{gas}}}\right) = -\frac{e\Delta V}{2kT} \tag{3.34}$$

Reactions at the surface consuming negative charge decreases the concentration of holes by electron transfer and therefore the upward band bending decreases.

This shows that the sensor signal for the same change in the surface charge in the case of *p*-type SMOXs is just the square root of the sensor signal in the case of *n*-type SMOXs. Thus, this explains the well-known lower sensor signals of *p*-type SMOXs due to a different transduction function (Eq. (3.33)).

If the surface charge becomes positive, the conduction mechanism changes and the space charge layer gets more resistive. In the outer part of the grains, a depletion layer of holes exists while the inner part is still unaffected. In this case, the conduction is determined by the height of the back to back double Schottky barriers at the grain boundaries.

3.4.3 Review of the modeling assumptions in the case of SnO_2

All past modeling of sensing reducing gases with SnO_2 based gas sensors assumes that in realistic operation conditions, that is, in air and in the presence of humidity:

- The reception takes place by the reaction between the reducing gases and the preadsorbed oxygen ions.
- The transduction is dominated by the depletion layer. Its extent and impact in the overall resistance of the sensing layer depends on the electronic properties of the respective SMOX and on the morphology of the layer.

In [359], the dependence of the sensor signal on the CO concentration in dry and humid conditions was derived (for details see [359]):

$$\text{for dry conditions}: \quad S = \left(1 + \frac{k_{\text{react}}}{k_{\text{des}}} \cdot p_{CO}^{\beta}\right)^{\frac{1}{(\alpha+\delta)}} \tag{3.35}$$

And

$$\text{for humid conditions}: \quad S = \left(1 + \frac{\frac{k_{\text{react}}}{k_{\text{des}}} \cdot p_{CO}^{\beta}}{\kappa \cdot p_{H_2O}}\right)^{\frac{1}{(\alpha+\gamma)}} \tag{3.36}$$

The derivation is based on the assumptions that the conduction is controlled by a depletion layer and that ionosorbed oxygen species are the reaction partners for both reducing gases and water vapor. Accordingly, the following surface reactions describe what happens in the presence of both humidity and CO:

$$\frac{1}{2}O_2^{gas} + \alpha \cdot e^- + S \rightleftharpoons O_S^{-\alpha} \tag{3.37}$$

$$CO^{gas} + O_S^{-\alpha} \rightarrow CO_2^{gas} + \alpha \cdot e^- + S \tag{3.38}$$

$$H_2O^{gas} + 2 \cdot Sn_{Sn} + O_S^{-\alpha} \rightleftharpoons 2 \cdot (SnOH) + \alpha \cdot e^- + S \tag{3.39}$$

$O_S^{-\alpha}$ is a chemisorbed oxygen species with $\alpha = 1$ for singly ionized oxygen species or 2 for doubly ionized oxygen species; e- represents an electron in the CB at the surface; S is a surface adsorption site.

However, operando measurements on SnO_2-based gas sensors [312] demonstrated that it is possible to switch the conduction mechanism from depletion layer controlled to accumulation layer controlled, even in application relevant conditions. These findings were explained by postulating a direct reaction of the reducing gas with the metal oxide surface forming donor species. Spectroscopic evidence of the nature of the donor species was presented by DRIFTS measurements. The results showed a reaction of reducing gases with lattice oxygen indicating that oxygen vacancies most probably act as donor species [257]. Due to the results of the latest findings, the models have to be extended for these surface reactions and for both conduction mechanism types.

The healing of vacancies leading to the formation of lattice oxygen can be described by:

$$\frac{1}{2}O_2^{gas} + 2\,e' + V_O^{\bullet\bullet} \rightleftharpoons O_O \tag{3.40}$$

The reaction of CO with lattice oxygen ions is given by:

$$CO + O_O \rightarrow CO_2^{gas} + 2e' + V_O^{\bullet\bullet} \tag{3.41}$$

For the flat-band situation, the concentration of vacancies at the surface $\left[V_{surf}^{\bullet\bullet}\right]$ and the vacancy concentration in the bulk are equal. If the vacancy concentration at the surface is lower than in the bulk of the material, depletion layers are formed. In this case, negative surface charge is present. For a higher surface vacancy concentration in comparison to the bulk value, positive surface charge and an accumulation layer of electrons is formed. This means:

$$\text{flatband situation:} \quad [V_O^{\bullet\bullet}] = \left[V_{surf}^{\bullet\bullet}\right] = a \cdot N_D \tag{3.42}$$

$$\text{depletion layers:}\quad [V_O^{\bullet\bullet}] < \left[V_{surf}^{\bullet\bullet}\right] \tag{3.43}$$

$$\text{accumulation layers:}\quad [V_O^{\bullet\bullet}] > \left[V_{surf}^{\bullet\bullet}\right] \tag{3.44}$$

Whereas a is the lattice constant and Eq. (3.42) corresponds to large grains, for which the geometry can be neglected. N_D is the donor concentration in the bulk.

The rate equation for the vacancy formation is:

$$\frac{d\left[V_O^{\bullet\bullet}\right]}{dt} = -k_{ads}\cdot p_{O_2}^{\frac{1}{2}}\cdot \left[V_O^{\bullet\bullet}\right]\cdot n_S^2 + k_{des}\cdot [O_O] + k_{react}\cdot p_{CO}\cdot [O_O] \tag{3.45}$$

With the total concentration of sites $[S_t]$:

$$\left[V_O^{\bullet\bullet}\right] + [O_O] = [S_t] \tag{3.46}$$

yields in the steady state:

$$k_{\text{ads}}\cdot p_{O_2}^{\frac{1}{2}}\cdot \frac{\left[V_O^{\bullet\bullet}\right]}{[S_t] - \left[V_O^{\bullet\bullet}\right]}\cdot n_S^2 = (k_{\text{des}} + k_{\text{react}}\cdot p_{CO}) \tag{3.47}$$

For the 1D approach (large grains), the electroneutrality equation in the case of depletion layers is:

$$2\left(\left[V_O^{\bullet\bullet}\right] - \left[V_{Ofl}^{\bullet\bullet}\right]\right) = -n_b x_0 \tag{3.48}$$

Solving for the vacancy concentration:

$$\left[V_O^{\bullet\bullet}\right] = \left[V_{Ofl}^{\bullet\bullet}\right] - \frac{n_b x_0}{2} \tag{3.49}$$

Substitution of the solution of the Poisson equation and the Boltzmann distribution with the definition of L_D results in:

$$\left[V_O^{\bullet\bullet}\right] = \left[V_{Ofl}^{\bullet\bullet}\right] - \frac{n_b \cdot L_D\sqrt{2ln\left(\frac{n_b}{n_S}\right)}}{2} \tag{3.50}$$

In the case of accumulation layers, the electroneutrality equation is:

$$2\left(\left[V_O^{\bullet\bullet}\right] - \left[V_{Ofl}^{\bullet\bullet}\right]\right) = \left(\tilde{n}_s - n_b\right)x_0 \tag{3.51}$$

$$\left[V_O^{\bullet\bullet}\right] = \frac{\left(\tilde{n}_s - n_b\right)x_0}{2} + \left[V_{Ofl}^{\bullet\bullet}\right] \tag{3.52}$$

Consequently,

$$\tilde{n}_S = \sqrt{n_S \cdot n_b} \tag{3.53}$$

Combining the equations yields:

$$\left[V_O^{\bullet\bullet}\right] = \frac{\left(\sqrt{n_S \cdot n_b} - n_b\right) \cdot \sqrt{2} \cdot L_D \cdot \left(1 - exp\left(\frac{1}{2} ln\left(\frac{n_b}{n_S}\right)\right)\right)}{2} + \left[V_{0fl}^{\bullet\bullet}\right] \tag{3.54}$$

Combining Eq. (3.47) with Eqs. (3.50) or (3.54), respectively, one obtains in the case of a depletion layer:

$$\frac{n_S^2}{\left(\frac{2[S_t]}{2\left[V_{Ofl}^{\bullet\bullet}\right] - n_b \cdot L_D \sqrt{2ln\left(\frac{n_b}{n_S}\right)}}\right)^{-1}} = \frac{(k_{\text{des}} + k_{\text{react}} \cdot p_{CO})}{k_{\text{ads}} \cdot p_{O_2}^{\frac{1}{2}}} \tag{3.55}$$

and in the case of an accumulation layer:

$$\left(\frac{(n_S - n_b) \cdot \sqrt{2} \cdot L_D + 2\left[V_{Ofl}^{\bullet\bullet}\right] - 2[S_t]}{2[S_t]}\right) \cdot n_S^2 = \frac{(k_{\text{des}} + k_{\text{react}} \cdot p_{CO})}{k_{\text{ads}} \cdot p_{O_2}^{\frac{1}{2}}} \tag{3.56}$$

Numerical simulations presented in [360] indicate that

$$\frac{\left[V_O^{\bullet\bullet}\right]}{[S_t] - \left[V_O^{\bullet\bullet}\right]} \tag{3.57}$$

can be approximated by $m \cdot n_S^\delta$ with δ taking values between 0 and 0.4 depending on the material parameters.

This means for the case of a depletion layer:

$$n_S = \left(\frac{(k_{\text{des}} + k_{\text{react}} \cdot p_{CO})}{k_{\text{ads}} \cdot p_{O_2}^{\frac{1}{2}}}\right)^{\frac{1}{2+\delta}} \tag{3.58}$$

For high concentrations of CO:

$$\text{for depletion layer:} \quad G \sim p_{CO}^{\frac{1}{(2+\delta)}} \tag{3.59}$$

If n_S dominates on the left side of Eq. (3.56) and $[S_t]$ is large in comparison to $\left[V_{surf}^{\bullet\bullet}\right]$, for the conduction mechanism controlled by the accumulation layer one obtains:

$$n_S \sim p_{CO}^{\frac{1}{2+1}} \tag{3.60}$$

Keeping in mind that $G \sim \sqrt{n_s}$ in this case, will result in:

$$\text{for accumulation layer:} \quad G \sim p_{CO}^{\frac{1}{2(2+1)}} \tag{3.61}$$

A comparison of Eqs. (3.60) and (3.61) indicates a large influence of the conduction mechanism on the dependence of the conductance on the CO concentration. It shows that the sensitivity will be dramatically decreased if the conduction mechanism switches to accumulation layer controlled.

The calculations presented here are valid for very strict boundary conditions. The grains/crystallites were considered large enough for a 1-dim model to be valid and for the case of depletion, the very strong Schottky model limitations (no free charges in the depletion layer) were considered fulfilled. Real materials will offer more variety and, accordingly, the modeling will have to be adapted. Any such effort needs to start with a thorough operando investigation of the properties of the sensing layer to determine the surface and bulk properties. For example, DCR/WF measurements could help to determine the position of the Fermi level in the bulk (and based on it n_b and L_D) and the presence of an initial/intrinsic band bending as well as the possible change of the conduction mechanism during sensing. This will allow for choosing the correct type of modeling approach from the semiconductor viewpoint that may ask for numerical simulations. DRIFTS investigations in combination with DFT calculations could help identify the correct surface chemistry and allow for the formulation of the respective quasi-chemical equations. All of that needs to be completed are thorough structural and compositional investigations that would correlate electronic properties and specific surface chemistry with synthesis.

References

[1] J. Maier, Physical Chemistry of Ionic Materials: Ions and Electrons in Solids, John Wiley & Sons, Ltd, New York, 2004. Available from: https://doi.org/10.1002/0470020229.

[2] C. Kittel, Introduction to Solid State Physiscs, eighth ed., John Wiley and Sons Ltd, New York, 2004.

[3] P. Atkins, J. de Paula, Atkins' Physical Chemistry, tenth ed, Oxford University Press, New York, 2014.
[4] J. Koryta, J. Dvorak, L. Kava, Principles of Electrochemistry, JohnWiley & Sons, Inc., Chichester, New York, Brisbane, Toronto, Singapore, 1993.
[5] F.A. Kröger, H.J. Vink, Relations between the concentrations of imperfections in crystalline solids, Solid State Phys. – Adv. Res. Appl., Elsevier, 1956, pp. 307–435. Available from: https://doi.org/10.1016/S0081-1947(08)60135-6.
[6] F.A. Kröger, The Chemistry of Imperfect Crystals, first ed., North-Holland, Amsterdam, 1964.
[7] T. Sahm, Flame spray synthesis of tin dioxide nanoparticles for gas sensing, Sens. Actuators B Chem. 98 (2004) 148–153. Available from: https://doi.org/10.1016/j.snb.2003.10.003.
[8] L. Mädler, A. Roessler, S.E. Pratsinis, T. Sahm, A. Gurlo, N. Barsan, et al., Direct formation of highly porous gas-sensing films by in situ thermophoretic deposition of flame-made Pt/SnO2 nanoparticles, Sens. Actuators B Chem. 114 (2006) 283–295. Available from: https://doi.org/10.1016/j.snb.2005.05.014.
[9] R.S. Popovic, Hall Effect Devices, second ed, CRC Press, Hoboken, 2003.
[10] H. Frei, G. Groetzinger, Liberation of electrical energy during the fusion of electrets, Phys. Zeitschrift. 37 (1936) 720–724.
[11] D.V. Lang, Deep-level transient spectroscopy: a new method to characterize traps in semiconductors, J. Appl. Phys. 45 (1974) 3023–3032. Available from: https://doi.org/10.1063/1.1663719.
[12] J.I. Pankove, Optical Processes in Semiconductors, second ed, Dover Publications, New York, 1971.
[13] C. Brian Smith, Fundamentals of Fourier Transform Infrared Spectroscopy, second ed, CRC Press, Hoboken, 2011.
[14] P.R. Griffith, JamesA. De Haseth, Fourier Transform Infrared Spectrometry, Wiley, New York, 2007.
[15] B. Stuart, Infrared Spectroscopy: Fundamentals and Applications, Wiley, New York, 2004.
[16] D.A. Long, The Raman Effect, John Wiley & Sons, Ltd, New York, 2002.
[17] D.A. Long, Raman Spectroscopy, second ed, McGraw Hill International Book Company, London, 1977.
[18] E. Smith, G. Dent, Modern Raman Spectroscopy—A Practical Approach, John Wiley & Sons, Ltd, Chichester, 2005.
[19] F. Reinert, S. Hüfner, Photoemission spectroscopy - from early days to recent applications, New J. Phys. 7 (2005). Available from: https://doi.org/10.1088/1367-2630/7/1/097.
[20] S. Hüfner, Photoelectron Spectroscopy. Principles and Applications, third ed., Springer-Verlag Berlin Heidelberg, Berlin, 2003.
[21] P.Y. Yu, M. Cardona, Fundamentals of Semiconductors, Springer-Verlag Berlin Heidelberg, Berlin, 2010. Available from: https://doi.org/10.1007/978-3-642-00710-1.
[22] M. Cardona, L. Ley, Photoemission in Solids I: General Principles, Springer-Verlag Berlin Heidelberg, Berlin, 1978.
[23] M. Cardona, L. Ley, Photoemission in Solids II: Case Studies, Springer-Verlag Berlin Heidelberg, *Berlin*, 1979.
[24] G. Zwicker, K. Jacobi, Experimental band structure of ZnO, Solid State Commun. 54 (1985) 701–704. Available from: https://doi.org/10.1016/0038-1098(85)90591-5.
[25] J.A. Weil, J.R. Bolton, Electron Paramagnetic Resonance: Elementary Theory and Practical Applications, second ed., Wiley-Interscience, New York, 2007.

[26] J. Wertz, Electron Spin Resonance: Elementary Theory and Practical Applications, second ed., Springer-Verlag Berlin Heidelberg, *Berlin*, 1986.
[27] B.D. Patterson, Muonium states in semiconductors, Rev. Mod. Phys. 60 (1988) 69–159. Available from: https://doi.org/10.1103/RevModPhys.60.69.
[28] S.F.J. Cox, Muonium as a model for interstitial hydrogen in the semiconducting and semimetallic elements, Reports Prog. Phys. 72 (2009) 116501. Available from: https://doi.org/10.1088/0034-4885/72/11/116501.
[29] A. Benninghoven, F.G. Rudenauer, H.W. Werner, Secondary Ion Mass Spectrometry: Basic Concepts, Instrumental Aspects, Applications and Trends, John Wiley and Sons, New York, 1987.
[30] W.K. Chu, Backscattering Spectrometry, Academic Press Inc, New York, 1978.
[31] C. van Bokhoven, Jeroen A. Lamberti, X-Ray Absorption and X-Ray Emission Spectroscopy: Theory and Applications, Wiley, Weinheim, 2016.
[32] D.C. Koningsberger, R. Prins, X-ray Absorption: Principles, Applications, Techniques of EXAFS, SEXAFS and XANES, 1987.
[33] M.D. McCluskey, E.E. Haller, Dopants and Defects in Semiconductors, first ed., CRC Press, Taylor-Francis Group, Boca Raton, London, New York, 2012.
[34] Fredrik Seitz, The Modern Theory of Solids, first ed., McGraw-Hill Book Company, Inc, New York, 1940.
[35] N.F. Mott, R.W. Gurney, Electronic Processes in Ionic Crystals, second ed., Calderon Press, Oxford, 1948.
[36] H.H. Baumbach, C. Wagner, Die elektrische Leitfähigkeit von Zinkoxyd und Cadmiumoxyd, Z. Phys. Chem. 22 (1933) 199.
[37] J.A. Marley, R.C. Dockerty, Electrical properties of stannic oxide single crystals, Phys. Rev. 140 (1965). Available from: https://doi.org/10.1103/PhysRev.140.A304.
[38] Z.M. Jarzebski, Preparation and physical properties of transparent conducting oxide films, Phys. Stat. Sol. 71 (1982) 13–41.
[39] H. Mizoguchi, P.M. Woodward, Electronic structure studies of main group oxides possess NG edge-sharing octahedra implications for the design of transparent conducting oxides, Chem. Mater. 16 (2004) 5233–5248. Available from: https://doi.org/10.1021/cm049249w.
[40] J. Maier, W. Göpel, Investigations tin (IV) oxide * of the bulk defect chemistry of polycrystalline, J. Solid State Chem. 302 (1988) 293–302.
[41] W. Göpel, K. Schierbaum, H.-D. Wiemhöfer, J. Maier, Defect chemistry of tin(iv)-oxide in bulk and boundary layers, Solid State Ionics. 32–33 (1989) 440–443. Available from: https://doi.org/10.1016/0167-2738(89)90253-1.
[42] F.D. Richardson, J.H.E. Jeffes, The thermodynamics of substances of interest in iron and steel making from 0°C to 2400°C: I-oxides, J. Iron Steel Inst. 160 (1948) 261.
[43] N.W. Ashcroft, N.D. Mermin, Solid State Physics, Hartcourt College Publisher, Fort Worth, Philadelphia, 1976.
[44] S.M. Sze, N.K. Kwok, Physics of Semiconductor Devices, third ed., John Wiley & Sons, Inc, New York, 2006.
[45] S. Samson, C.G. Fonstad, Defect structure and electronic donor levels in stannic oxide crystals, J. Appl. Phys. 44 (1973) 4618–4621. Available from: https://doi.org/10.1063/1.1662011.
[46] M. Nagasawa, S. Shionoya, Exciton structure in optical absorption of SnO_2 crystals, Phys. Lett. 22 (1966) 409–410. Available from: https://doi.org/10.1016/0031-9163(66)91202-9.

[47] C.G. Fonstad, R.H. Rediker, Electrical properties of high-quality stannic oxide crystals, J. Appl. Phys. 42 (1971) 2911–2918. Available from: https://doi.org/10.1063/1.1660648.
[48] Y. Mizokawa, S. Nakamura, ESR and electric conductance studies of the fine-powdered SnO_2, Jpn. J. Appl. Phys. 14 (1975) 779.
[49] H.J. van Daal, Polar optical-mode scattering of electrons in SnO_2, Solid State Commun. 6 (1968) 5–9. Available from: https://doi.org/10.1016/0038-1098(68)90325-6.
[50] D.J.M. Bevan, J.P. Shelton, J.S. Anderson, Properties of some simple oxides and spinels at high temperatures, J. Chem. Soc. (1948) 1729. Available from: https://doi.org/10.1039/jr9480001729.
[51] E.E. Hahn, Some electrical properties of zinc oxide semiconductor, J. Appl. Phys. 22 (1951) 855–863. Available from: https://doi.org/10.1063/1.1700063.
[52] D.C. Look, D.C. Reynolds, J.R. Sizelove, R.L. Jones, C.W. Litton, G. Cantwell, et al., Electrical properties of bulk ZnO, Solid State Commun. 105 (1998) 399–401. Available from: https://doi.org/10.1016/S0038-1098(97)10145-4.
[53] D.C. Look, J.W. Hemsky, J.R. Sizelove, Residual native shallow donor in ZnO, Phys. Rev. Lett. 82 (1999) 2552–2555. Available from: https://doi.org/10.1103/PhysRevLett.82.2552.
[54] P.H. Kasai, Electron Spin resonance studies of donors and acceptors in ZnO, Phys. Rev. 130 (1963) 989–995. Available from: https://doi.org/10.1103/PhysRev.130.989.
[55] D.C. Look, G.C. Farlow, P. Reunchan, S. Limpijumnong, S.B. Zhang, K. Nordlund, Evidence for native-defect donors in n-type ZnO, Phys. Rev. Lett. 95 (2005) 225502. Available from: https://doi.org/10.1103/PhysRevLett.95.225502.
[56] L.E. Halliburton, N.C. Giles, N.Y. Garces, M. Luo, C. Xu, L. Bai, et al., Production of native donors in ZnO by annealing at high temperature in Zn vapor, Appl. Phys. Lett. 87 (2005) 172108. Available from: https://doi.org/10.1063/1.2117630.
[57] A. Pöppl, G. Völkel, ESR and photo-ESR investigations of zinc Vacancies and interstitial oxygen ions in undoped ZnO ceramics, Phys. Status Solidi 125 (1991) 571–581. Available from: https://doi.org/10.1002/pssa.2211250218.
[58] K. Vanheusden, C.H. Seager, W.L. Warren, D.R. Tallant, J.A. Voigt, Correlation between photoluminescence and oxygen vacancies in ZnO phosphors, Appl. Phys. Lett. 68 (1996) 403–405. Available from: https://doi.org/10.1063/1.116699.
[59] D.M. Hofmann, D. Pfisterer, J. Sann, B.K. Meyer, R. Tena-Zaera, V. Munoz-Sanjose, et al., Properties of the oxygen vacancy in ZnO, Appl. Phys. A Mater. Sci. Process. 88 (2007) 147–151. Available from: https://doi.org/10.1007/s00339-007-3956-2.
[60] P.H. Miller, The electrical conductivity of zinc oxide, Phys. Rev. 60 (1941) 890–895. Available from: https://doi.org/10.1103/PhysRev.60.890.
[61] C. Gonzalez, D. Galland, A. Herve, Interactions hyperfines du centre F + dans ZnO, Phys. Status Solidi 72 (1975) 309–320. Available from: https://doi.org/10.1002/pssb.2220720134.
[62] L.S. Vlasenko, G.D. Watkins, Optical detection of electron paramagnetic resonance in room-temperature electron-irradiated ZnO, Phys. Rev. B. 71 (2005) 125210. Available from: https://doi.org/10.1103/PhysRevB.71.125210.
[63] K. Leutwein, J. Schneider, Defects in neutron-irradiated ZnO (I), Zeitschrift F?r Naturforsch. A. 26 (1971). Available from: https://doi.org/10.1515/zna-1971-0727.
[64] Q.X. Zhao, P. Klason, M. Willander, H.M. Zhong, W. Lu, J.H. Yang, Deep-level emissions influenced by O and Zn implantations in ZnO, Appl. Phys. Lett. 87 (2005) 1–3. Available from: https://doi.org/10.1063/1.2135880.

[65] A. Janotti, C.G. Van de Walle, Native point defects in ZnO, Phys. Rev. B. 76 (2007) 165202. Available from: https://doi.org/10.1103/PhysRevB.76.165202.
[66] M.D. McCluskey, S.J. Jokela, Defects in ZnO, J. Appl. Phys. 106 (2009) 71101. Available from: https://doi.org/10.1063/1.3216464.
[67] J.H.W. De Wit, Electrical properties of In_2O_3, J. Solid State Chem. 8 (1973) 142–149. Available from: https://doi.org/10.1016/0022-4596(73)90007-8.
[68] J.H.W. De Wit, The high temperature behavior of In_2O_3, J. Solid State Chem. 13 (1975) 192–200. Available from: https://doi.org/10.1016/0022-4596(75)90118-8.
[69] J.H.W. De Wit, Structural aspects and defect chemistry in In_2O_3, J. Solid State Chem. 20 (1977) 143–148. Available from: https://doi.org/10.1016/0022-4596(77)90061-5.
[70] J.H.W. De Wit, G. Van Unen, M. Lahey, Electron concentration and mobility in In_2O_3, J. Phys. Chem. Solids. 38 (1977) 819–824. Available from: https://doi.org/10.1016/0022-3697(77)90117-2.
[71] G. Rupprecht, Untersuchungen der elektrischen und lichtelektrischen Leitfähigkeit dünner Indiumoxydschichten, Zeitschrift Für Phys. 139 (1954) 504–517. Available from: https://doi.org/10.1007/BF01374559.
[72] A.J. Rosenberg, Oxidation of intermetallic compounds. II. Interrupted oxidation of InSb, J. Phys. Chem. 64 (1960) 1143–1150. Available from: https://doi.org/10.1021/j100838a011.
[73] L.N. Cojocaru, I.D. Alecu, Electrical properties of ß-Ga_2O_3, Z. Phys. Chem. 84 (1973) 325–331. Available from: https://doi.org/10.1524/zpch.1973.84.5_6.325.
[74] T. Harwig, G.J. Wubs, G.J. Dirksen, Electrical-properties of beta-Ga_2O_3 single-crystals, Solid State Commun. 18 (1976) 1223–1225. Available from: https://doi.org/10.1016/0038-1098(76)90944-3.
[75] T. Harwig, J. Schoonman, Electrical properties of β-Ga_2O_3 single crystals. II, J. Solid State Chem. 23 (1978) 205–211. Available from: https://doi.org/10.1016/0022-4596(78)90066-X.
[76] K. Irmscher, Z. Galazka, M. Pietsch, R. Uecker, R. Fornari, Electrical properties of beta-Ga_2O_3 single crystals grown by the Czochralski method, J. Appl. Phys. 110 (2011). Available from: https://doi.org/10.1063/1.3642962.
[77] H.J. Kümmerer, G. Denninger, The oxygen vacancy in Ga_2O_3: a double resonance investigation, Magn. Reson. Chem. 43 (2005) S145–S152. Available from: https://doi.org/10.1002/mrc.1682.
[78] B.L. Crowder, M.J. Sienko, Some solid-state studies of tungsten trioxide and their significance to tungsten bronze theory, J. Chem. Phys. 38 (1963) 1576–1583. Available from: https://doi.org/10.1063/1.1776924.
[79] J.M. Berak, M.J. Sienko, Effect of oxygen-deficiency on electrical transport properties of tungsten trioxide crystals, J. Solid State Chem. 2 (1970) 109–133. Available from: https://doi.org/10.1016/0022-4596(70)90040-X.
[80] M. Gillet, C. Lemire, E. Gillet, K. Aguir, The role of surface oxygen vacancies upon WO_3 conductivity, Surf. Sci. 532–535 (2003) 519–525. Available from: https://doi.org/10.1016/S0039-6028(03)00477-1.
[81] M. Gillet, K. Aguir, C. Lemire, E. Gillet, K. Schierbaum, The structure and electrical conductivity of vacuum-annealed WO_3 thin films, Thin Solid Films. 467 (2004) 239–246. Available from: https://doi.org/10.1016/j.tsf.2004.04.018.
[82] J.S. Anderson, B.G. Hyde, On the possible role of dislocations in generating ordered and disordered shear structures, J. Phys. Chem. Solids. 28 (1967) 1393–1408. Available from: https://doi.org/10.1016/0022-3697(67)90268-5.

[83] L. Ottaviano, L. Lozzi, M. Passacantando, S. Santucci, On the spatially resolved electronic structure of polycrystalline WO_3 films investigated with scanning tunnelling spectroscopy, Surf. Sci. 475 (2001) 73–82.

[84] C.J. Kevane, Oxygen vacancies and electrical conduction in metal oxides, Phys. Rev. 133 (1964) A1431–A1436. Available from: https://doi.org/10.1103/PhysRev.133.A1431.

[85] W. Göpel, G. Rocker, R. Feierabend, Intrinsic defects of $TiO_2(110)$: interaction with chemisorbed O_2, H_2, CO, and CO_2, Phys. Rev. B. 28 (1983) 3427–3438. Available from: https://doi.org/10.1103/PhysRevB.28.3427.

[86] J. Yahia, Dependence of the electrical conductivity and thermoelectric power of pure and aluminum-doped rutile on equilibrium oxygen pressure and temperature, Phys. Rev. 130 (1963) 1711–1719. Available from: https://doi.org/10.1103/PhysRev.130.1711.

[87] R.G. Breckenridge, W.R. Hosler, Electrical properties of titanium dioxide semiconductors, Phys. Rev. 91 (1953) 793–802. Available from: https://doi.org/10.1103/PhysRev.91.793.

[88] E. Iguchi, K. Yajima, T. Asahina, Y. Kanamori, Resistivities of reduced rutile (TiO_2) from 300°K to exhaustion range, J. Phys. Chem. Solids. 35 (1974) 597–599. Available from: https://doi.org/10.1016/S0022-3697(74)80016-8.

[89] R.R. Hasiguti, E. Yagi, Electrical conductivity below 3 K of slightly reduced oxygen-deficient rutile TiO_{2-x}, Phys. Rev. B. 49 (1994) 7251–7256. Available from: https://doi.org/10.1103/PhysRevB.49.7251.

[90] E. Yagi, R.R. Hasiguti, M. Aono, Electronic conduction above 4 K of slightly reduced oxygen-deficient rutile TiO_{2-x}, Phys. Rev. B. 54 (1996) 7945–7956. Available from: https://doi.org/10.1103/PhysRevB.54.7945.

[91] M.K. Nowotny, L.R. Sheppard, T. Bak, J. Nowotny, Defect Chemistry of titanium dioxide. application of defect engineering in processing of TiO_2-based photocatalysts, J. Phys. Chem. C. 112 (2008) 5275–5300. Available from: https://doi.org/10.1021/jp077275m.

[92] L. Forro, O. Chauvet, D. Emin, L. Zuppiroli, H. Berger, F. Lévy, High mobility n-type charge carriers in large single crystals of anatase (TiO_2), J. Appl. Phys. 75 (1994) 633–635. Available from: https://doi.org/10.1063/1.355801.

[93] D. Eder, R. Kramer, Stoichiometry of "titanium suboxide", Phys. Chem. Chem. Phys. 5 (2003) 1314–1319. Available from: https://doi.org/10.1039/b210004e.

[94] G. Frank, H. Köstlin, Electrical properties and defect model of tin-doped indium oxide layers, Appl. Phys. A Solids Surfaces. 27 (1982) 197–206. Available from: https://doi.org/10.1007/BF00619080.

[95] J.H. Hwang, D.D. Edwards, D.R. Kammler, T.O. Mason, Point defects and electrical properties of Sn-doped In-based transparent conducting oxides, Solid State Ionics. 129 (2000) 135–144. Available from: https://doi.org/10.1016/S0167-2738(99)00321-5.

[96] S.P. Harvey, T.O. Mason, Y. Gassenbauer, R. Schafranek, A. Klein, Surface versus bulk electronic/defect structures of transparent conducting oxides: I. Indium oxide and ITO, J. Phys. D. Appl. Phys. 39 (2006) 3959–3968. Available from: https://doi.org/10.1088/0022-3727/39/18/006.

[97] S.P. Harvey, T.O. Mason, C. Körber, A. Klein, Bulk defect chemistry and surface electronic behavior of Zn,Sn codoped In_2O_3 transparent conducting oxides, Phys. Chem. Chem. Phys. 11 (2009) 3099–3104. Available from: https://doi.org/10.1039/b822149a.

[98] T.O. Mason, G.B. Gonzalez, D.R. Kammler, N. Mansourian-Hadavi, B.J. Ingram, Defect chemistry and physical properties of transparent conducting oxides in the CdO-In_2O_3-SnO_2 system, Thin Solid Films. 411 (2002) 106–114. Available from: https://doi.org/10.1016/S0040-6090(02)00197-9.

[99] J.-H. Park, K. Natesan, Electronic transport in thermally grown Cr_2O_3, Oxid. Met. 33 (1990) 31–54. Available from: https://doi.org/10.1007/BF00665668.

[100] W.C. Hagel, Electrical conductivity of li-substituted Cr_2O_3, J. Appl. Phys. 36 (1965) 2586–2587. Available from: https://doi.org/10.1063/1.1714536.

[101] A. Holt, P. Kofstad, Electrical conductivity and defect structure of Cr_2O_3. II. Reduced temperatures ($< \sim 1000°C$), Solid State Ionics. 69 (1994) 137–143. Available from: https://doi.org/10.1016/0167-2738(94)90402-2.

[102] R.F.G. Gardner, F. Sweett, D.W. Tanner, The electrical properties of alpha ferric oxide - II. Ferric oxide of high purrity, J. Phys. Chem. Solids. 24 (1963) 1183–1196. Available from: https://doi.org/10.1016/0022-3697(63)90235-X.

[103] J.A. Glasscock, P.R.F. Barnes, I.C. Plumb, A. Bendavid, P.J. Martin, Structural, optical and electrical properties of undoped polycrystalline hematite thin films produced using filtered arc deposition, Thin Solid Films. 516 (2008) 1716–1724. Available from: https://doi.org/10.1016/j.tsf.2007.05.020.

[104] R.H. Chang, J.B. Wagner, Direct-current conductivity and iron tracer diffusion in hematite at high temperatures, J. Am. Ceram. Soc. 55 (1972) 211–213. Available from: https://doi.org/10.1111/j.1151-2916.1972.tb11261.x.

[105] K. Hoshino, N.L. Peterson, Cation self-diffusion and impurity diffusion in Fe_2O_3, J. Phys. Chem. Solids. 46 (1985) 1247–1254. Available from: https://doi.org/10.1016/0022-3697(85)90127-1.

[106] A. Atkinson, R.I. Taylor, Diffusion of ^{55}Fe in Fe_2O_3 single crystals, J. Phys. Chem. Solids. 46 (1985) 469–475. Available from: https://doi.org/10.1016/0022-3697(85)90114-3.

[107] R. Dieckmann, Point defects and transport in haematite (ε-Fe_2O_3), Philos. Mag. A. 68 (1993) 725–745. Available from: https://doi.org/10.1080/01418619308213994.

[108] Y. Ling, G. Wang, J. Reddy, C. Wang, J.Z. Zhang, Y. Li, The influence of oxygen content on the thermal activation of hematite nanowires, Angew. Chemie Int. Ed. 51 (2012) 4074–4079. Available from: https://doi.org/10.1002/anie.201107467.

[109] C.-S. Cheng, M. Serizawa, H. Sakata, T. Hirayama, Electrical conductivity of Co_3O_4 films prepared by chemical vapour deposition, Mater. Chem. Phys. 53 (1998) 225–230. Available from: https://doi.org/10.1016/S0254-0584(98)00044-3.

[110] S. Suda, S. Fujitsu, K. Koumoto, H. Yanagida, The effect of atmosphere and doping on electrical conductivity of CuO, Jpn. J. Appl. Phys. 31 (1992) 2488.

[111] K. Watanabe, T. Ohsawa, I. Sakaguchi, O. Bierwagen, M.E. White, M.-Y. Tsai, et al., Investigation of charge compensation in indium-doped tin dioxide by hydrogen insertion via annealing under humid conditions, Appl. Phys. Lett. 104 (2014) 132110. Available from: https://doi.org/10.1063/1.4870425.

[112] D.M. Hofmann, A. Hofstaetter, F. Leiter, H. Zhou, F. Henecker, B.K. Meyer, et al., Hydrogen: a relevant shallow donor in zinc oxide, Phys. Rev. Lett. 88 (2002) 45504. Available from: https://doi.org/10.1103/PhysRevLett.88.045504.

[113] D.-G. Kim, S. Lee, G.-H. Lee, S.-C. Kwon, Effects of hydrogen gas on properties of tin-doped indium oxide films deposited by radio frequency magnetron sputtering method, Thin Solid Films. 515 (2007) 6949–6952. Available from: https://doi.org/10.1016/j.tsf.2007.02.056.

[114] U. Diebold, The surface science of titanium dioxide, Surf. Sci. Rep. 48 (2003) 53–229. Available from: https://doi.org/10.1016/S0167-5729(02)00100-0.

[115] S.F.J. Cox, E.A. Davis, S.P. Cottrell, P.J.C. King, J.S. Lord, J.M. Gil, et al., Experimental confirmation of the predicted shallow donor hydrogen state in zinc oxide, Phys. Rev. Lett. 86 (2001) 2601–2604. Available from: https://doi.org/10.1103/PhysRevLett.86.2601.

[116] S.F.J. Cox, E.A. Davis, P.J.C. King, J.M. Gil, H.V. Alberto, R.C. Vilão, et al., Shallow versus deep hydrogen states in ZnO and HgO, J. Phys. Condens. Matter. 13 (2001) 9001.
[117] P.D.C. King, R.L. Lichti, Y.G. Celebi, J.M. Gil, R.C. Vilão, H.V. Alberto, et al., Shallow donor state of hydrogen in In_2O_3 and SnO_2: implications for conductivity in transparent conducting oxides, Phys. Rev. B. 80 (2009) 81201. Available from: https://doi.org/10.1103/PhysRevB.80.081201.
[118] P.D.C. King, I. McKenzie, T.D. Veal, Observation of shallow-donor muonium in Ga_2O_3: evidence for hydrogen-induced conductivity, Appl. Phys. Lett. 96 (2010) 62110. Available from: https://doi.org/10.1063/1.3309694.
[119] E.V. Lavrov, F. Börrnert, J. Weber, Photoconductivity and infrared absorption study of hydrogen-related shallow donors in ZnO, Phys. Rev. B. 72 (2005) 85212. Available from: https://doi.org/10.1103/PhysRevB.72.085212.
[120] E.V. Lavrov, F. Herklotz, J. Weber, Identification of two hydrogen donors in ZnO, Phys. Rev. B. 79 (2009) 165210. Available from: https://doi.org/10.1103/PhysRevB.79.165210.
[121] M.D. McCluskey, M.C. Tarun, S.T. Teklemichael, Hydrogen in oxide semiconductors, J. Mater. Res. 27 (2012) 2190–2198. Available from: https://doi.org/10.1557/jmr.2012.137.
[122] G. Brauer, W. Anwand, W. Skorupa, J. Kuriplach, O. Melikhova, C. Moisson, et al., Defects in virgin and N + -implanted ZnO single crystals studied by positron annihilation, Hall effect, and deep-level transient spectroscopy, Phys. Rev. B - Condens. Matter Mater. Phys 74 (2006) 1–10. Available from: https://doi.org/10.1103/PhysRevB.74.045208.
[123] M.C. Tarun, M.Z. Iqbal, M.D. McCluskey, Nitrogen is a deep acceptor in ZnO, AIP Adv. 1 (2011) 22105. Available from: https://doi.org/10.1063/1.3582819.
[124] M.J. Alam, D.C. Cameron, Investigation of annealing effects on sol-gel deposited indium tin oxide thin films in different atmospheres, Thin Solid Films. 420–421 (2002) 76–82. Available from: https://doi.org/10.1016/S0040-6090(02)00737-X.
[125] M.E. White, O. Bierwagen, M.Y. Tsai, J.S. Speck, Electron transport properties of antimony doped SnO_2 single crystalline thin films grown by plasma-assisted molecular beam epitaxy, J. Appl. Phys. 106 (2009) 93704. Available from: https://doi.org/10.1063/1.3254241.
[126] J. Rebholz, P. Bonanati, C. Jaeschke, M. Hübner, L. Mädler, U. Weimar, et al., Conduction mechanism in undoped and antimony doped SnO_2 based FSP gas sensors, Sens. Actuators, B Chem. 188 (2013) 631–636. Available from: https://doi.org/10.1016/j.snb.2013.07.063.
[127] B.K. Meyer, H. Alves, D.M. Hofmann, W. Kriegseis, D. Forster, F. Bertram, et al., Bound exciton and donor-acceptor pair recombinations in ZnO, Phys, Status Solidi Basic Res. 241 (2004) 231–260. Available from: https://doi.org/10.1002/pssb.200301962.
[128] D. de Cogan, G.A. Lonergan, Electrical conduction in Fe_2O_3 and Cr_2O_3, Solid State Commun. 15 (1974) 1517–1519. Available from: https://doi.org/10.1016/0038-1098(74)90929-6.
[129] K.H. Kim, S.H. Lee, J.S. Choi, Electrical conductivity of pure and doped alpha-ferric oxides, J. Phys. Chem. Solids. 46 (1985) 331–338. Available from: https://doi.org/10.1016/0022-3697(85)90175-1.
[130] M. Hübner, N. Barsan, U. Weimar, Influences of Al, Pd and Pt additives on the conduction mechanism as well as the surface and bulk properties of SnO_2 based

polycrystalline thick film gas sensors, Sens. Actuators B Chem. 171–172 (2012) 172–180. Available from: https://doi.org/10.1016/j.snb.2012.02.080.
[131] W.R. Gardner, G.C. Danielson, Electrical resistivity and hall coefficient of sodium tungsten bronze, Phys. Rev. 93 (1954) 46–51. Available from: https://doi.org/10.1103/PhysRev.93.46.
[132] P.D.C. King, T.D. Veal, D.J. Payne, A. Bourlange, R.G. Egdell, C.F. McConville, Surface electron accumulation and the charge neutrality level in In_2O_3, Phys. Rev. Lett. 101 (2008) 116808. Available from: https://doi.org/10.1103/PhysRevLett.101.116808.
[133] P.D.C. King, T.D. Veal, F. Fuchs, C.Y. Wang, D.J. Payne, A. Bourlange, et al., Band gap, electronic structure, and surface electron accumulation of cubic and rhombohedral In_2O_3, Phys. Rev. B. 79 (2009) 205211. Available from: https://doi.org/10.1103/PhysRevB.79.205211.
[134] J. Tersoff, Schottky barriers and semiconductor band structures, Phys. Rev. B. 32 (1985) 6968–6971. Available from: https://doi.org/10.1103/PhysRevB.32.6968.
[135] T. Nagata, O. Bierwagen, M.E. White, M.Y. Tsai, Y. Yamashita, H. Yoshikawa, et al., XPS study of Sb-/In-doping and surface pinning effects on the Fermi level in SnO_2 (101) thin films, Appl. Phys. Lett. 98 (2011) 232107. Available from: https://doi.org/10.1063/1.3596449.
[136] O. Bierwagen, T. Nagata, M.E. White, M.-Y. Tsai, J.S. Speck, Electron transport in semiconducting SnO_2: intentional bulk donors and acceptors, the interface, and the surface, J. Mater. Res. 27 (2012) 2232–2236. Available from: https://doi.org/10.1557/jmr.2012.172.
[137] C. Kilic, A. Zunger, Origins of coexistence of conductivity and transparency in SnO_2, Phys. Rev. Lett. 88 (2002) 955011–955014. Available from: https://doi.org/10.1103/PhysRevLett.88.095501.
[138] P.D.C. King, T.D. Veal, Conductivity in transparent oxide semiconductors, J. Phys. Condens. Matter. 23 (2011) 334214. Available from: https://doi.org/10.1088/0953-8984/23/33/334214.
[139] S. Lany, A. Zunger, Assessment of correction methods for the band-gap problem and for finite-size effects in supercell defect calculations: case studies for ZnO and GaAs, Phys. Rev. B. 78 (2008) 235104. Available from: https://doi.org/10.1103/PhysRevB.78.235104.
[140] C. Freysoldt, B. Grabowski, T. Hickel, J. Neugebauer, G. Kresse, A. Janotti, et al., First-principles calculations for point defects in solids, Rev. Mod. Phys. 86 (2014) 253–305. Available from: https://doi.org/10.1103/RevModPhys.86.253.
[141] P. Hohenberg, W. Kohn, Inhomogeneous electron gas, Phys. Rev. 136 (1964) B864–B871. Available from: https://doi.org/10.1103/PhysRev.136.B864.
[142] W. Kohn, L.J. Sham, Self-consistent equations including exchange and correlation effects, Phys. Rev. 140 (1965) A1133–A1138. Available from: https://doi.org/10.1103/PhysRev.140.A1133.
[143] U. von Barth, Basic density-functional theory - an overview, Phys. Scr. T109 (2004) 9–39. Available from: https://doi.org/10.1238/Physica.Topical.109a00009.
[144] R.O. Jones, O. Gunnarsson, The density functional formalism, its applications and prospects, Rev. Mod. Phys. 61 (1989) 689–746. Available from: https://doi.org/10.1103/RevModPhys.61.689.
[145] D.C. Langreth, J.P. Perdew, Theory of nonuniform electronic systems. I. Analysis of the gradient approximation and a generalization that works, Phys. Rev. B. 21 (1980) 5469–5493. Available from: https://doi.org/10.1103/PhysRevB.21.5469.

[146] D.C. Langreth, M.J. Mehl, Beyond the local-density approximation in calculations of ground-state electronic properties, Phys. Rev. B. 28 (1983) 1809–1834. Available from: https://doi.org/10.1103/PhysRevB.28.1809.
[147] J.P. Perdew, S. Kurth, A. Zupan, P. Blaha, Accurate density functional with correct formal properties: a step beyond the generalized gradient approximation, Phys. Rev. Lett. 82 (1999) 2544–2547. Available from: https://doi.org/10.1103/PhysRevLett.82.2544.
[148] A.D. Becke, A new mixing of Hartree–Fock and local density-functional theories, J. Chem. Phys. 98 (1993) 1372–1377. Available from: https://doi.org/10.1063/1.464304.
[149] J. Heyd, G.E. Scuseria, M. Ernzerhof, Hybrid functionals based on a screened Coulomb potential, J. Chem. Phys. 118 (2003) 8207–8215. Available from: https://doi.org/10.1063/1.1564060.
[150] J.P. Perdew, M. Ernzerhof, K. Burke, Rationale for mixing exact exchange with density functional approximations, J. Chem. Phys. 105 (1996) 9982–9985. Available from: https://doi.org/10.1063/1.472933.
[151] V.I. Anisimov, F. Aryasetiawan, A.I. Lichtenstein, First-principles calculations of the electronic structure and spectra of strongly correlated systems: the LDA + U method, J. Phys. Condens. Matter. 9 (1997) 767.
[152] D.M. Ceperley, B.J. Alder, Ground state of the electron gas by a stochastic method, Phys. Rev. Lett. 45 (1980) 566–569. Available from: https://doi.org/10.1103/PhysRevLett.45.566.
[153] A.D. Becke, Density-functional thermochemistry. III. The role of exact exchange, J. Chem. Phys. 98 (1993) 5648–5652. Available from: https://doi.org/10.1063/1.464913.
[154] C. Lee, W. Yang, R.G. Parr, Development of the Colle-Salvetti correlation-energy formula into a functional of the electron density, Phys. Rev. B. 37 (1988) 785–789. Available from: https://doi.org/10.1103/PhysRevB.37.785.
[155] VASP, n.d. https://www.vasp.at.
[156] DFT codes and website, n.d. http://dft.sandia.gov/Quest/DFT_codes.html.
[157] S. Lany, A. Zunger, Dopability, intrinsic conductivity, and nonstoichiometry of transparent conducting oxides, Phys. Rev. Lett. 98 (2007) 2–5. Available from: https://doi.org/10.1103/PhysRevLett.98.045501.
[158] A.K. Singh, A. Janotti, M. Scheffler, C.G. Van De Walle, Sources of electrical conductivity in SnO_2, Phys. Rev. Lett. 101 (2008) 1–4. Available from: https://doi.org/10.1103/PhysRevLett.101.055502.
[159] A. Janotti, C.G. Van de Walle, Oxygen vacancies in ZnO, Appl. Phys. Lett. 87 (2005) 122102. Available from: https://doi.org/10.1063/1.2053360.
[160] A. Janotti, C.G. Van De Walle, New insights into the role of native point defects in ZnO, J. Cryst. Growth. 287 (2006) 58–65. Available from: https://doi.org/10.1016/j.jcrysgro.2005.10.043.
[161] F. Oba, M. Choi, A. Togo, A. Seko, I. Tanaka, Native defects in oxide semiconductors: a density functional approach, J. Physics. Condens. Matter. 22 (2010) 384211. Available from: https://doi.org/10.1088/0953-8984/22/38/384211.
[162] F. Trani, M. Causaç, D. Ninno, G. Cantele, V. Barone, Density functional study of oxygen vacancies at the SnO_2 surface and subsurface sites, Phys. Rev. B. 77 (2008) 245410. Available from: https://doi.org/10.1103/PhysRevB.77.245410.
[163] I. Tanaka, F. Oba, K. Tatsumi, M. Kunisu, M. Nakano, H. Adachi, Theoretical formation energy of oxygen-vacancies in oxides, Mater. Trans. 43 (2002) 1426. Available from: https://doi.org/10.1021/1429.

[164] W.C. Mackrodt, E.-A. Simson, N.M. Harrison, An ab initio Hartree-Fock study of the electron-excess gap states in oxygen-deficient rutile TiO_2, Surf. Sci. 384 (1997) 192–200. Available from: https://doi.org/10.1016/S0039-6028(97)00219-7.
[165] G. Mattioli, F. Filippone, P. Alippi, A. Amore Bonapasta, Ab initio study of the electronic states induced by oxygen vacancies in rutile and anatase TiO_2, Phys. Rev. B. 78 (2008) 241201. Available from: https://doi.org/10.1103/PhysRevB.78.241201.
[166] P. Reunchan, X. Zhou, S. Limpijumnong, A. Janotti, C.G. Van De Walle, Vacancy defects in indium oxide: an ab-initio study, Curr. Appl. Phys. 11 (2011) S296–S300. Available from: https://doi.org/10.1016/j.cap.2011.03.051.
[167] F. Oba, S.R. Nishitani, S. Isotani, H. Adachi, I. Tanaka, Energetics of native defects in ZnO, J. Appl. Phys. 90 (2001) 824–828. Available from: https://doi.org/10.1063/1.1380994.
[168] J. Robertson, Defect levels of SnO_2, Phys. Rev. B. 30 (1984) 3520–3522. Available from: https://doi.org/10.1103/PhysRevB.30.3520.
[169] T.S. Rantala, V. Lantto, T.T. Rantala, A cluster approach for modelling of surface characteristics of stannic oxide, Phys. Scr. 54 (1994) 252–255.
[170] S.B. Zhang, S.-H. Wei, A. Zunger, Intrinsic n -type versus p -type doping asymmetry and the defect physics of ZnO, Phys. Rev. B. 63 (2001) 75205. Available from: https://doi.org/10.1103/PhysRevB.63.075205.
[171] K. Palandage, G.W. Fernando, Role of point defects on conductivity, magnetism and optical properties in In_2O_3, Phys. Lett. Sect. A Gen. At. Solid State Phys. 374 (2010) 2879–2885. Available from: https://doi.org/10.1016/j.physleta.2010.05.003.
[172] J. Lee, S. Han, Thermodynamics of native point defects in alpha-Fe_2O_3: an ab initio study, Phys. Chem. Chem. Phys. 15 (2013) 18906. Available from: https://doi.org/10.1039/c3cp53311e.
[173] F. Lebreau, M.M. Islam, B. Diawara, P. Marcus, Structural, magnetic, electronic, defect, and diffusion properties of Cr_2O_3: a DFT + U study, J. Phys. Chem. C. 118 (2014) 18133–18145. Available from: https://doi.org/10.1021/jp5039943.
[174] D. Wu, Q. Zhang, M. Tao, LSDA + U study of cupric oxide: electronic structure and native point defects, Phys. Rev. B - Condens. Matter Mater. Phys 73 (2006) 1–6. Available from: https://doi.org/10.1103/PhysRevB.73.235206.
[175] Y. Peng, Z. Zhang, T. Viet Pham, Y. Zhao, P. Wu, J. Wang, Density functional theory analysis of dopants in cupric oxide, J. Appl. Phys. 111 (2012) 103708. Available from: https://doi.org/10.1063/1.4719059.
[176] S. Lany, A. Zunger, Anion vacancies as a source of persistent photoconductivity in II-VI and chalcopyrite semiconductors, Phys. Rev. B. 72 (2005) 35215. Available from: https://doi.org/10.1103/PhysRevB.72.035215.
[177] P.M. Mooney, Deep donor levels (D X centers) in III-V semiconductors, J. Appl. Phys. 67 (1990) R1–R26. Available from: https://doi.org/10.1063/1.345628.
[178] J.M. Smith, W.E. Vehse, ESR of electron irradiated ZnO confirmation of the F + center, Phys. Lett. A. 31 (1970) 147–148. Available from: https://doi.org/10.1016/0375-9601(70)90199-4.
[179] M.C. Tarun, F.A. Selim, M.D. McCluskey, Persistent photoconductivity in strontium titanate, Phys. Rev. Lett. 111 (2013) 187403. Available from: https://doi.org/10.1103/PhysRevLett.111.187403.
[180] T.S. Bjørheim, S. Stølen, T. Norby, Ab initio studies of hydrogen and acceptor defects in rutile TiO_2, Phys. Chem. Chem. Phys. 12 (2010) 6817–6825. Available from: https://doi.org/10.1039/b925823j.

[181] J.B. Varley, A. Janotti, A.K. Singh, C.G. Van de Walle, Hydrogen interactions with acceptor impurities in SnO_2: first-principles calculations, Phys. Rev. B. 79 (2009) 245206. Available from: https://doi.org/10.1103/PhysRevB.79.245206.
[182] J.B. Varley, H. Peelaers, A. Janotti, C.G. Van de Walle, Hydrogenated cation vacancies in semiconducting oxides, J. Phys. Condens. Matter. 23 (2011) 334212. Available from: https://doi.org/10.1088/0953-8984/23/33/334212.
[183] A. Janotti, C.G. Van de Walle, Hydrogen multicentre bonds, Nat. Mater. 6 (2007) 44–47. Available from: https://doi.org/10.1038/nmat1795.
[184] J.B. Varley, A. Janotti, C.G. Van De Walle, Group-V impurities in SnO_2 from first-principles calculations, Phys. Rev. B - Condens. Matter Mater. Phys. 81 (2010) 1–7. Available from: https://doi.org/10.1103/PhysRevB.81.245216.
[185] A. Zunger, Practical doping principles, Appl. Phys. Lett. 83 (2003) 57–59. Available from: https://doi.org/10.1063/1.1584074.
[186] H. Raebiger, S. Lany, A. Zunger, Electronic structure, donor and acceptor transitions, and magnetism of 3d impurities in In_2O_3 and ZnO, Phys. Rev. B. 79 (2009) 165202. Available from: https://doi.org/10.1103/PhysRevB.79.165202.
[187] X.H. Zhou, Q.H. Hu, Y. Fu, First-principles LDA + U studies of the In-doped ZnO transparent conductive oxide, J. Appl. Phys. 104 (2008). Available from: https://doi.org/10.1063/1.2978324.
[188] S. Lany, A. Zunger, Polaronic hole localization and multiple hole binding of acceptors in oxide wide-gap semiconductors, Phys. Rev. B - Condens. Matter Mater. Phys. 80 (2009) 1–5. Available from: https://doi.org/10.1103/PhysRevB.80.085202.
[189] H. Peng, J. Li, S.-S. Li, J.-B. Xia, First-principles study of the electronic structures and magnetic properties of 3d transition metal-doped anatase TiO_2, J. Phys. Condens. Matter. 20 (2008) 125207. Available from: https://doi.org/10.1088/0953-8984/20/12/125207.
[190] A. Hjelm, C.G. Granqvist, J.M. Wills, Electronic structure and optical properties of WO_3, Phys. Rev. B. 54 (1996) 2436–2445. Available from: https://doi.org/10.1103/PhysRevB.54.2436.
[191] B. Ingham, S.C. Hendy, S.V. Chong, J.L. Tallon, Density-functional studies of tungsten trioxide, tungsten bronzes, and related systems, Phys. Rev. B. 72 (2005) 75109. Available from: https://doi.org/10.1103/PhysRevB.72.075109.
[192] H. Eschrig, Metal-Insulator Transition, second ed., Taylor & Francis, London, 1990.
[193] J.P. Perdew, R.G. Parr, M. Levy, J.L. Balduz, Density-functional theory for fractional particle number: derivative discontinuities of the energy, Phys. Rev. Lett. 49 (1982) 1691–1694. Available from: https://doi.org/10.1103/PhysRevLett.49.1691.
[194] J.P. Perdew, M. Levy, Physical content of the exact Kohn-Sham orbital energies: band gaps and derivative discontinuities, Phys. Rev. Lett. 51 (1983) 1884–1887. Available from: https://doi.org/10.1103/PhysRevLett.51.1884.
[195] J.P. Perdew, Density functional theory and the band gap problem, Int. J. Quantum Chem. 28 (2009) 497–523. Available from: https://doi.org/10.1002/qua.560280846.
[196] J.P. Perdew, A. Zunger, Self-interaction correction to density-functional approximations for many-electron systems, Phys. Rev. B. 23 (1981) 5048–5079. Available from: https://doi.org/10.1103/PhysRevB.23.5048.
[197] R.W. Godby, M. Schlüter, L.J. Sham, Accurate exchange-correlation potential for silicon and its discontinuity on addition of an electron, Phys. Rev. Lett. 56 (1986) 2415–2418. Available from: https://doi.org/10.1103/PhysRevLett.56.2415.
[198] C.W.M. Castleton, A. Höglund, S. Mirbt, Density functional theory calculations of defect energies using supercells, Model. Simul. Mater. Sci. Eng. 17 (2009) 84003. Available from: https://doi.org/10.1088/0965-0393/17/8/084003.

[199] H.J. Monkhorst, J.D. Pack, Special points for Brillouin-zone integrations, Phys. Rev. B. 13 (1976) 5188–5192. Available from: https://doi.org/10.1103/PhysRevB.13.5188.
[200] A. Alkauskas, M.D. McCluskey, C.G. Van de Walle, Tutorial: defects in semiconductors—combining experiment and theory, J. Appl. Phys. 119 (2016) 181101. Available from: https://doi.org/10.1063/1.4948245.
[201] J.M. Crowley, J. Tahir-Kheli, W.A. Goddard, Resolution of the band gap prediction problem for materials design, J. Phys. Chem. Lett. 7 (2016) 1198–1203. Available from: https://doi.org/10.1021/acs.jpclett.5b02870.
[202] D. Bagayoko, Understanding density functional theory (DFT) and completing it in practice, AIP Adv. 4 (2014) 127104. Available from: https://doi.org/10.1063/1.4903408.
[203] G. Heiland, Zum Einfluß von adsorbiertem Sauerstoff auf die elektrische Leitfähigkeit von Zinkoxydkristallen, Zeitschrift Für Phys. 138 (1954) 459–464. Available from: https://doi.org/10.1007/BF01340692.
[204] A. Bielanski, J. Deren, J. Haber, Electric conductivty and catalytic activity of semiconduction oxide catalysts, Nature 179 (1957) 668–669. Available from: https://doi.org/10.1038/179668a0.
[205] N. Barsan, M. Schweizer-Berberich, W. Göpel, Fundamental and practical aspects in the design of nanoscaled SnO2 gas sensors: a status report, Fresenius J. Anal. Chem. 365 (1999) 287–304.
[206] W. Göpel, Solid-state chemical sensors: atomistic models and research trends, Sens. Actuators. 16 (1989) 167–193.
[207] W. Göpel, K.-D. Schierbaum, SnO_2 sensors: current status and future prospects, Sens. Actuators B Chem (1995) 1–12.
[208] G. Heiland, Homogeneous semiconducting gas sensors, Sens. Actuators. 2 (1982) 343–361.
[209] D. Kohl, Surface processes in the dectection of reducing gases with SnO_2-based devices, Sens. Actuators. 18 (1989) 71–113.
[210] R. Jaaniso, O.K. Tan, Semiconductor Gas Sensors, first ed., Woodhead Publishing Series in Electric and Optical Materials, Cambridge, 2013.
[211] N. Barsan, C. Simion, T. Heine, S. Pokhrel, U. Weimar, Modeling of sensing and transduction for p-type semiconducting metal oxide based gas sensors, J. Electroceramics. 25 (2010) 11–19. Available from: https://doi.org/10.1007/s10832-009-9583-x.
[212] N. Barsan, U. Weimar, Conduction model of metal oxide gas sensors, J. Electroceramics. 7 (2001) 143–167. Available from: https://doi.org/10.1023/A:1014405811371.
[213] N. Barsan, D. Koziej, U. Weimar, Metal oxide-based gas sensor research: how to? Sens. Actuators B Chem. 121 (2007) 18–35. Available from: https://doi.org/10.1016/j.snb.2006.09.047.
[214] A. Gurlo, R. Riedel, In situ and operando spectroscopy for assessing mechanisms of gas sensing, Angew. Chemie Interantional Ed. 46 (2007) 3826–3848. Available from: https://doi.org/10.1002/anie.200602597.
[215] B.M. Weckhuysen, Determining the active site in a catalytic process: operando spectroscopy is more than a buzzword, Phys. Chem. Chem. Phys. 5 (2003) 4351–4360. Available from: https://doi.org/10.1039/b309650p.
[216] A. Urakawa, Trends and advances in Operando methodology, Curr. Opin. Chem. Eng. 12 (2016) 31–36. Available from: https://doi.org/10.1016/j.coche.2016.02.002.

[217] M.A. Bañares, Operando methodology: combination of in situ spectroscopy and simultaneous activity measurements under catalytic reaction conditions, Catal. Today. 100 (2005) 71–77. Available from: https://doi.org/10.1016/j.cattod.2004.12.017.
[218] V.E. Heinrich, P.A. Cox, Surface Science of Metal Oxides, first ed., Cambridge University Press, Cambridge, 1994.
[219] M. Batzill, Surface Science studies of gas sensing materials: SnO_2, Sensors 6 (2006) 1345–1366.
[220] M. Batzill, U. Diebold, The surface and materials science of tin oxide, Prog. Surf. Sci. 79 (2005) 47–154. Available from: https://doi.org/10.1016/j.progsurf.2005.09.002.
[221] M. Batzill, U. Diebold, Surface studies of gas sensing metal oxides, Phys. Chem. Chem. Phys. 9 (2007) 2307–2318. Available from: https://doi.org/10.1039/b617710g.
[222] S. Lenaerts, J. Roggen, G. Maes, FT-IR characterization of tin dioxide gas sensor materials under working conditions, Spectrochim. Acta. 51 (1995) 883–894.
[223] J.J. Benítez, M.A. Centeno, C. Lois dit Pcard, O. Merdrignac, Y. Lauernt, J.A. Odriozola, In situ diffuse reflectance infrared spectroscopy (DRIFTS) study of the reversibility of CdGeON sensors towards oxygen, Sens. Actuators B Chem. 31 (1996) 197–202.
[224] S. Harbeck, A. Szatvanyi, N. Barsan, U. Weimar, V. Hoffmann, DRIFT studies of thick film un-doped and Pd-doped SnO_2 sensors: temperature changes effect and CO detection mechanism in the presence of water vapour, Thin Solid Films. 436 (2003) 76–83. Available from: https://doi.org/10.1016/S0040-6090(03)00512-1.
[225] M. Gaidi, M. Labeau, B. Chenevier, J.L. Hazemann, In-situ EXAFS analysis of the local environment of Pt particles incorporated in thin films of SnO_2 semi-conductor oxide used as gas-sensors, Sens. Actuators B Chem. 48 (1998) 277–284. Available from: https://doi.org/10.1016/S0925-4005(98)00056-2.
[226] O.V. Safonova, T. Neisius, A. Ryzhikov, B. Chenevier, A.M. Gaskov, M. Labeau, Characterization of the H2 sensing mechanism of Pd-promoted SnO_2 by XAS in operando conditions, Chem. Commun. (Camb) (2005) 5202–5204. Available from: https://doi.org/10.1039/b509826b.
[227] M. Hübner, D. Koziej, M. Bauer, N. Barsan, K.O. Kvashnina, M.D. Rossell, et al., The structure and behavior of platinum in SnO_2-based sensors under working conditions, Angew. Chem. Int. Ed. Engl. 50 (2011) 2841–2844. Available from: https://doi.org/10.1002/anie.201004499.
[228] D. Koziej, M. Hübner, N. Barsan, U. Weimar, M. Sikora, J.-D. Grunwaldt, Operando X-ray absorption spectroscopy studies on Pd-SnO_2 based sensors, Phys. Chem. Chem. Phys. 11 (2009) 8620–8625. Available from: https://doi.org/10.1039/b906829e.
[229] D. Degler, N. Barz, U. Dettinger, H. Peisert, T. Chassé, U. Weimar, et al., Extending the toolbox for gas sensor research: operando UV/vis diffuse reflectance spectroscopy on SnO2-based gas sensors, Sens. Actuators, B Chem. 224 (2016) 256–259. Available from: https://doi.org/10.1016/j.snb.2015.10.040.
[230] S. Sänze, A. Gurlo, C. Hess, Monitoring gas sensors at work: operando Raman-FTIR study of ethanol detection by indium oxide, Angew. Chem. Int. Ed. Engl. 52 (2013) 3607–3610. Available from: https://doi.org/10.1002/anie.201207258.
[231] S. Sänze, C. Hess, Ethanol gas sensing by indium oxide: an operando spectroscopic Raman-FTIR study, J. Phys. Chem. C. 118 (2014) 25603–25613. Available from: https://doi.org/10.1021/jp509068s.
[232] O. Hirsch, K.O. Kvashnina, L. Luo, M.J. Süess, P. Glatzel, D. Koziej, High-energy resolution X-ray absorption and emission spectroscopy reveals insight into unique

selectivity of La-based nanoparticles for CO2, Proc. Natl. Acad. Sci. 112 (2015) 15803–15808. Available from: https://doi.org/10.1073/pnas.1516192113.
[233] M. Che, J.C. Vedrine, Characterization of Solid Materials and Heterogeneous Catalysts: From Structure to Surface Reactivity, first ed., Wiley VCH, Weinheim, 2012.
[234] S. de Monredon, A. Cellot, F. Ribot, C. Sanchez, L. Armelao, L. Gueneau, et al., Synthesis and characterization of crystalline tin oxide nanoparticles, J. Mater. Chem. 12 (2002) 2396–2400. Available from: https://doi.org/10.1039/B203049G.
[235] G. Neri, A. Bonavita, G. Micali, G. Rizzo, E. Callone, G. Carturan, Resistive CO gas sensors based on In_2O_3 and $InSnO_x$ nanopowders synthesized via starch-aided sol-gel process for automotive applications, Sens. Actuators, B Chem. 132 (2008) 224–233. Available from: https://doi.org/10.1016/j.snb.2008.01.030.
[236] L.F. Gladden, M. Lutecki, J. McGregor, Characterization of solid materials and heterogeneous catalysts - nuclear magnetic resonance spectroscopy, in: M. Che, J.C. Vedrine (Eds.), Charact. Solid Mater. Heterog. Catal. From Struct. to Surf. React., first ed., Wiley VCH, Weinheim, 2012, pp. 289–342.
[237] M. Hunger, In situ flow MAS NMR spectroscopy: state of the art and applications in heterogeneous catalysis, Prog. Nucl. Magn. Reson. Spectrosc. 53 (2008) 105–127. Available from: https://doi.org/10.1016/j.pnmrs.2007.08.001.
[238] E.A. Konstantinova, I.S. Pentegov, A. Marikutsa, M. Rumyantseva, A. Gaskov, P.K. Kashkarov, EPR study of nanocrystalline tin dioxide, Phys. Status Solidi 8 (2011) 1957–1960. Available from: https://doi.org/10.1002/pssc.201000140.
[239] D. Amalric-Popescu, J.-M. Herrmann, A. Ensuque, F. Bozon-Verduraz, Nanosized tin dioxide: spectroscopic (UV-VIS, NIR, EPR) and electrical conductivity studies, Phys. Chem. Chem. Phys. 3 (2001) 2522–2530. Available from: https://doi.org/10.1039/b100553g.
[240] M. D'Arienzo, D. Cristofori, R. Scotti, F. Morazzoni, New insights into the SnO_2 sensing mechanism based on the properties of shape controlled tin oxide nanoparticles, Chem. Mater. 25 (2013) 3675–3686. Available from: https://doi.org/10.1021/cm401895x.
[241] A. Marikutsa, M. Rumyantseva, E. Konstantinova, T. Shatalova, A. Gaskov, Active sites on nanocrystalline tin dioxide surface: effect of palladium and ruthenium oxides clusters, J. Phys. Chem. C. 118 (2014) 21541–21549. Available from: https://doi.org/10.1021/jp5071902.
[242] M. D'Arienzo, M. Redaelli, B. Di Credico, S. Polizzi, R. Scotti, F. Morazzoni, New insights into the sensing mechanism of shape controlled ZnO particles, RSC Adv. 6 (2016) 52987–52997. Available from: https://doi.org/10.1039/C6RA09824J.
[243] R. Summitt, N.F. Borrelli, Infrared absorption in single crystal stannic oxide, J. Phys. Chem. Solids. 26 (1965) 921–925.
[244] E.W. Thornton, P.G. Harrison, Tin oxide surfaces - part 1.-surface hydroxyl groups and the chemisorption of carbon dioxide and carbon monoxide on tin(IV) oxide, J. Chem. Soc. Faraday Trans. 1 Phys. Chem. Condens. Phases. 71 (1974) 461–472.
[245] T.A. Gundrizer, A.A. Davydov, IR spectra of oxygen adsorbed on SnO_2, React. Kinet. Catal. Lett. 3 (1975) 63–70.
[246] P.G. Harrison, E.W. Thornton, Tin oxide surfaces part 4. - infrared study of the adsorption of oxygen and carbon monoxide + oxygen mixtures on tin(IV) oxide, and the adsorption of carbon dioxide on ammoniapretreated Tin(IV) oxide, J. Chem. Soc. Faraday Trans. 1 Phys. Chem. Condens. Phases. 71 (1977) 2597–2603.

[247] P.G. Harrison, A. Guest, Tin oxide surface part 17. - an infrared and thermogravimetric analysis of the thermal dehydration of tin(IV) oxide gel, J. Chem. Soc. Faraday Trans. 1 Phys. Chem. Condens. Phases. 83 (1987) 3383–3397.

[248] P.G. Harrison, A. Guest, Tin oxide surfaces part 18 - infrared study of the adsorption of very low levels (2- 50 ppm) of carbon monoxide in air on to Tin(IV) oxide gel, J. Chem. Soc. Faraday Trans. 1 Phys. Chem. Condens. Phases. 85 (1989) 1897–1906.

[249] G. Ghiotti, A. Chiorino, W.X. Pan, L. Marchese, A comparative study of SnO_2, and CrO_x/SnO_2 powders: their microstructures and their different response to CO, NO, H_2, and CH_4, atmospheres, Sens. Actuators B Chem. 7 (1992) 691–699.

[250] G. Ghiotti, A. Chiorino, W.X. Pan, Surface chemistry and electronic effects of H_2 (D_2) on pure SnO_2 and Cr-doped SnO_2, Sens. Actuators B Chem. 16 (1993) 367–371.

[251] P.G. Harrison, M.J. Willett, The mechanism of operation of tin(iv) oxide carbon monoxide sensors, Nature 332 (1988) 337–339. Available from: https://doi.org/10.1038/332337a0.

[252] M.-I. Baraton, L. Merhari, Influence of the particle size on the surface reacticity and gas sensing properties of SnO_2 nanopowders, Mater. Trans. 42 (2001) 1616–1622.

[253] M.-I. Baraton, L. Merhari, Nanoparticles-based chemical gas sensors for outdoor air quality monitoring microstations, Mater. Sci. Eng. B. 112 (2004) 206–213. Available from: https://doi.org/10.1016/j.mseb.2004.05.033.

[254] R. Pohle, M. Fleischer, H. Meixner, In situ infrared emission spectroscopic study of H_2O and hydrogen-containing gases on Ga_2O_3 gas sensors, Sens. Actuators B Chem. 68 (2000) 151–156.

[255] R. Pohle, M. Fleischer, H. Meixner, Infrared emission spectroscopic study of the adsorption of oxygen on gas sensors based on polycrystalline metal oxide films, Sens. Actuators, B Chem. 78 (2001) 133–137. Available from: https://doi.org/10.1016/S0925-4005(01)00803-6.

[256] S. Emiroglu, N. Barsan, U. Weimar, V. Hoffmann, In situ diffuse reflectance infrared spectroscopy study of CO adsorption on SnO_2, Thin Solid Films. 391 (2001) 176–185.

[257] D. Degler, S. Wicker, U. Weimar, N. Barsan, Identifying the active oxygen species in SnO_2 based gas sensing materials: an operando IR spectrsocopy study, J. Phys. Chem. C. 119 (2015) 11792–11799. Available from: https://doi.org/10.1021/acs.jpcc.5b04082.

[258] K. Grossmann, R.G. Pavelko, N. Barsan, U. Weimar, Chemical Interplay of H_2, water vapor and oxygenat the surface of SnO_2 based gas sensors – an operando investigation utilizing deuterated gases, Sens. Actuators B Chem. 166–167 (2012) 787–793. Available from: https://doi.org/10.1016/j.snb.2012.03.075.

[259] K. Großmann, S. Wicker, U. Weimar, N. Barsan, Impact of Pt additives on the surface reactions between SnO_2, water vapour, CO and H_2; an operando investigation, Phys. Chem. Chem. Phys. 15 (2013) 19151. Available from: https://doi.org/10.1039/c3cp52782d.

[260] D. Koziej, K. Thomas, N. Barsan, F. Thibault-Starzyk, U. Weimar, Influence of annealing temperature on the CO sensing mechanism for tin dioxide based sensors–Operando studies, Catal. Today. 126 (2007) 211–218. Available from: https://doi.org/10.1016/j.cattod.2007.03.011.

[261] D. Koziej, N. Barsan, K. Shimanoe, N. Yamazoe, J. Szuber, U. Weimar, Spectroscopic insights into CO sensing of undoped and palladium doped tin dioxide sensors derived from hydrothermally treated tin oxide sol, Sens. Actuators B Chem. 118 (2006) 98–104. Available from: https://doi.org/10.1016/j.snb.2006.04.014.

[262] D. Koziej, N. Barsan, V. Hoffmann, J. Szuber, U. Weimar, Complementary phenomenological and spectroscopic studies of propane sensing with tin dioxide based sensors, Sens. Actuators B Chem. 108 (2005) 75–83. Available from: https://doi.org/10.1016/j.snb.2004.12.076.

[263] R.G. Pavelko, H. Daly, M. Hu, C. Hardacre, E. Llobet, Time-resolved DRIFTS, MS, and resistance study of SnO_2 materials: the role of surface hydroxyl groups in formation of donor states, J. Phys. Chem. C. 117 (2013) 4158–4167. Available from: https://doi.org/10.1021/jp312532u.

[264] T. Simons, P. Chen, D. Rauch, R. Moos, U. Simon, Sensing catalytic conversion: simultaneous DRIFT and impedance spectroscopy for in situ monitoring of NH_3–SCR on zeolites, Sens. Actuators B Chem. 224 (2016) 492–499.

[265] M. Hübner, C.E. Simion, A. Haensch, N. Barsan, U. Weimar, CO sensing mechanism with WO_3 based gas sensors, Sens. Actuators B Chem. 151 (2010) 103–106. Available from: https://doi.org/10.1016/j.snb.2010.09.040.

[266] M. Boulova, A. Gaskov, G. Lucazeau, Tungsten oxide reactivity versus CH_4, CO and NO_2 molecules studied by Raman spectroscopy, Sens. Actuators B. 81 (2001) 99–106.

[267] T. Pagnier, M. Boulova, A. Galerie, A. Gaskov, G. Lucazeau, In situ coupled raman and impedance measurements of the reactivity of nanocrystalline SnO_2 versus H_2S, J. Solid State Chem. 143 (1999) 86–94.

[268] F. Fan, Z. Feng, L. Can, Characterization of Solid materials and heterogeneous catalysts - Raman and UV-Raman spectroscopies, in: M. Che, J.C. Védrine (Eds.), Charact. Solid Mater. Heterog. Catal. vol. 1 2, first ed., Wiley VCH, Weinheim, Germany, 2012, pp. 49–81.

[269] F.C. Jentoft, in: M. Che, J.C. Védrine (Eds.), Characterization of Solid Materials and Heterogeneous Catalysts - Electronic Spectroscopy: Ultra Violet-visible and Near IR Spectroscopies, Wiley VCH, Weinheim, Germany, 2012, pp. 89–140.

[270] C. Malagù, M.C. Carotta, S. Morandi, S. Gherardi, G. Ghiotti, A. Giberti, et al., Surface barrier modulation and diffuse reflectance spectroscopy of MoO_3–WO_3 thick films, Sens. Actuators B Chem. 118 (2006) 94–97. Available from: https://doi.org/10.1016/j.snb.2006.04.009.

[271] S. Morandi, E. Comini, G. Faglia, G. Ghiotti, Cr–Sn oxide thin films: electrical and spectroscopic characterisation with CO, NO_2, NH_3 and ethanol, Sens. Actuators B Chem. 118 (2006) 142–148. Available from: https://doi.org/10.1016/j.snb.2006.04.052.

[272] S. Morandi, G. Ghiotti, A. Chiorino, E. Comini, FT-IR and UV-Vis-NIR characterisation of pure and mixed MoO_3 and WO_3 thin films, Thin Solid Films. 490 (2005) 74–80. Available from: https://doi.org/10.1016/j.tsf.2005.04.020.

[273] S. Morandi, G. Ghiotti, A. Chiorino, B. Bonelli, E. Comini, G. Sberveglieri, MoO_3–WO_3 mixed oxide powder and thin films for gas sensing devices: a spectroscopic characterisation, Sens. Actuators B Chem. 111–112 (2005) 28–35. Available from: https://doi.org/10.1016/j.snb.2005.06.037.

[274] D. Koziej, Revealing complexity of nanoparticle synthesis in solution by in situ hard X-ray spectroscopy—today and beyond, Chem. Mater. 28 (2016) 2478–2490. Available from: https://doi.org/10.1021/acs.chemmater.6b00486.

[275] M. Bauer, H. Bertagnolli, X-ray absorbation spectroscopy – the method and its applications, Bunsen-Magazin 9 (2007) 216–231. Available from: https://doi.org/10.1002/9783527636839.ch7.

[276] H.W. Pereira de Carvalho, D. Degler, N. Barsan, J.-D. Grunwaldt, Chapter 27: Sensors, in: Y. Iwasawa, K. Asakura, M. Tada (Eds.), XAFS Tech. Catal. Nanomater. Surfaces, Springer International Publsihing, 2017, pp. 383–396.

[277] C. Geantet, C. Pichon, Characterization of solid materials and heterogeneous catalysts - X-ray absorption spectroscopy, in: M. Che, J.C. Védrine (Eds.), Charact. Solid Mater. Heterog. Catal. vol. 1 Fo 2, Wiley VCH, Weinheim, 2012, pp. 511−536.

[278] C.J. Belle, G.E. Wesch, S. Neumeier, M.J. Lozano-Rodríguez, A.C. Scheinost, U. Simon, Volume-doped cobalt titanates for ethanol sensing: an impedance and X-ray absorption spectroscopy study, Sens. Actuators B Chem. 192 (2014) 60−69. Available from: https://doi.org/10.1016/j.snb.2013.10.094.

[279] S.R. Davis, A.V. Chadwick, J.D. Wright, A combined EXAFS and diffraction study of pure and doped nanocrystalline tin oxide, J. Phys. Chem. B. 5647 (1997) 9901−9908. Available from: https://doi.org/10.1021/jp971756w.

[280] P. Serrini, V. Briois, M.C. Horrillo, A. Traverse, L. Manes, Chemical composition and crystalline structure of SnO_2 thin films used as gas sensors, Thin Solid Films. 304 (1997) 113−122. Available from: https://doi.org/10.1016/S0040-6090(97)00219-8.

[281] D.P. Volanti, A.A. Felix, P.H. Suman, E. Longo, J.A. Varela, M.O. Orlandi, Monitoring a CuO gas sensor at work: an advanced in situ X-ray absorption spectroscopy study, Phys. Chem. Chem. Phys. 17 (2015) 18761−18767. Available from: https://doi.org/10.1039/C5CP02150B.

[282] S.R. Davis, A.V. Chadwick, J.D. Wright, The effects of crystalline growth and dopant migration on the carbon monoxide sensing characteristics of nanocrystalline tin oxide based sensor materials, J. Mater. Chem. 8 (1998) 2065−2071.

[283] F. Gyger, A. Sackmann, M. Hübner, P. Bockstaller, D. Gerthsen, H. Lichtenberg, et al., Pd@SnO_2 and SnO2 @Pd Core@Shell nanocomposite sensors, Part. Part. Syst. Charact. 31 (2014) 591−596. Available from: https://doi.org/10.1002/ppsc.201300241.

[284] M. Hübner, D. Koziej, J.-D. Grunwaldt, U. Weimar, N. Barsan, An Au clusters related spill-over sensitization mechanism in SnO_2-based gas sensors identified by operando HERFD-XAS, work function changes, DC resistance and catalytic conversion studies, Phys. Chem. Chem. Phys. 14 (2012) 13249−13254. Available from: https://doi.org/10.1039/c2cp41349c.

[285] L. Mädler, T. Sahm, A. Gurlo, J.-D. Grunwaldt, N. Barsan, U. Weimar, et al., Sensing low concentrations of CO using flame-spray-made Pt/SnO_2 nanoparticles, J. Nanoparticle Res. 8 (2006) 783−796. Available from: https://doi.org/10.1007/s11051-005-9029-6.

[286] A. Marikutsa, M. Rumyantseva, D. Frolov, I. Morozov, A. Boltalin, A. Fedorova, et al., Role of PdO_x and RuO_y clusters in oxygen exchange between nanocrystalline tin dioxide and the gas phase, J. Chem. Phys. (2013) 23858−23867.

[287] N. Murata, T. Suzuki, M. Kobayashi, F. Togoh, K. Asakura, Characterization of Pt-doped SnO_2 catalyst for a high-performance micro gas sensor, Phys. Chem. Chem. Phys. 15 (2013) 17938. Available from: https://doi.org/10.1039/c3cp52490f.

[288] T. Wagner, M. Bauer, T. Sauerwald, D. Kohl, M. Tiemann, X-ray absorption near-edge spectroscopy investigation of the oxidation state of Pd species in nanoporous SnO_2 gas sensors for methane detection, Thin Solid Films. 520 (2011) 909−912. Available from: https://doi.org/10.1016/j.tsf.2011.04.187.

[289] D. Degler, H.W. Pereira de Carvalho, K.O. Kvashnina, J.-D. Grunwaldt, U. Weimar, N. Barsan, Structure and chemistry of surface-doped Pt:SnO_2 gas sensing materials, RSC Adv. 6 (2016) 28149−28155. Available from: https://doi.org/10.1039/C5RA26302F.

[290] T. Wada, N. Murata, T. Suzuki, H. Uehara, H. Nitani, Y. Niwa, et al., Improvement of a real gas-sensor for the origin of methane selectivity degradation by μ-XAFS

investigation, Nano-Micro Lett. 7 (2015) 255–260. Available from: https://doi.org/10.1007/s40820-015-0035-7.
[291] P. Glatzel, T.C. Weng, K. Kvashnina, J. Swarbrick, M. Sikora, E. Gallo, et al., Reflections on hard x-ray photon-in/photon-out spectroscopy for electronic structure studies, J. Elec. Spect. Rel. Phen. 188 (2013) 17–25. Available from: https://doi.org/10.1016/j.elspec.2012.09.004.
[292] M. Gaidi, B. Chenevier, M. Labeau, J.L. Hazemann, In situ simultaneous XAS and electrical characterizations of Pt-doped tin oxide thin film deposited by pyrosol method for gas sensors application, Sens. Actuators, B Chem. 120 (2006) 313–315. Available from: https://doi.org/10.1016/j.snb.2006.02.021.
[293] M. Labeau, B. Gautheron, F. Cellier, M. Vallet-Regi, E. Garcia, J.M. Gonzalez Calbet, Pt nanoparticles dispersed on SnO_2 thin films: a microstructural study, J. Solid State Chem. 102 (1993) 434–439.
[294] N. Yamazoe, New approaches for improving semiconductor gas sensors, Sens. Actuators B Chem. 5 (1991) 7–19. Available from: https://doi.org/10.1016/0925-4005(91)80213-4.
[295] L. Stievano, F. Wagner, Characterization of Solid Materials and Heterogeneous Catalysts - Mössbauer Spectroscopy, in: M. Che, J. Védrine (Eds.), Charact. Solid Mater. Heterog. Catal. vol. 1 2, first ed., Weinheim, (Germany), 2012: pp. 407–452.
[296] O.V. Safonova, I. Bezverkhy, P. Fabrichnyi, M. Rumyantseva, A. Gaskov, Mechanism of sensing CO in nitrogen by nanocrystalline SnO_2 and SnO_2(Pd) studied by Mössbauer spectroscopy and conductance measurements, J. Mater. Chem. 12 (2002) 1174–1178. Available from: https://doi.org/10.1039/b109219g.
[297] S.S. Sharma, Y. Ujihira, Mössbauer studies on tin-bismuth oxide CO selective gas sensor, J. Appl. Phys. 71 (1992) 2000–2005.
[298] J.W. Niemantsverdriet, W.N. Delgass, In situ Mössbauer spectroscopy in catalysis, Top. Catal. 8 (1999) 133–140.
[299] M. Bauer, N. Barsan, K. Ingrisch, A. Zeppenfeld, I. Denk, B. Schumann, et al., Influence of measuring voltage and geometry of the sensing layer on the characteristics of thick film SnO_2 gas sensors, in: Proc. 11th Eur. Microelectron. Conf., 1997.
[300] U. Hoefer, K. Steiner, E. Wagner, Contact and sheet resistance of SnO_2 thin films from transmission-line model measurements, Sens. Actuators B Chem. 26 (1995) 59–63. Available from: https://doi.org/10.1016/0925-4005(94)01557-X.
[301] S. Rank, The Influence of the Electrode Material on the Sensor Characteristics of SnO_2 Thick Film Gas Sensors, Eberhard Karls Universität Tübingen, 2014.
[302] N. Barsan, U. Weimar, Understanding the fundamental principles of metal oxide based gas sensors; the example of CO sensing with SnO_2 sensors in the presence of humidity, J. Phys. Condens. Matter. 15 (2003) R813–R839.
[303] A. Oprea, N. Bârsan, U. Weimar, Work function changes in gas sensitive materials: fundamentals and applications, Sens. Actuators, B Chem. 142 (2009) 470–493. Available from: https://doi.org/10.1016/j.snb.2009.06.043.
[304] N. Barsan, M. Hübner, U. Weimar, Conduction mechanisms in SnO_2 based polycrystalline thick film gas sensors exposed to CO and H_2 in different oxygen backgrounds, Sens. Actuators B Chem. 157 (2011) 510–517. Available from: https://doi.org/10.1016/j.snb.2011.05.011.
[305] K.-I. Choi, M. Hübner, A. Haensch, H.-J. Kim, U. Weimar, N. Barsan, et al., Ambivalent effect of Ni loading on gas sensing performance in SnO_2 based gas sensor, Sens. Actuators B Chem. 183 (2013) 401–410. Available from: https://doi.org/10.1016/j.snb.2013.04.007.

[306] M. Hübner, C. Simion, A. Tomescu-Stănoiu, S. Pokhrel, N. Barsan, U. Weimar, Influence of humidity on CO sensing with p-type CuO thick film gas sensors, Sens. Actuators B Chem. 153 (2011) 347–353. Available from: https://doi.org/10.1016/j.snb.2010.10.046.
[307] S. Pokhrel, C. Simion, V. Quemener, N. Barsan, U. Weimar, Investigations of conduction mechanism in Cr_2O_3 gas sensing thick films by ac impedance spectroscopy and work function changes measurements, Sens. Actuators B Chem. 133 (2008) 78–83. Available from: https://doi.org/10.1016/j.snb.2008.01.054.
[308] A. Marikutsa, M. Rumyantseva, A. Gaskov, Specific interaction of PdO_x- and RuO_y-modified tin dioxide with CO and NH_3 gases: kelvin probe and DRIFT studies, J. Phys. Chem. C. 119 (2015) 24342–24350. Available from: https://doi.org/10.1021/acs.jpcc.5b02532.
[309] J. Mizsei, J. Harsányi, Resistivity and work function measurements on Pd-doped SnO_2 sensor surface, Sens. Actuators. 4 (1983) 397–402.
[310] K.-D. Schierbaum, U. Weimar, W. Göpel, R. Kowalkowski, Conductance, work function and catalytic activity of SnO_2-based gas sensors, Sens. Actuators B. Chem. 3 (1991) 205–214. Available from: https://doi.org/10.1016/0925-4005(91)80007-7.
[311] T. Doll, J. Lechner, I. Eisele, K.-D. Schierbaum, W. Göpel, Ozone detection in the ppb range with work function sensors operating at room temperature, Sens. Actuators B Chem. 34 (1996) 506–510. Available from: https://doi.org/10.1016/S0925-4005(96)01941-7.
[312] N. Barsan, J. Rebholz, U. Weimar, Conduction mechanism switch for SnO_2 based sensors during operation in application relevant conditions; implications for modeling of sensing, Sens. Actuators B Chem. 207 (2015) 455–459. Available from: https://doi.org/10.1016/j.snb.2014.10.016.
[313] A. Oprea, N. Barsan, U. Weimar, Characterization of granular metal oxide semiconductor gas sensitive layers by using Hall effect based approaches, J. Phys. D. Appl. Phys. 40 (2007) 7217–7237. Available from: https://doi.org/10.1088/0022-3727/40/23/S05.
[314] A. Oprea, E. Moretton, N. Barsan, W.J. Becker, J. Wöllenstein, U. Weimar, Conduction model of SnO2 thin films based on conductance and Hall effect measurements, J. Appl. Phys. 100 (2006) 1–10. Available from: https://doi.org/10.1063/1.2229802.
[315] A. Oprea, A. Gurlo, N. Barsan, U. Weimar, Transport and gas sensing properties of In_2O_3 nanocrystalline thick films: a Hall effect based approach, Sens. Actuators, B Chem. 139 (2009) 322–328. Available from: https://doi.org/10.1016/j.snb.2009.03.002.
[316] J. Kappler, A. Tomescu-Stănoiu, N. Barsan, U. Weimar, CO consumption of Pd doped SnO_2 based sensors, Thin Solid Films. 391 (2001) 186–191. Available from: https://doi.org/10.1016/S0040-6090(01)00980-4.
[317] M. Hübner, R.G. Pavelko, N. Barsan, U. Weimar, Influence of oxygen backgrounds on hydrogen sensing with SnO_2 nanomaterials, Sens. Actuators B Chem. 154 (2011) 264–269. Available from: https://doi.org/10.1016/j.snb.2010.01.049.
[318] D. Degler, Spectroscopic Insights in the Gas Detection Mechanism of Tin Dioxide Based Gas Sensors, Eberhard Karls University of Tübingen, 2017.
[319] S.R. Morrison, Selectivity in semiconductor gas sensors, Sens. Actuators. 12 (1987) 425–440. Available from: https://doi.org/10.1016/0250-6874(87)80061-6.
[320] N. Yamazoe, Y. Kurokawa, T. Seiyama, Effects of additives on semiconductor gas sensors, Sens. Actuators B Chem. 4 (1983) 283–289. Available from: https://doi.org/10.1016/0250-6874(83)85034-3.

[321] D. Degler, S. Rank, S. Müller, H.W. Pereira de Carvalho, J. Grunwaldt, U. Weimar, et al., Gold-loaded tin dioxide gas sensing materials: mechanistic insights and the role of gold dispersion, ACS Sensors. 1 (2016) 1322–1329. Available from: https://doi.org/10.1021/acssensors.6b00477.

[322] A. Cabot, J. Arbiol, J.R. Morante, U. Weimar, N. Barsan, W. Göpel, Analysis of the noble metal catalytic additives introduced by impregnation of as obtained SnO_2 sol – gel nanocrystals for gas sensors, Sens. Actuators B Chem. 70 (2000) 87–100. Available from: https://doi.org/10.1016/S0925-4005(00)00565-7.

[323] A. Diéguez, A. Vilà, A. Cabot, A. Romano-Rodriguez, J.R. Morante, J. Kappler, et al., Influence on the gas sensor performances of the metal chemical states introduced by impregnation of calcinated SnO_2 sol-gel nanocrystals, Sens. Actuators B Chem. 68 (2000) 94–99. Available from: https://doi.org/10.1016/S0925-4005(00)00467-6.

[324] D. Degler, S.A. Müller, D. Doronkin, D. Wang, J.-D. Grunwaldt, U. Weimar, et al., The Role of the Oxide-Oxide Interface for Gas Sensing with PtO_x-loaded SnO_2, 2017.

[325] A. Urakawa, T. Bürgi, A. Baiker, Sensitivity enhancement and dynamic behavior analysis by modulation excitation spectroscopy: principle and application in heterogeneous catalysis, Chem. Eng. Sci. 63 (2008) 4902–4909. Available from: https://doi.org/10.1016/j.ces.2007.06.009.

[326] C. Ledesma, J. Yang, D. Chen, A. Holmen, Recent approaches in mechanistic and kinetic studies of catalytic reactions using SSITKA technique, ACS Catal. 4 (2014) 4527–4547. Available from: https://doi.org/10.1021/cs501264f.

[327] A. Staerz, C. Berthold, T. Russ, S. Wicker, U. Weimar, N. Barsan, The oxidizing effect of humidity on WO_3 based sensors, Sens. Actuators B Chem (2016) 0–14. Available from: https://doi.org/10.1016/j.snb.2016.06.072.

[328] H.L. Han, G. Melaet, S. Alayoglu, G.A. Somorjai, In Situ Microscopy and spectroscopy applied to surfaces at work, ChemCatChem. 7 (2015) 3625–3638. Available from: https://doi.org/10.1002/cctc.201500642.

[329] A. Kolmakov, L. Gregoratti, M. Kiskinova, S. Günther, Recent approaches for bridging the pressure gap in photoelectron microspectroscopy, Top. Catal. 59 (2016) 448–468. Available from: https://doi.org/10.1007/s11244-015-0519-1.

[330] J.-M. Ducéré, A. Hemeryck, A. Estève, M.D. Rouhani, G. Landa, P. Ménini, et al., A computational chemist approach to gas sensors: modeling the response of SnO_2 to CO, O_2, and H_2O gases, J. Comput. Chem. 33 (2011) 247–258. Available from: https://doi.org/10.1002/jcc.21959.

[331] Y. Kim, I. Hwang, S. Kim, C. Lee, J. Lee, Chemical CuO nanowire gas sensors for air quality control in automotive cabin, Sens. Actuators B Chem. 135 (2008) 298–303. https://doi.org/10.1016/j.snb.2008.08.026.

[332] M. Akiyama, J. Tamaki, N. Miura, N. Yamazoe, Tungsten oxide-based semiconductor sensor highly sensitive to NO and NO_2, Chem. Lett. (1991) 1611–1614. Available from: https://doi.org/10.1246/cl.1991.1611.

[333] M. Ivanovskaya, A. Gurlo, P. Bogdanov, Mechanism of O_3 and NO_2 detection and selectivity of In_2O_3 sensors, Sens. Actuators B Chem. 77 (2001) 264–267.

[334] O. In, C. Wo, Sensors for the nitrogen oxides, NO_2, NO and N_2O, based on In_2O_3 and WO_3 nanowires, Appl. Phys. A 246 (2006) 241–246. Available from: https://doi.org/10.1007/s00339-006-3707-9.

[335] I. Manassidis, J. Goniakowski, L.N. Kantorovich, M.J. Gillan, The structure of the stoichiometric and reduced SnO_2 (110) surface, Surf. Sci. 339 (1995) 258–271. Available from: https://doi.org/10.1016/0039-6028(95)00677-X.

[336] M. Batzill, W. Bergermayer, I. Tanaka, U. Diebold, Tuning the chemical functionality of a gas sensitive material: water adsorption on SnO_2 (101), Surf. Sci. 600 (2006) 1−4. Available from: https://doi.org/10.1016/j.susc.2005.11.034.

[337] M. Batzill, K. Katsiev, J.M. Burst, U. Diebold, A.M. Chaka, B. Delley, Gas-phase-dependent properties of SnO (110), (100), and (101) single-crystal surfaces: structure, composition, and electronic properties, Phys. Rev. B 2 (2005) 1−20. Available from: https://doi.org/10.1103/PhysRevB.72.165414.

[338] A. Gurlo, M. Ivanovskaya, N. Ba, Grain size control in nanocrystalline In_2O_3 semiconductor gas sensors, Sens. Actuators B Chem. 44 (1997) 327−333.

[339] S. Wicker, M. Guiltat, U. Weimar, A. Hémeryck, N. Barsan, Ambient Humidity Influence on CO Detection with SnO_2 Gas Sensing Materials – a Combined DRIFTS / DFT Investigation, J. Phys. Chem. C 121 (2017) 25064–25073. Available from: https:// doi.org /10.1021/acs.jpcc.7b06253.

[340] S.P. Bates, Full-coverage adsorption of water on SnO_2 (110): the stabilisation of the molecular species, Surf. Sci. 512 (2002) 29−36.

[341] P.J.D. Lindan, Water chemistry at the SnO_2 (110) surface: the role of inter-molecular interactions and surface geometry, Chem. Phys. Lett. 328 (2000) 325−329.

[342] J. Goniakowski, M.J. Gillan, The adsorption of H_2O on TiO_2 and SnO_2 (110) studied by first-principles calculations, Surf. Sci. 350 (1996) 145−158.

[343] G. Santarossa, K. Hahn, A. Baiker, Free Energy and Electronic Properties of Water Adsorption on the SnO_2 (110) Surface, Langmuir 2 (2013).

[344] N. Kumar, P.R.C. Kent, A.V. Bandura, J.D. Kubicki, D.J. Wesolowski, D.R. Cole, et al., Faster proton transfer dynamics of water on SnO_2 compared to TiO_2, J. Chem. Phys. 134 (2011) 44706. Available from: https://doi.org/10.1063/1.3509386.

[345] H. Xu, R.Q. Zhang, A.M.C. Ng, A.B. Djuri, H.T. Chan, W.K. Chan, et al., Splitting Water on Metal Oxide Surfaces, J. Phys. Chem. C (2011) 19710−19715.

[346] X. Wang, H. Qin, Y. Chen, J. Hu, Sensing Mechanism of SnO_2 (110) Surface to CO: Density Functional Theory Calculations, J. Phys. Chem. C 2 (2014).

[347] G. Henkelman, B.P. Uberuaga, H. Jonsson, A climbing image nudged elastic band method for finding saddle points and minimum energy paths, J. Chem. Phys. 113 (2000).

[348] N.O. Sensitivity, E.A. Metals, M. Sciences, S. Kyushu, Promotion of NO sensitivity of WO_3 element with the addition of noble metals, J. Surf. Sci. Japan. 14 (1993) 295−300.

[349] W. Tang, E. Sanville, G. Henkelman, A. Grid-Based Bader, Analysis algorithm without lattice bias, J. Phys. Condens. Matter. 21 (2009) 84204. Available from: https://doi.org/10.1088/0953-8984/21/8/084204.

[350] P.K. Clifford, D.T. Tuma, Characteristics of Semiconductor Gas Sensors I. Steady State Gas Response, Sens. Actuators 3 (1983) 255−281.

[351] P.K. Clifford, D.T. Tuma, Characteristics of semiconductor gas sensors II. Transient response to temperature change, Sens. Actuators. 3 (1982) 255−281. Available from: https://doi.org/10.1016/0250-6874(82)80027-9.

[352] S.R. Morrison, The Chemical Physics of Surfaces, Plenum Press, New York, USA, 1977. Available from: https://doi.org/10.1063/1.2995077.

[353] J.F. Mcaleer, P.T. Moseley, J.W. Norris, D.E. Williams, O.O.X. Ora, Dioxide Gas Sensors, J. Chem. Soc. Faraday Trans. 1, 83 (1987) 1323.

[354] A. Rothschild, Y. Komem, The effect of grain size on the sensitivity of nanocrystalline metal-oxide gas sensors, J. Appl. Phys. 95 (2004) 6374. Available from: https://doi.org/10.1063/1.1728314.

[355] N. Yamazoe, K. Shimanoe, Roles of shape and size of component crystals in semiconductor gas sensors II. Response to NO_2 and H_2, J. Electrochem. Soc. 155 (2008) 93–98. Available from: https://doi.org/10.1149/1.2832655.

[356] N. Yamazoe, K. Shimanoe, Roles of shape and size of component crystals in semiconductor gas sensors I. Response to oxygen, J. Electrochem. Soc. 155 (2008) 85–91. Available from: https://doi.org/10.1149/1.2832655.

[357] N. Barsan, Transduction in semiconducting metal oxide based gas sensors - Implications of the conduction mechanism, Procedia Eng. 25 (2011) 100–103. Available from: https://doi.org/10.1016/j.proeng.2011.12.025.

[358] J. Rebholz, U. Weimar, N. Barsan, Influence of conduction mechanism changes on the sensor performance of SMOX based gas sensors, Procedia Eng. 87 (2014) 20–23. Available from: https://doi.org/10.1016/j.proeng.2014.11.256.

[359] J. Rebholz, P. Bonanati, U. Weimar, Grain shape influence on semiconducting metal oxide based gas sensor performance: modeling versus experiment, 2014, pp. 3977–3983. https://doi.org/10.1007/s00216-013-7502-0.

[360] J.M. Rebholz, Influence of Conduction Mechanism Changes and Related Effects on the Sensing Performance of Metal Oxide Based Gas Sensors, Eberhard Karls University of Tübingen, 2016.

Technological realization of semiconducting metal oxide–based gas sensors

4

Jong-Heun Lee
Department of Materials Science and Engineering, Korea University, Seoul, Korea

Oxide semiconductor chemiresistors have been widely investigated and used to detect various harmful, toxic, and explosive gases because of their unique advantages such as excellent sensitivity, cost effectiveness, simple sensing mechanism, and facile integration [1–4]. Gas-sensing characteristics such as gas response, selectivity, stability, and response/recovery speed are closely dependent upon transducer designs, sensing materials, and the deposition of sensing materials on transducers.

In the early stages of research in this field, bulky porous pellets were used as sensors. These have gradually evolved to thick films on alumina substrates and thick films on transducers based on microelectromechanical systems (MEMS). The significant miniaturization of sensors provides various advantages such as power consumption low enough to allow battery operation, rapid sensor heating due to small thermal mass, a new mode of sensor operation using dynamic sensing transients, integration of sensors or sensor arrays into other devices such as smart phones, and wireless sensor networks assisted by the Internet of Things. More recently, gas sensors formed on plastic foils are being considered for wearable devices [5]. Thus, in the first part of this chapter, the advantages, limitations, and future perspectives of various transducers for oxide semiconductor gas sensors will be critically reviewed and discussed.

The gas-sensing reaction in oxide semiconductors consists of the diffusion of the analyte gas from the ambient atmosphere toward the surface of the oxide semiconductor and the charge-transferring interaction between the analyte gas and the oxide surface. The former highlights the importance of gas transfer via micro-, meso-, and nanoporous sensing films, while the latter is dominated by the gas adsorption, surface reaction, and change in charge carrier concentration near the surface of the oxide semiconductor [1–4]. All the above parameters that determine the gas-sensing characteristics are dependent upon the size, crystallinity, composition, porosity, and morphology of oxide nanostructures including nanorods (1D), nanowires (1D), nanosheets (2D), nanocubes (3D), and their assembled hierarchical, hollow, and yolk-shell structures. Accordingly, in the second part, current and new strategies for synthesizing oxide nanostructures for high-performance gas sensors will be investigated.

Gas Sensors Based on Conducting Metal Oxides. DOI: https://doi.org/10.1016/B978-0-12-811224-3.00004-4

Sensing films consisting of nanobuilding blocks differ in the size, volume, and distribution of pores, as well as connectivity between nanostructures according to the packed configuration, which may vary the gas transfer from the ambient atmosphere to the sensing surface and the conduction along the nanostructures. This suggests that not only the nanostructures themselves, but also their consolidated configuration, play key roles in gas-sensing characteristics such as gas response and sensor resistance. Therefore, in the last part of this chapter, various methods to coat sensing materials on transducers will be reviewed, and their advantages and limitations for gas sensor applications will be discussed.

4.1 Transducer technology evolution from pellets to microhotplates

4.1.1 Thick films on Al_2O_3 tubes and substrates

First-generation oxide semiconductor gas sensors were prepared in the form of porous pellets (ZnO and SnO_2) by the sintering of a powder compact. Then, the slurry of sensing materials was coated onto an alumina ceramic tube and subsequently sintered at an elevated temperature to form a thick sensing film. The operation temperature was controlled by a metal-coil heater within the tube (Fig. 4.1A). This type of sensor is still available in the marketplace [6]. In the 1990s, gas sensor companies started to produce gas sensors on a large scale by screen printing the sensing film onto an alumina substrate with the patterns of a sensing electrode and heater (Fig. 4.1B). This significantly improved the fabrication productivity and quality of sensors [7]. This sensor structure is still dominant in the marketplace (Fig. 4.1B).

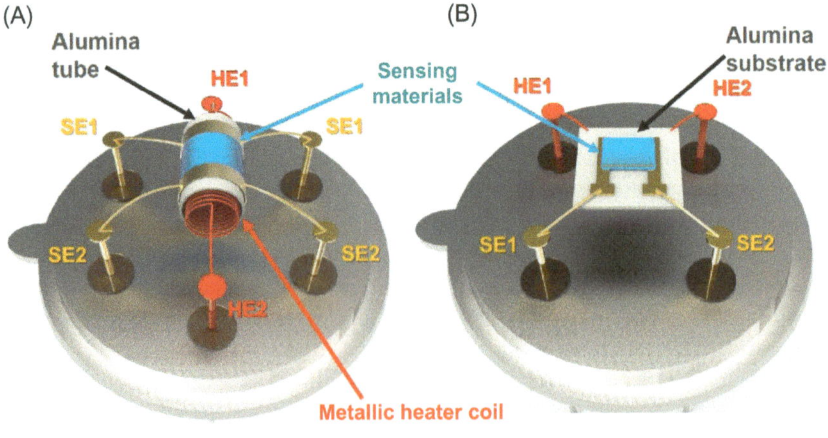

Figure 4.1 Schematic structure of (A) tube-type and (B) substrate-type oxide semiconductor gas sensors (*SE*, sensor electrode; *HE*, heater electrode).

There are several reasons for the use of alumina tubes or substrates as supporting materials for sensors. High thermal conductivity, low thermal expansion coefficients, and high mechanical strength of supporting materials are required to achieve uniform temperature distribution in the sensing area and to prevent mechanical failure by thermal shock. Moreover, the materials should be cost effective and electrically insulating. From this perspective, alumina is one of the best available supporting materials. In addition, the high thermal stability of alumina, due to its high melting temperature (2072°C), is advantageous during the heat treatment of printed metallic heater patterns and sensing materials at elevated temperatures. In contrast, high thermal conductivity can decrease the thermal efficiency of the heater by the active dissipation of heat or can damage the neighboring sensor circuits. Accordingly, a heat-insulating design of the substrate can be considered to obtain high and uniform sensor temperatures without significant heat loss or thermal damage to the circuits.

4.1.2 MEMS-based sensors

The MEMS technology enables the integration of gas sensors, sensor arrays, and microheaters into a tiny chip. The decrease of sensor size has various benefits for gas sensors, such as (1) low power consumption due to small thermal mass [8], (2) integration of sensors and sensor arrays into a miniaturized chip [9,10], (3) mass production via a batch process [11], and (4) quick warm-up or change of sensing temperatures via rapid sensor heating [12].

Briand [13] wrote a comprehensive review of micromachined semiconductor gas sensors. He summarized the power consumption of microhotplate designs reported in the literature from 1990 to 2008. The power consumption required to achieve a sensor temperature of 300°C ranged from 125 to 6 mW, which tended to decrease with the progress of fabrication technology. More recently, the Figaro company announced MEMS-based oxide semiconductor gas sensors with a power consumption of 15 mW as a commercial product [14]. The integration of gas sensors or electronic noses based on sensor arrays into mobile phones, as well as the wireless communication of sensor signals via the Internet of Things and sensor networks, are very important to modern technology, both of which require microsensors with power consumption low enough to be operated by a battery for extended periods of time.

Simon et al. published an extensive review on micromachined metal oxide gas sensors. They illustrated typical sensor structures with (A) a closed membrane, and (B) a suspended membrane fabricated by micromachining (Fig. 4.2) [15]. The back side of a silicon substrate is generally etched out by a KOH or EDP (ethylene diamine, pyrocatechol, and water) solution after forming the heater pattern, dielectric layer, electrodes, and sensing layers. Thin membranes and air pockets block the thermal conduction from heated sensors to substrates and, thus, high temperatures can be accomplished at minimum power consumption. Note that the heat conduction along the thin membrane can be further reduced by employing a suspended design (Fig. 4.2B). The main parameters that determine the power consumption in

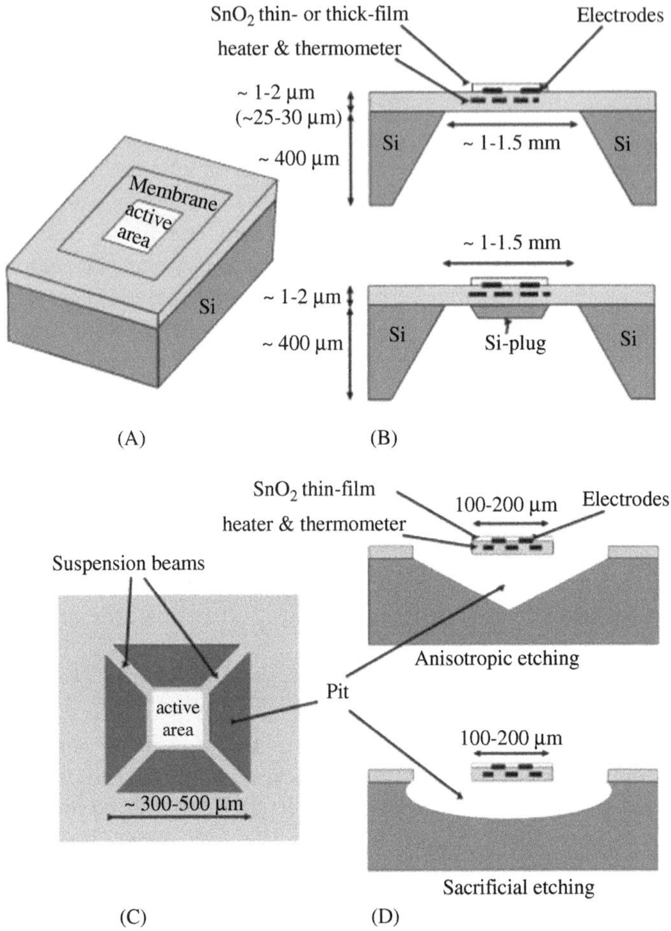

Figure 4.2 Schematic of (A, B) closed membrane-type gas sensors and (C, D) suspended membrane-type gas sensors.
Source: Reprinted with permission from I. Simon, N. Bârsan, M. Bauer, U. Weimar, Micromachined metal oxide gas sensors: opportunities to improve sensor performance, Sensors Actuators B Chem. 73 (2001) 1–26. doi:10.1016/S0925-4005(00)00639-0.

MEMS-based gas sensors are the (A) sensing area, (B) thickness, (C) thermal conductivity, and (D) design of the membrane. Simon et al. [15] provided an insightful discussion on the thermal design of MEMS-based sensors considering heat conduction through the membrane as well as radiation and convection of heat in the ambient atmosphere.

The temperature in the sensing area of the membrane is significantly different from that in adjacent Si substrates and the ambient atmosphere, which may cause an uneven temperature distribution. Thus, a uniform distribution of the sensor temperature requires precise design of the heater patterns [16]. Uniformity of the sensor temperature can also be improved by modifying the sensor structures. Briand et al. [17] fabricated an isolated Si layer underneath the membrane (lower figure in Fig. 4.2B) by two successive KOH etching steps using different masking processes and investigated the influence of an isolated Si layer on the uniformity of temperature in the sensing area by finite element modeling [18]. They suggested that uniform sensor distribution can be achieved by controlling the thickness and thermal conductivity of the isolated Si layer, as well as by tuning the heating area and operation temperature.

Rapid thermal response in MEMS-based gas sensors provides a new mode of sensor operation. For example, temperature-programmed measurements of gas-sensing characteristics can be used to detect a single analyte and to discriminate complex chemicals containing multiple analytes [10,19,20]. Meier et al. [19] deposited SnO_2 and TiO_2 sensing layers on suspended membrane-type sensing platforms (Fig. 4.3A) and measured dynamic sensing transients while heating the sensors to 15 different temperatures (between 50 and 450°C) within a short period of time (15 seconds) via the application of programmed pulse heater voltage (Fig. 4.3B). Even a single sensor showed different gas-sensing patterns to multianalytes at different sensing temperatures [21,22] because defect formation, adsorption/desorption of gases, and the reaction between gas and adsorbent depends closely on temperature. Thus, high-throughput collection of diverse sensing data via temperature-programmed measurements provides a new method to detect and analyze gases within a wide gas-sensing library. It should be noted, however, that the short duration at sensing temperature is often insufficient to reach the steady state in the gas-sensing reaction because it takes substantial time for the diffusion of analyte gases into the sensing area—the reaction between analyte and adsorbed oxygen during the gas-sensing reaction—and the in-diffusion, dissociation, and ionization of

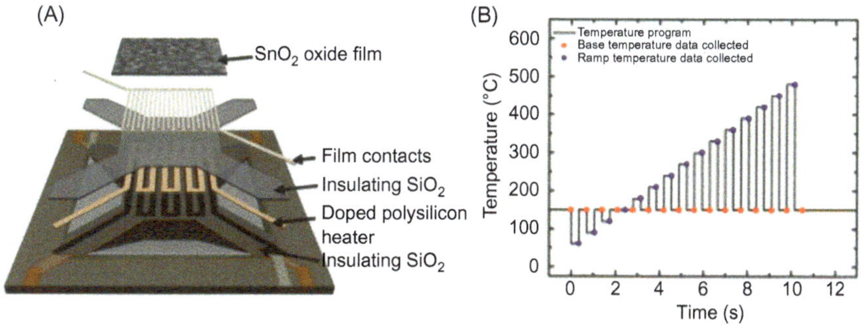

Figure 4.3 (A) Schematic of SnO_2 microsensor element and (B) temperature-programmed sensing operation profile.
Source: Reprinted with permission from D.C. Meier, J.K. Evju, Z. Boger, B. Raman, K.D. Benkstein, C.J. Martinez, et al., The potential for and challenges of detecting chemical hazards with temperature-programmed microsensors, Sensors Actuators, B Chem. 121 (2007) 282–294. doi:10.1016/j.snb.2006.09.050.

oxygen during recovery. Thus, the analysis of dynamic sensing transients assisted by advanced signal processing is necessary for obtaining meaningful interpretations.

The microheaters can be generally classified into buried and coplanar designs. If the sensing area is similar to the heating area, the sensor temperature at the core of the heater will usually be higher than that at the edges, which establishes nonuniform distribution of the sensing temperature. If the coplanar design is used, as shown in Fig. 4.4A, the temperature distribution of the sensing area can become more uniform. Elmi et al. [23] demonstrated that the sensor can be heated to 400°C with ultralow-power consumption (8.9 mW) using a coplanar heater design. However, in the coplanar design, it is better to passivate the heater layer to avoid any interaction between the heater and sensing layer (Fig. 4.4B). In principle, the

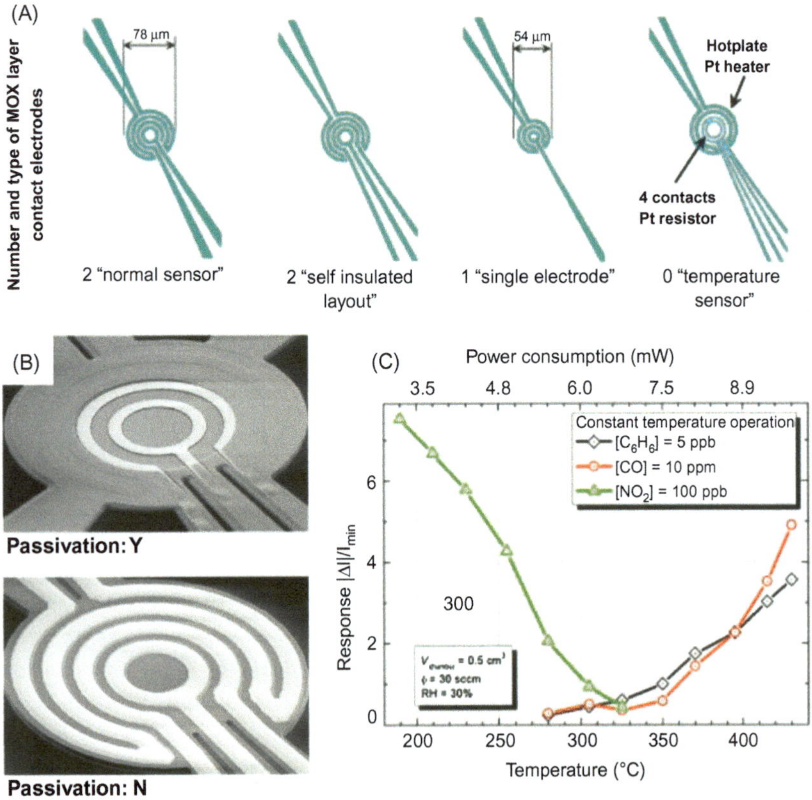

Figure 4.4 (A) Various coplanar designs of microheaters; (B) microheaters with passivation (upper) and without passivation (lower); and (C) responses to various gases in relation to the temperature and power consumption.
Source: Reprinted with permission from I. Elmi, S. Zampolli, E. Cozzani, F. Mancarella, G.C. Cardinali, Development of ultra-low-power consumption MOX sensors with ppb-level VOC detection capabilities for emerging applications, Sensors Actuators, B Chem. 135 (2008) 342–351. doi:10.1016/j.snb.2008.09.002.

operation of sensors at a low temperature is advantageous to decrease power consumption, to reach sensor temperature quickly, and to decrease thermal shock. However, microheaters that can be heated to high sensing temperatures are often necessary to detect gases with low reactivity and to refresh the sensor surface by heat treatment. For example, responses to the less reactive benzene and CO increase with increasing sensor temperature (Fig. 4.4C), while the response to NO_2 is high at low temperature (200°C).

4.1.3 Sensors on plastic foils

Plastic foils can also be considered as supporting materials. The major benefits of plastic foils are their flexible (foldable and wearable) design, light weight, transparency, and mass productivity via roll-to-roll printing processes, which can open new applications. In general, because oxide semiconductor gas sensors are operated at elevated temperatures (150–400°C), polyimide (PI), with a relatively high thermal stability and high glass transition temperature, have been widely used as supporting materials [24–26] (Fig. 4.5). However, the operation of metal oxide semiconductor sensors at >400°C seemed difficult using PI substrates [27].

4.2 Synthesis of sensitive materials: synthetic strategies and gas-sensing characteristics

4.2.1 Historic perspective

The gas response of oxide semiconductor chemiresistors is primarily determined by the size, connecting configuration, and agglomeration of the nanostructures of sensing materials. Initially, coarse commercial SnO_2 powders were used [28] as sensing materials. Later, many synthetic routes, including precipitation [29], sputtering [30], spray chemical vapor deposition [31], and thermal oxidation [32], were employed for preparing gas-sensing materials. After the pioneering work by Xu et al. on the enhancement of the gas response using nanosized particles [33], researchers focused

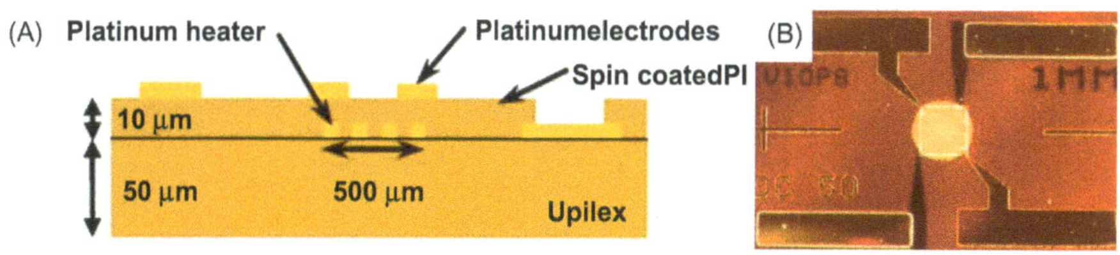

Figure 4.5 (A) Schematic diagram of PI sensor substrates with heater and electrodes; (B) image of metal oxide gas sensors on a PI microhotplate made on a PI sheet as substrates. *Source:* Reprinted with permission from D. Briand, S. Colin, J. Courbat, S. Raible, J. Kappler, N.F. de Rooij, Integration of MOX gas sensors on polyimide hotplates, Sensors Actuators, B Chem. 130 (2008) 430–435. doi:10.1016/j.snb.2007.09.013.

on decreasing the particle size to several nanometers. For example, Baik et al. [34] succeeded in preparing SnO_2 films with a crystallite size of ~6 nm, even after heat treatment at 600°C, using an SnO_2 nanosol solution prepared by hydrothermal treatment of *Sn*-containing gel precursors.

In smaller nanostructures, more effective electron depletion or hole accumulation occurs near the surface, resulting in higher gas responses. However, small particles are more prone to the formation of large and dense secondary particles because the van der Waals attraction is inversely proportional to the particle size. The agglomeration hampers the in-diffusion of analyte gases to the entire sensing surface, resulting in the deterioration of the gas response. Accordingly, highly gas-accessible nanostructures whose small size has not been sacrificed are necessary for enhancing the gas response. The following chemical and physical approaches have been employed or are being explored to synthesize oxide nanostructures or their assembled nanoarchitectures for use as gas-sensing materials exhibiting high gas responses.

4.2.2 Hydrothermal/solvothermal self-assembly reaction

Highly crystalline nanoparticles, nanorods, nanoplates, and nanocubes can be prepared by employing a solvothermal or hydrothermal reaction of a solution containing metal precursors, directing agents, and/or a surfactant. The evaporation of a solvent above its boiling point in a small and closed volume results in a high pressure that facilitates the growth of highly crystalline nanocrystals, even at relatively low reaction temperatures (typically 120–250°C).

Nanorods or nanoplates are obtained when a seed layer with a preferred orientation [35] is present, or a specific lattice plane is covered with polymeric surfactants or anionic adsorbates [36] during the growth of nanocrystals. Both of these anisotropic nanostructures with high aspect ratios are more advantageous for suppressing agglomeration between nanostructures compared to their nanoparticle counterparts, resulting in an increase in gas accessibility in sensing reactions. In addition, the gas response can be significantly enhanced when a specific crystallographic plane exhibiting a strong interaction with analyte gases becomes abundant [37–39].

Nanobuilding blocks such as nanoparticles, nanorods, nanowires (NWs), nanosheets, and nanocubes are often spontaneously assembled into collective structures of higher dimensions with periodic pores, which are called hierarchical nanostructures. The gas accessibility can be maximized without sacrificing electron depletion or hole accumulation near the surface of the nanobuilding blocks. Thus, a high gas response and rapid responding speed can be accomplished simultaneously. Hierarchical nanostructures can be classified according to the dimension and morphology of the nanobuilding block assembled structures (Fig. 4.6) [4]. Particularly, when the polymeric segment is formed at the center and the inorganic segment is assembled on the outer surface during the self-assembly process, hollow nanostructures can be obtained by thermally decomposing the cores [40–42]. Among several physicochemical routes, the solvothermal/hydrothermal reaction is the most effective chemical route to prepare highly crystalline, hierarchical, and hollow

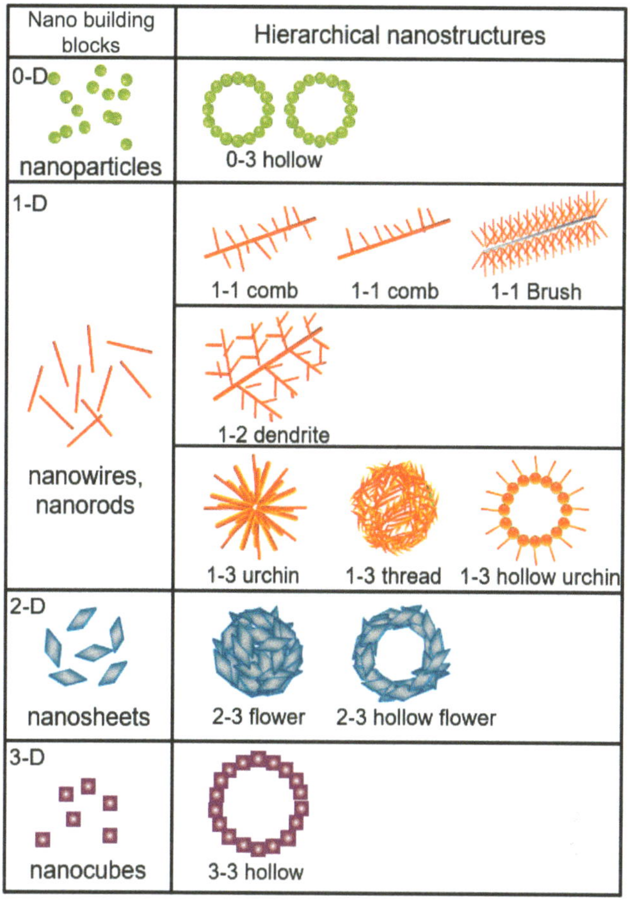

Figure 4.6 Nomenclature of hierarchical structures according to the dimensions of the corresponding nanobuilding blocks (the former number) and the consequent hierarchical structures (the latter number).
Source: Reprinted with permission from J.H. Lee, Gas sensors using hierarchical and hollow oxide nanostructures: Overview, Sensors Actuators, B Chem. 140 (2009) 319–336. doi:10.1016/j.snb.2009.04.026.

nanostructures with periodic and well-defined pores. The effect of hierarchical oxide nanostructures on gas-sensing characteristics are discussed in detail in a review article by the chapter author [4].

4.2.3 Template-based synthesis: tubular nanofibers, hemispheres, hollow spheres, and inverse opals

Agglomeration of nanoparticles significantly hampers the transport of analyte gases to the sensing surface, thereby deteriorating the gas response. The use of sacrificial templates for the formation of well-defined porous structures is an effective strategy for increasing the gas accessibility of nanostructures. Although hollow spheres can be prepared using template-free solvothermal/hydrothermal reactions, the thickness and porosity of shells can be more precisely tuned using sacrificial templates. Various sacrificial templates made of polymers—for example, polystyrene (PS) and polymethylmethacrylate (PMMA) [43]—metals [44], and SiO_2 [45,46] have been proposed for preparing tubular nanofibers, hollow spheres, and inverse opals that can then be removed by thermal decomposition and dissolution in acidic or alkaline solutions.

Tubular nanofibers: Oxide nanofibers with solid inner structures are commonly prepared by electrospinning a precursor solution containing a metal and polymer, followed by thermal decomposition of the polymeric component. Tubular oxide nanofibers can be prepared by coating oxide or metal precursors on pure polymeric sacrificial nanofibers via atomic layer deposition (ALD), sputtering, and subsequent heat treatment (Fig. 4.7A). SnO_2 [47,48], ZnO [49,50], WO_3 [51], and $InGaZnO_4$ [52] tubular nanofibers have been prepared using various sacrificial nanofibers made of polyacrylonitrile (PAN) [47], polyvinyl pyrrolidone (PVP)/PMMA [48,51], PVP [52], and polyvinyl alcohol (PVA) [52]. Tubular structures with uncoated parts or hemi-tubular structures are often obtained when sputtering is used, whereas uniform tubular structures are obtained with the highly conformal ALD technique. The inner or outer part of the tubular structures can be decorated by adding steps to coat the catalyst (Fig. 4.7B). For example, Cho et al. [53] prepared NiO nanofibers by uniformly decorating Pt catalysts on the inner, outer, or both surfaces of NiO nanotubes and reported that the gas-sensing characteristics of the tubular oxide nanofibers depended on the location and configuration of the noble metal catalyst.

Coaxial electrospinning is a viable alternative for preparing tubular nanofibers. In this method, the precursor solution is spun via an outer spinneret, while the polymeric solutions (paraffin or mineral oil) are spun via an inner spinneret (Fig. 4.7C). Thermal decomposition of the sacrificial core leads to the formation of tubular

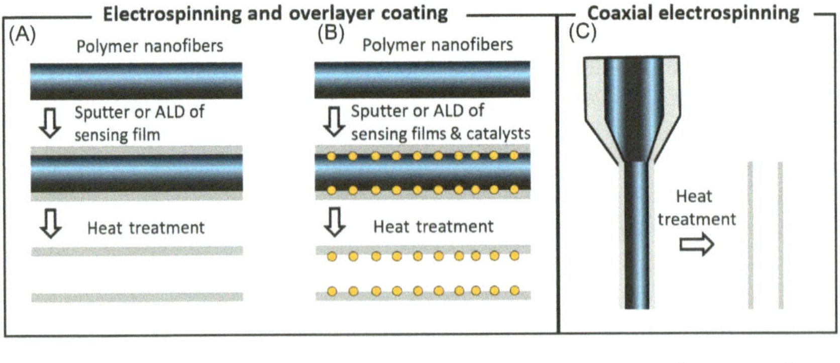

Figure 4.7 Schematic of the preparation of tubular oxide nanofibers: (A, B) pure and catalyst-loaded tubular oxide nanofibers prepared using polymeric nanofiber templates; (C) oxide nanofibers prepared by coaxial electrospinning.

nanofibers. The key experimental parameters used for controlling the diameter, thickness, and porosity of the shell layers are the compositions of the precursor solutions (solvent, polymer, and metal ion). SnO_2 [54], TiO_2 [55], and In_2O_3 tubular nanofibers [56] have been successfully prepared by coaxial electrospinning. The obtained gas response increased upon introducing hollow inner structures within nanofibers [54] and with a decrease in the diameter of the nanotubes [56].

Embossed structures with hemispheres: Close-packed arrays of polymeric (PMMA, PS) spheres have been used as templates for preparing oxide films with embossed structures. Kim et al. prepared embossed structures of SnO_2 [57], TiO_2 [58], NiO [59], and $CaCu_3Ti_4O_{12}$ [60] and demonstrated that the gas responses of the embossed sensing films were significantly higher than those of the corresponding dense films. The high gas response could be attributed to the gas-sensing reaction occurring at the upper and lower parts of the sensing film due to the macroporosity of the films. Moon et al. [61,62] employed plasma etching to control the connection between template PS spheres in a monolayer configuration. On increasing the O_2 plasma etching time, the close-linked configuration of PS balls gradually changed to an isolated configuration (upper diagrams in Fig. 4.8A,B).

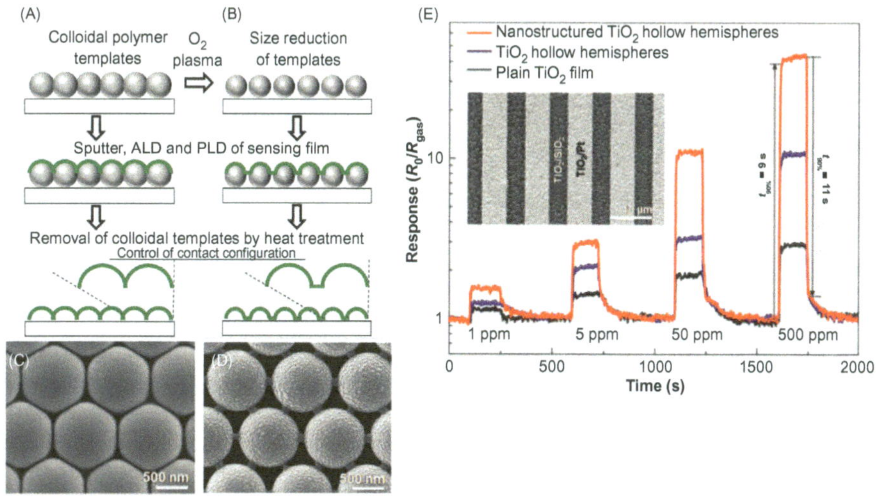

Figure 4.8 Schematic for the preparation of (A) embossed TiO_2 nanostructures of closely connected TiO_2 hemispheres (denoted as "TiO_2 hollow hemispheres" in [e]) and (B) TiO_2 hemispheres connected by nanobridges (denoted as "nanostructured TiO_2 hollow hemispheres" in [e]) prepared using the closely packed array of PS spheres and O_2 plasma treatment; SEM images of (C) closely connected TiO_2 hemispheres and (D) TiO_2 hemispheres connected by nanobridges; (E) sensing transients of sensors to 1, 5, 50, and 500 ppm CO at 250°C; (C, D).

Source: Reprinted and modified with permission from H.G. Moon, Y.S. Shim, H.W. Jang, J.S. Kim, K.J. Choi, C.Y. Kang, et al., Highly sensitive CO sensors based on cross-linked TiO_2 hollow hemispheres, Sensors Actuators, B Chem. 149 (2010) 116–121. doi:10.1016/j.snb.2010.06.014[62].

After coating with TiO_2 by sputtering (lower diagrams in Fig. 4.8A,B) and subsequent heat treatment, closely connected TiO_2 hemispheres (Fig. 4.8C) and TiO_2 hemispheres connected by nanobridges (Fig. 4.8D) were obtained. The highest gas response was obtained with the latter, while the former exhibited a low gas response (Fig. 4.8E), which could be attributed to the high chemiresistive variations occurring at the nanoscale bridge.

Inverse opals: 2D inverse opals are important nanoarchitectures for gas sensor applications (Fig. 4.9A–C). To prepare these structures, the closely packed array of PS spheres formed on a glass substrate is dipped into the desired precursor solution and then transferred to a substrate with electrodes [63]. The precursor solution enters the interstitials among the PS spheres in the monolayer by capillary force. Heat treatment causes the removal of the PS templates and conversion of the precursor to the corresponding oxide, thereby forming a metal oxide sensor with a 2D inverse opal structure. Dai et al. [65] reported that 2D SnO_2 inverse opals exhibit high and ultrafast gas responses (response time <1 seconds) compared to the dense SnO_2 films. Lee et al. [66] reported that the C_2H_5OH response of 2D Co_3O_4 inverse opals was significantly higher than that of the dense Co_3O_4 thin films. These results could be attributed to the high gas accessibility of inverse opal structures and the

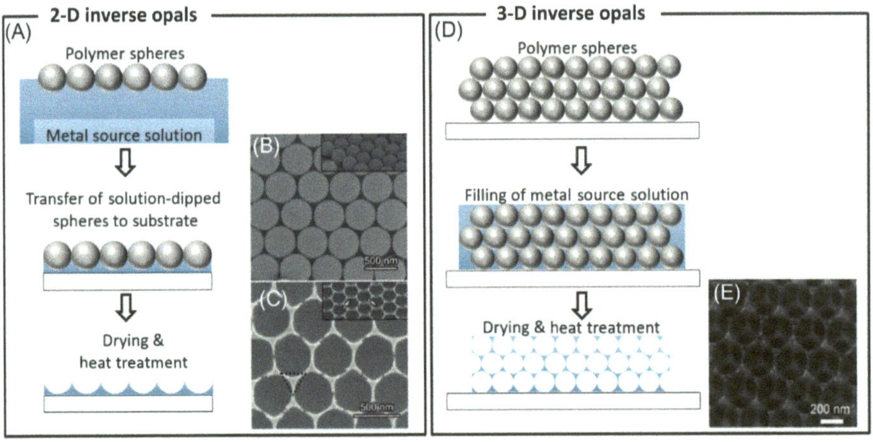

Figure 4.9 (A) Schematic of the preparation of 2D inverse opals; SEM images of (B) the monolayer of PS microspheres on a Si substrate and (C) 2D inverse opals of Fe_2O_3 [63]; (D) schematic of the preparation of 3D inverse opals; (E) SEM image of 3D inverse opals of ZnO-CuO [64]; (B, C, E).

Source: Modified with permission from Z. Dai, C.-S. Lee, Y. Tian, I.-D. Kim, J.-H. Lee, Highly reversible switching from P- to N-type NO2 sensing in a monolayer Fe2O3 inverse opal film and the associated P–N transition phase diagram, J. Mater. Chem. A. 3 (2015) 3372–3381. doi:10.1039/C4TA05438E; Y. Xie, R. Xing, Q. Li, L. Xu, H. Song, Three-dimensional ordered ZnO – CuO inverse opals toward low concentration acetone detection for exhaled breath sensing, Sensors Actuators B. Chem. 211 (2015) 255–262. doi:10.1016/j.snb.2015.01.086.

high chemiresistive variation achieved at the thinnest neck region. Ternary oxide inverse opal structures [67] can be prepared using a mixed precursor solution.

3D closely packed arrays of polymer spheres have been used for obtaining 3D inverse opals for gas sensor applications (Fig. 4.9D,E). For example, SnO_2 [68,69], Pt-SnO_2 [69], Zn-SnO_2 [70], $LaFeO_3$ [71], and ZnO-CuO [64] have been prepared for use as gas sensors. Wang et al. [72] reported the response of 3D SnO_2 inverse opals to 500 ppm methanol to be 2.1 times higher than that of random macroporous SnO_2 structures prepared by mixing PS microspheres and SnO_2 precursor sols and 5.6 times higher than that of the SnO_2 thin film prepared using SnO_2 precursor sols. Xing et al. [73] reported that the gas response of SnO_2 inverse opals can be tuned by varying the diameter of the template PMMA spheres. These results demonstrate that highly periodic arrays of macropores are effective for enhancing the gas response, and the gas-sensing characteristics can be tuned by controlling the porosity of the structures.

4.2.4 Spray pyrolysis: dense, hollow, yolk-shell, and hierarchically porous spheres

Ultrasonic spray pyrolysis is a synthetic route used for preparing spherical oxide powders by employing a thermal reaction of spherical droplets containing metal precursors [74]. Generally, the reaction involves the evaporation of solvent and thermal decomposition of precursors. The droplet size depends on the surface tension of the solution and the resonant frequency of the ultrasonic transducer. Ultrasonic transducers with a resonant frequency of 1.67 MHz are generally used for generating droplets with a diameter of $\sim 3\ \mu m$ when water is used as the solvent. After spray pyrolysis, submicrometer-sized spheres are obtained. The main advantages of the ultrasonic spray pyrolysis technique are the spherical morphology, narrow particle size distribution, compositional homogeneity, and high controllability of composition of the obtained products, as well as their suitability for successive processing.

Rapid evaporation of the solvent might lead to supersaturation of the solute near the surface of the droplets, which often results in the production of a hollow morphology. In contrast, if the solubility of the solute is very high, and neither the solvent boiling nor the rapid decomposition of precursors occurs, solid spheres can be obtained by ultrasonic spray pyrolysis. The morphology of products also depends on parameters such as the solubility of the solute, solvent evaporation, temperature dependence of solute solubility, permeability, and plasticity of the shell formed, as well as the presence of organic components [74].

To date, various hollow spheres of WO_3 [75], In_2O_3 [76], CuO-SnO_2 [77], Rh-SnO_2 [78], Rh-WO_3 [79], and In_2O_3-decorated NiO [80] have been prepared by ultrasonic spray pyrolysis for gas-sensing applications. The composition of these sensors can be controlled without restriction when all the source materials can be dissolved in solution. Thus, spray pyrolysis is a promising and versatile synthetic route for preparing various compositions of hollow spheres with high gas accessibility.

The porosity of the spheres can be further controlled using polymeric templates. For example, Shimizu et al. reported the preparation of macroporous SnO_2 [81], In_2O_3 [82], and MoO_3-doped SnO_2 [83] by employing spray pyrolysis of droplets containing metal precursors and PMMA template spheres. The macroporosity was controlled by varying the size/content of the macroporous PMMA templates and by heat treatment (Fig. 4.10). The gas response of the macroporous spheres was found to be higher than that of dense spheres. The spherical macropores cannot be connected to each other when present in low amounts. Even when present in high amounts, the spherical macropores often make contact with a narrow contact area. If the diameters of the connecting channels are smaller than 2 nm, sluggish surface diffusion becomes the rate-determining step of gas transport to the sensor surface.

One of the chapter authors [84] prepared trimodal porous SnO_2 spheres (pore sizes $\approx$3, 20, and 100 nm) by employing ultrasonic spray pyrolysis of droplets containing source materials, as well as PS spheres (diameter = 100 nm) and carbon nanotubes (diameter = 20 nm), as templates. The macropores (100 nm) were well interconnected, exhibiting an elongated configuration of 1D pores (diameter = 20 nm) achieved by the decomposition of the carbon nanotube template. The maximum gas response (R_a/R_g) of the trimodal porous SnO_2 spheres to 5 ppm C_2H_5OH was as high as 316.5, which was significantly higher than that of the bimodal porous SnO_2 spheres (pore diameters $\approx$ 3 and 20 nm) prepared using only the PS templates (R_a/R_g = 117.6) and the dense SnO_2 spheres prepared without templates (R_a/R_g = 34.2) (Fig. 4.11). When interconnecting pores with sizes ranging from 2 to 50 nm are present (i.e., mesopores), Knudsen diffusion occurs along the pores. Under these conditions, gas diffusion coefficients are proportional to the diameter of the pores [85]. Thus, the high gas response of the trimodal porous SnO_2

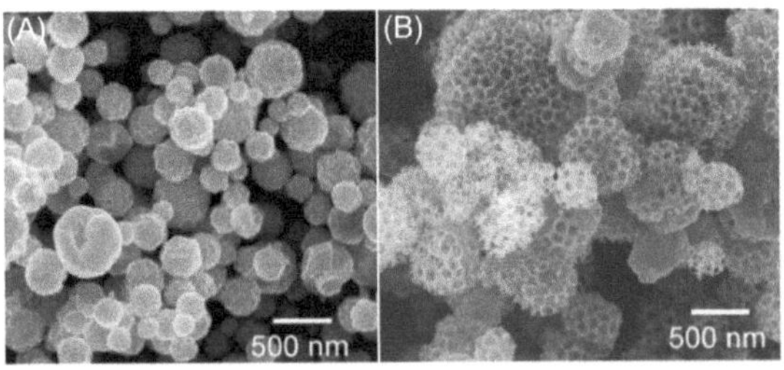

Figure 4.10 SEM images of (A) dense In_2O_3 spheres and (B) macroporous In_2O_3 spheres prepared by spray pyrolysis of droplets without and with PMMA precursor templates, respectively.

Source: Reprinted with permission from T. Hyodo, H. Inoue, H. Motomura, K. Matsuo, T. Hashishin, J. Tamaki, et al., NO2 sensing properties of macroporous In_2O_3-based powders fabricated by utilizing ultrasonic spray pyrolysis employing polymethylmethacrylate microspheres as a template, Sensors Actuators, B Chem. 151 (2010) 265–273. doi:10.1016/j.snb.2010.09.002.

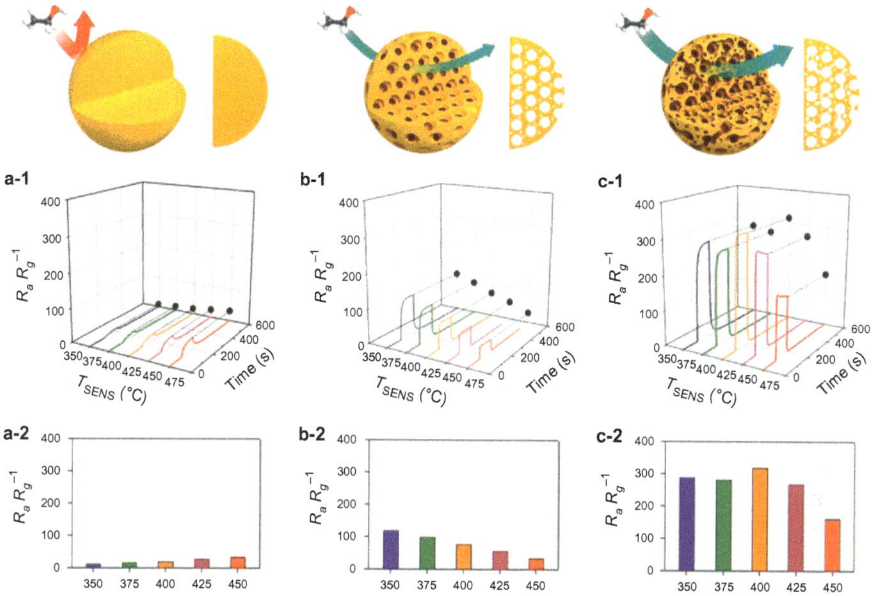

Figure 4.11 Gas-sensing characteristics of (A) dense, (B) bimodal porous, and (C) trimodal porous SnO_2 spheres to 5 ppm of ethanol; (A-1, B-1, C-1) gas response (R_a/R_g) transients as a function of time and sensing temperature; (A-2, B-2, C-2) gas responses of the three sensors.
Source: Adapted with permission from J.W. Yoon, S.H. Choi, J.S. Kim, H.W. Jang, Y.C. Kang, J.H. Lee, Trimodally porous SnO2nanospheres with three dimensional interconnectivity and size tunability: A one-pot synthetic route and potential application as an extremely sensitive ethanol detector, NPG Asia Mater. 8 (2016) 1–8. doi:10.1038/am.2016.16.

spheres can be attributed to the presence of well interconnected macropores and the enhanced Knudsen diffusion occurring along the mesopores with a diameter of 20 nm.

4.2.5 Core-shell nanostructures (nanowires, nanofibers, spheres): modulation of the electron depletion layer

The gas response can be enhanced by coating shell layers with different thicknesses on the oxide nanostructures. When core nanostructures are completely covered with a shell layer and the conduction across the core-shell hetero-interface is difficult, the chemiresistive variation is dominated by the shell layer. Studies focusing on controlling or enhancing the gas-sensing characteristics of core-shell nanostructures are summarized in Table 4.1 [86–104]. Core(*p*)-shell(*n*), core(*n*)-shell(*n*), and core (*n*)-shell(*p*) nanostructures have all been investigated as gas sensors. When a thin *n*-shell layer is formed on *p*-core nanostructures, electron depletion layers are formed both at the surface of the shell (by the adsorption of negatively charged

Table 4.1 Gas sensors obtained using core-shell nanostructures (nanowires, nanofibers, spheres) [85–103]

Synthetic routes	Nanostructures	Core	Shell	References
Hydrothermal reaction	Spheres	Co_3O_4	SnO_2	[86]
Hydrothermal reaction	Spheres	SnO_2	$ZnSnO_3$	[87]
Galvanic replacement	Spheres	Co_3O_4	SnO_2	[88]
Hydrothermal (c) + Hydrothermal (s)	Nanorods	α-Fe_2O_3	SnO_2	[89]
Hydrothermal (c) + Solution coating (s)	Nanorods	α-MoO_3	TiO_2	[90]
Hydrolysis (c) + Hydrolysis (s)	Nanorods	Fe_2O_3	ZnO	[91]
Hydrothermal (c) + Hydrothermal (s)	Nanospindles	Fe_2O_3	ZnO	[92]
Coaxial electrospinning	Nanofibers	α-Fe_2O_3	NiO	[93]
Coaxial electrospinning	Nanofibers	Co_3O_4	α-Fe_2O_3	[94]
Coaxial electrospinning	Nanofibers	NiO	ZnO	[95]
Electrospinning (c) + ALD (s)	Nanofibers	SnO_2	ZnO	[96–98]
Electrospinning (c) + ALD (s)	Nanofibers	TiO_2	ZnO	[99]
Vapor growth(c) + Vapor growth(s)	Nanowires	ZnO	SnO_2	[100]
Vapor growth(c) + Vapor growth(s)	Nanowires	In_2O_3	ZnO	[101]
Vapor growth(c) + Vapor-phase cation substitution (s)	Nanowires	ZnO	Cr_2O_3	[102]
Vapor growth(c) + Vapor-phase cation substitution (s)	Nanowires	ZnO	$ZnMnO_4$	[103]
Vapor growth(c) + Vapor-phase cation substitution (s)	Nanowires	ZnO	$ZnCo_2O_4$	[104]

oxygen) and at the interface between the *p*-core and the *n*-shells. Accordingly, when the shell is sufficiently thin, full electron depletion can be obtained, resulting in an increase in the gas response.

Formation of *n*–*n* hetero-interfaces can also be employed for modulating the electron depletion layer of the chemiresistive shell layer. Katoch et al. [97] prepared SnO_2 nanofibers and coated them with ZnO shell layers of different thicknesses by ALD. They reported that the response of the SnO_2-ZnO core-shell nanofibers to 1 ppm CO increased with increasing shell thicknesses up to 20 nm, and decreased rapidly with further increases in the shell layer thickness up to 120 nm. The effective electron depletion of the thin shell layer owing to oxygen adsorption and the presence of $n-n$ hetero-interfaces could be responsible for the enhanced gas response, which is supported by the correlation between sensor resistance and gas response. Choi [98] reported a similar enhancement in the gas response and sensor resistance on coating a ZnO shell layer on crystalline SnO_2 core NWs. These results clearly indicate that the modulation of the electron depletion layer using core-shell nanostructures is an efficient method for enhancing the gas response. The key parameters determining the gas response of core-shell nanostructures are the thickness of the shell layer and modulation of the electron depletion layer or hole accumulation layer in the shell layer achieved with hetero-interfaces.

4.2.6 Oxide nanowires or spheres decorated with different oxides

The $p-n$ junction in Si-based semiconductors or III–V compound semiconductors has many useful applications including diodes, field effect transistors, photovoltaic cells, and light emitting diodes. The heterojunction between n-type (SnO_2, ZnO, WO_3, In_2O_3, Fe_2O_3, and TiO_2) and p-type (CuO, NiO, Co_3O_4, Cr_2O_3, and Mn_3O_4) oxide semiconductors can be tuned to control the gas-sensing characteristics [105]. Composites between p-type CuO and n-type oxide semiconductors (SnO_2, ZnO, In_2O_3, WO_3, MoO_3) [29,106–110] have been proposed for detecting H_2S in a selective manner because the resistive $p-n$ junction can be destroyed upon converting p-semiconducting CuO to metallic CuS. The high chemical affinity of CuO to H_2S is responsible for the selective gas detection.

The $p-n$ junction between oxide semiconductors can be used to control or enhance the gas response of oxide NW network sensors. The diameter of oxide semiconductor NWs grown by a vapor-phase reaction (typically >50 nm) is generally higher than twice the width of the electron depletion layer. Thus, it is difficult to achieve full electron depletion in pure n-type oxide semiconductor NWs. When p-type nanoclusters with a discrete configuration are uniformly coated on n-type oxide semiconductor NWs, the electron depletion layer beneath the p-type nanoclusters expands, and the chemiresistive variation of the oxide NW sensor is controlled more effectively upon exposure to analyte gases. Hence, decorating p-nanoclusters is an effective way to increase the gas response. The radial extension of the electron depletion layer in n-type oxide semiconductor NWs achieved from decorating the p-nanocluster resulted in a significant increase in the sensor resistance in air [102,111–113]. Gas selectivity can be achieved when p-type nanoclusters exhibit catalytic activity or chemical affinity toward a specific analyte gas. For example, the decoration of Cr_2O_3 nanoclusters on ZnO or SnO_2 NW sensors enhanced the selectivity to trimethylamine [102,113].

The $p-n$, $n-p$, $n-n$, and $p-p$ hetero-nanostructures of oxide semiconductors with different work functions have been explored with the aim of enhancing or controlling the gas response [114,115]. This strategy can be also applied to various other nanostructures such as spheres [116], hierarchical nanostructures [117], and NW arrays [118].

4.2.7 Catalyst-loaded microreactor

Oxide yolk-shell spheres consist of thin shells and movable cores, which can be used as attractive nanostructures in nanoreactors, drug delivery systems, and Li-ion batteries [119]. The major benefits of the yolk-shell morphology in oxide chemiresistors are the high gas accessibility and effective electron depletion or hole accumulation of thin shells. For gas sensor applications, yolk-shell spheres with the metal-in-oxide and oxide-in-oxide configurations have been explored. In metal@oxide gas sensors, the Au, Ag, Pt, and Pd have been used as core metals and TiO_2, SnO_2, and Cu_2O have been used as metal oxide sensing materials; for overview see Ref. [120]. Gas response and selectivity have been significantly improved by adding yolk noble metal catalysts [121–123]. Gyger et al. [121,124] prepared Pd@SnO_2 and SnO_2@Pd sensors and compared their gas-sensing characteristics. They reported that the CO response of Pd@SnO_2 sensor at <250°C is significantly higher than those of SnO_2@Pd and pure SnO_2 sensors. Substantial variation of gas-sensing characteristics without direct exposure of noble metal catalysts on the shell surfaces strongly suggests the possibility to enhance or tune the gas-sensing characteristics using micro- or nanogas reactor.

In general, the synthesis of yolk-shell spheres requires a multistep process, which hampers their widespread applications. More recently, Hong et al. [125] reported a facile one-pot route to prepare oxide yolk-shell spheres by employing ultrasonic spray pyrolysis, and they further demonstrated the preparation of multicompositional oxide yolk-shell spheres [126]. The simple and large-scale preparation of yolk-shell spheres promotes the application of these distinctive and unique nanostructures.

Hong et al. [127] prepared Pd-loaded SnO_2 hollow spheres by ultrasonic spray pyrolysis using Sn and Pd precursors and sucrose. The reactions in spray pyrolysis involve the evaporation of solvent, formation of Pd-C-Sn composite spheres by polymerization and carbonization of carbon, decomposition of the outermost Pd-C-Sn layer into a Pd-SnO_2 shell and the contraction of core Pd-C-Sn spheres (formation of Pd-C-Sn@Pd-SnO_2 yolk-shell), formation of a second Pd-SnO_2 shell layer and the contraction of core Pd-C-Sn spheres (formation of double-shelled Pd-C-Sn@Pd-SnO_2@Pd-SnO_2), as well as decomposition of the core Pd-C-Sn sphere into the Pd-SnO_2 core (formation of double-shelled Pd-SnO_2 yolk-shell spheres) (Fig. 4.12).

Sensors produced with dense and undoped SnO_2 yolk-shell spheres exhibited selective detection of ethanol at 350–450°C [127] (Fig. 4.13A,B). In contrast, 0.3 wt% Pd-loaded SnO_2 yolk-shell spheres exhibited selective detection of xylene and toluene at 350–400°C (Fig. 4.13C). In *n*-type oxide semiconductor gas sensors, the response to highly reactive ethanol is generally higher than that to the less reactive methylbenzene. Hence, the selective detection of xylene and toluene by employing catalyst-loaded yolk-shell structures is unusual and interesting. The selectivity to methylbenzene could be attributed to the in-diffusion of methylbenzene thorough thin, semipermeable shells into the microreactor and the subsequent reforming of methylbenzene into smaller and more reactive species assisted by

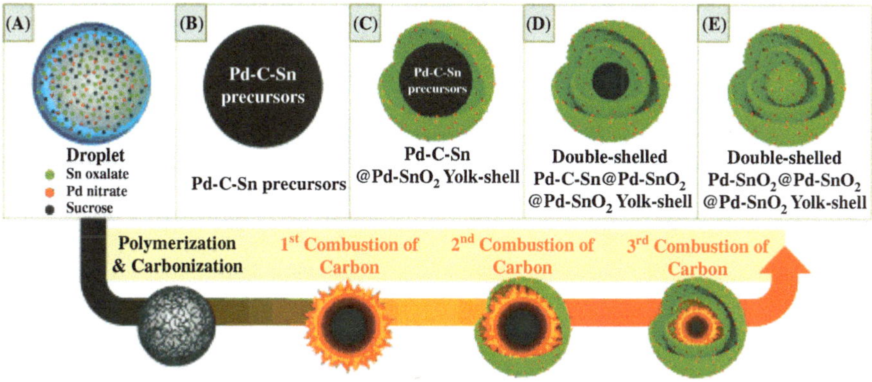

Figure 4.12 Schematic showing the formation of double-shelled Pd-SnO_2@Pd-SnO_2@Pd-SnO_2 yolk-shell spheres.
Source: Reprinted with permission from Y.J. Hong, J.W. Yoon, J.H. Lee, Y.C. Kang, One-pot synthesis of Pd-loaded SnO_2 yolk-shell nanostructures for ultraselective methyl benzene sensors, Chem. - A Eur. J. 20 (2014) 2737–2741. doi:10.1002/chem.201304502.

catalytic Pd nanoparticles at the inner wall of shells and yolks. This is a new strategy for achieving selective gas detection based on gas reforming using a microreactor.

Not only noble metal catalysts, but also catalytically active sensing materials, can be combined to enhance gas selectivity. For instance, Yoon et al. [128] prepared quintuple-shelled Pd-loaded Co_3O_4 yolk-shell spheres by the heat treatment of spray-dried gel spheres containing Co and Pd salts. A high selectivity to methylbenzene was achieved by promoting the methylbenzene reforming reaction via the synergistic combination of Pd catalysts and catalytically active Co_3O_4 sensing materials. To date, gas sensors with WO_3 [129], Ag-loaded SnO_2 [130], and Au-loaded Cr_2O_3 yolk-shell spheres [131] have been investigated to detect NO_2, H_2S, and trimethylamine, respectively, in a selective and sensitive manner. Considering the compositional controllability achieved with the spray pyrolysis and spray drying processes, this method can be used for designing selective gas sensors.

4.3 Design of sensing films for highly sensitive and selective gas detection

4.3.1 Slurry deposition, spin coating, Langmuir–Blodgett method, and dielectrophoretic positioning

Most oxide nanostructures can be dispersed in aqueous/organic solvents with ultrasonic treatment and can be subsequently coated on a substrate containing electrodes by drop coating [132,133]. Organic components are added to the slurry to control the viscosity, improve the dispersion of oxide nanostructures by forming an organic

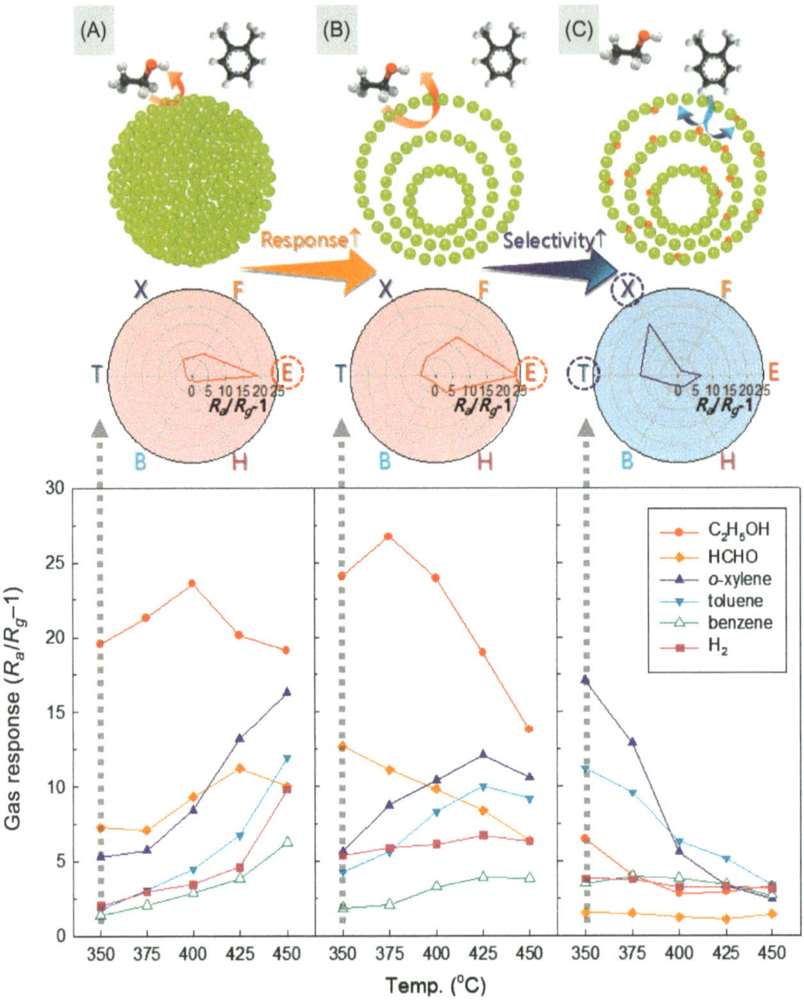

Figure 4.13 Gas responses (R_a/R_g-1) of (A) dense SnO_2 spheres, (B) SnO_2 yolk-shell spheres, and (C) Pd-loaded SnO_2 yolk-shell spheres to various analytic gases at 350−450°C (B: benzene; H: H_2; E: C_2H_5OH; F: HCHO; X: *o*-xylene; T: toluene).
Source: Reprinted with permission from Y.J. Hong, J.W. Yoon, J.H. Lee, Y.C. Kang, One-pot synthesis of Pd-loaded SnO2 yolk-shell nanostructures for ultraselective methyl benzene sensors, Chem. - A Eur. J. 20 (2014) 2737−2741. doi:10.1002/chem.201304502.

capping layer, or enhance the binding between the nanostructures or the contact between the nanostructures and the electrodes. Slurry deposition provides a facile route to fabricate gas-sensing films. The gas response closely depends on the thickness and porosity of the sensing film [85,134]. Accordingly, efforts should be focused on fabricating sensing films in a reproducible manner, which include controlling the solvent drying process and decomposition of the organic vehicle, if used.

Spin coating is a reproducible method for producing uniform sensing films by coating a sol solution containing nanosized oxide particles followed by heat treatment [34,135,136]. The thickness and porosity of the sensing film can be controlled by repeating the spin coating process, varying the spinning speed, changing the sol content, and tuning the viscosity of the sol solution [137–140].

When nanostructures are dispersed in a solvent that is immiscible with water, a thin slurry layer floating on top of the water can be obtained, which can then be coated onto a substrate by pulling up the vertically dipped substrate. This method is known as the Langmuir–Blodgett (LB) method [141,142]. In this method, the oxide surface can be made hydrophobic, and the film layer needs to be heat treated at elevated temperatures to remove the organic groups present on the surface of the oxide nanostructures. A metal precursor solution immiscible in water can also be used to produce LB films, which can be converted to an oxide film by transferring it to a substrate and then employing heat treatment [143]. In particular, the LB technique is suitable for preparing ultrathin sensing films (thickness <10 nm). This might result in an increase in gas accessibility; however, it might also deteriorate the connectivity of the sensing film. The packing density and thickness of the film can be controlled by varying the number of LB coating steps [141,143]. The LB technique can also be used for obtaining aligned coating of anisotropic nanostructures. Cheng et al. [142] demonstrated that single crystalline and ultrathin $W_{18}O_{29}$ NWs (diameter <2 nm) can be coated on large-area flat and patterned substrates in a highly aligned manner using the LB technique, and the sensor with 10 layers of aligned NWs exhibited a high response to hydrogen at room temperature.

Anisotropic oxide nanostructures such as NWs and nanorods can also be aligned in the slurry state by applying an AC electric field between two electrodes. This method is known as dielectrophoretic positioning [144] and has been applied for fabricating gas sensors with SnO_2 [145,146], CuO [147], ZnO [145], TiO_2 [145], and SnO_2-CuO NWs [148]. This is a cost effective and convenient route for fabricating gas sensors. Guilera et al. [145] successfully demonstrated that SnO_2, TiO_2, and ZnO NWs dispersed in isopropanol can be aligned perpendicular to the electrode edges by dielectrophoretic positioning (Fig. 4.14), and that the resultant sensors exhibited good gas-sensing characteristics. Even oxide nanoparticles form elongated particle chains along the electric field line because of the force exerted on the polarized oxide particles by the inhomogeneous electric field. Gas sensors have been fabricated by forming chains of ZnO [149], TiO_2 [150], and WO_3 [151] nanoparticles across two electrodes.

4.3.2 Screen printing

Screen printing is one of the most representative and facile methods used to form thick sensing films on flat substrates. In general, inks containing oxide nanostructures, solvents, binders, and plasticizers are formed by vigorous mixing and then coated through patterned meshes. During the coating of the film by the application

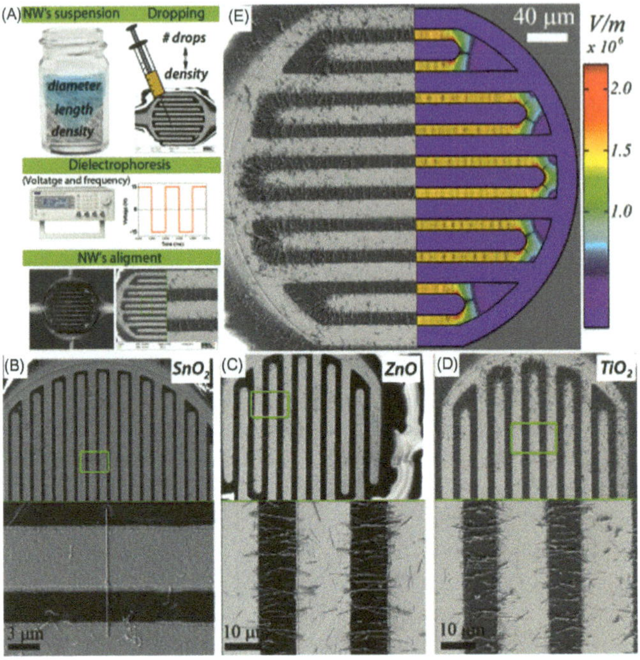

Figure 4.14 (A) Fabrication process steps for NW suspension preparation, drop transfer on the final substrate, and NW alignment with dielectrophoretic forces; (B) SnO_2, (C) ZnO, and (D) TiO_2 NWs obtained using this technique. (E) According to finite element simulations, the NW arrangement (in concentration and direction) follows the electric field distribution imposed by the electrode (IDE) geometry.
Source: Reproduced with permission from J. Guilera, C. Fàbrega, O. Casals, F. Hernández-Ramírez, S. Wang, S. Mathur, et al., Facile integration of ordered nanowires in functional devices, Sensors Actuators, B Chem. 221 (2015) 104–112. doi:10.1016/j.snb.2015.06.069.

of high shear stress (point B in Fig. 4.15) using a squeegee, low viscosity is necessary for the facile transfer of ink. However, after printing, high ink viscosity is required to level the surface and maintain the printed area (point A in Fig. 4.15). Thus, the preparation of a paste with pseudoplastic (shear thinning) rheological behaviors, as shown in Fig. 4.15, is necessary. The key advantages of screen printing are facile mass production, excellent reproducibility, and easy control of sensing film thickness. The main parameters used to control the thickness of sensing films are the viscosity of ink, mesh size, mesh tension, screen emulsion thickness, squeeze attack angle, and squeeze hardness. The organic contents must then be removed by heat treatment.

4.3.3 Inkjet printing

Inkjet printing is also a promising technique for coating metal oxide films onto electrodes. The main advantages of this technique are the facile coating of films

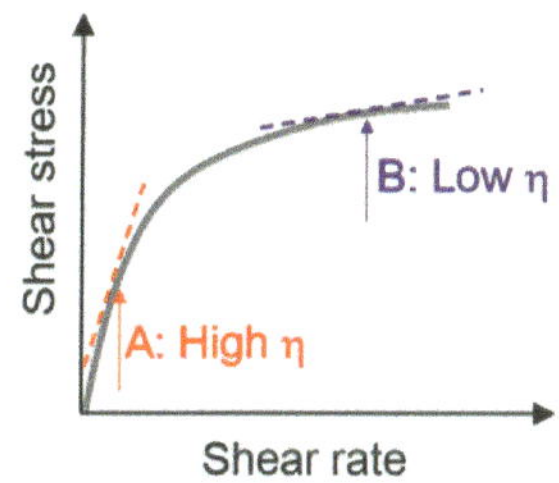

Figure 4.15 Rheological behaviors of inks for screen printing.

achieved without the use of a mask, highly controlled release of a small amount of slurry, formation of patterned films, high-throughput coating, and a process suitable for highly integrated devices [152,153]. Moreover, both flexible and hard substrates can be used, and the composition of the sensing/catalytic materials can be easily controlled when a multinozzle system is employed [154]. Gas sensors incorporating CuO [153], SnO_2 [155], ZnO [156], and WO_3 [157] nanostructures have been prepared by inkjet printing. The fabrication of an electronic nose using a sensor array has been successfully demonstrated by coating different organic sensing materials onto an array of electrodes [158]. The size of the particles should be chosen according to the diameter of the nozzle, and the key parameters that determine film formation are the viscosity, solid loading, organic component, rheological behaviors, and drying kinetics of the slurry. Drop-on-demand coating can also be used to fabricate gas sensors by employing precise inkjet deposition of tiny and dilute droplets (1–100 pL) on a designated area [159].

4.3.4 Near-field electrospinning

Nanofibers containing metal precursors can be prepared by electrospinning, which can then be converted to oxide semiconductor nanofibers. When nanofibers are collected at a distance of a few tens of centimeters, chaotic whipping of nanofibers would lead to the formation of randomly tangled nanofiber networks. Hence, the oxide nanofibers obtained after heat treatment are generally dispersed in a solvent, and the slurry is coated onto a substrate containing electrodes to fabricate the sensors. In contrast, when the electrode-to-collector distance is decreased to 0.5–3.0 mm, linear and stable liquid jets can be coated onto a substrate in a highly aligned manner [160,161]. This method is known as near-field (NF) electrospinning and has been successfully employed for the fabrication of gas sensors using a single polyaniline nanofiber field effect transistor [162]. Ruggieri et al. [163] prepared gas sensors using highly aligned nitrogen-doped TiO_2 nanofibers via NF electrospinning, and Wang et al. [164] demonstrated the detection of moisture and CH_4 using a single SnO_2 nanofiber gas sensor. The high gas accessibility and the formation of hetero-nanostructures between oxide nanofibers provide versatility in the gas sensor design.

4.3.5 *Flame spray pyrolysis*

In a flame spray pyrolysis (FSP) method, oxide thin films are directly deposited onto a substrate by evaporation and decomposition of sprayed liquid precursors at high temperatures using a flame to achieve the gas-phase reaction to form particles, followed by deposition of nanoparticles by thermophoresis [165–168]. The major advantages of this method are the single-step process, the controllability of primary particle size and film porosity, and the possibility of in situ deposition of noble metal catalysts. This process can also be used to deposit a catalytic filtering layer for controlling the gas selectivity. Sahm et al. [169] reported a Pd/SnO_2 film obtained by FSP that exhibited selective detection of ethanol with low cross-responses to CH_4 and CO, while the Pd/SnO_2 film with a Pd/Al_2O_3 catalytic filtering overlayer exhibited selective detection of CH_4 with negligible cross-responses to ethanol and CO (Fig. 4.16). Thus, FSP is useful in controlling the microstructure of the filtering layer and the uniform loading of noble metal catalysts that are essential for controlling the gas selectivity.

4.3.6 *Vapor deposition: sputtering, e-beam evaporation, GAD, ALD, and PLD*

4.3.6.1 Sputtering and e-beam evaporation

Sputtering forms a thin film by depositing the atoms detached from the targets due to the bombardment of the Ar^+ ion. Direct current (DC) sputtering is advantageous to deposit metal films with high conductivity. In contrast, radio frequency (RF) sputtering is generally used to deposit oxide semiconductors and offers reactive sputtering, which is the formation of a thin film by the reaction between the target material and reactant gas in the chamber [170]. The main advantages of sputtering as a method to form films for gas sensors are the facile control of composition, thickness, and catalyst loading. Accordingly, sputtering is a good candidate process to realize oxide semiconductor gas sensor arrays with different compositions and catalysts on a MEMS-based sensor platform [171]. To date, most sensing materials have been deposited by sputtering process [172–176]. However, the deposition of porous thin films and the tuning of the film microstructure is relatively difficult.

E-beam evaporation also involves physical vapor deposition, which uses a focused electron beam with high energy as a source to evaporate the target materials under high vacuum [177]. The evaporated materials then condense on the substrate materials. Not only metals (for electrodes, heaters, and catalysts), but also metal oxides (for chemiresistors) can be coated. Associated difficulties in conformal coating on complex shaped substrates is not an issue for the applications of gas sensors with flat geometry. The precise control of film thickness (a few nanometers per minute) is also advantageous for the reproducible formation of sensor films or catalytic layers. To date, e-beam evaporation has been used to deposit sensing materials such as SnO_2 [178], ZnO [179], and In_2O_3 [180] and to form catalytic layers of Au [181], Ag [182], and Co_3O_4 [183].

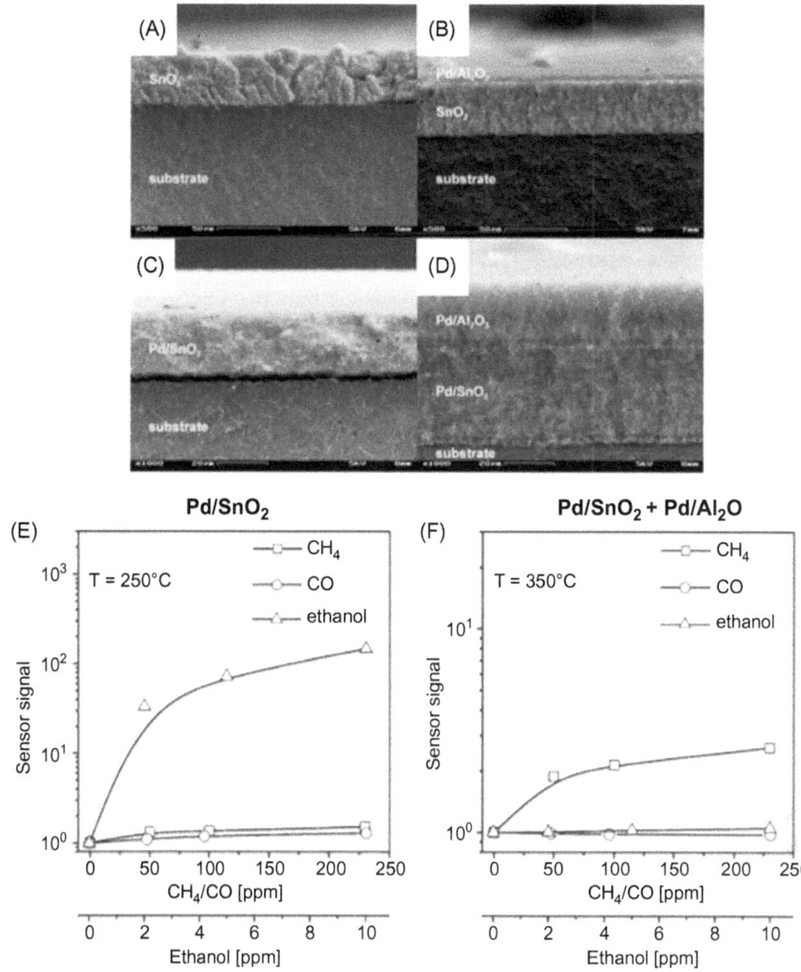

Figure 4.16 SEM images of cross-sections of sensors obtained with (A) SnO_2 layer, (B) Pd/ Al_2O_3 layer on the top of a SnO_2 layer, (C) Pd/SnO_2 layer, and (D) Pd/Al_2O_3 layer on top of a Pd/SnO_2 layer deposited on ceramic substrates; Sensor signal for CH_4 (squares), CO (circles), and ethanol (triangles). The measurements performed at (E) 250°C for Pd/SnO_2 and (F) 350°C for Pd/SnO_2 + Pd/Al_2O_3.
Source: Reprinted with permission from T. Sahm, W. Rong, N. Bârsan, L. Mädler, U. Weimar, Sensing of CH4, CO and ethanol with in situ nanoparticle aerosol-fabricated multilayer sensors, Sensors Actuators, B Chem. 127 (2007) 63–68. doi:10.1016/j.snb.2007.07.001.

4.3.6.2 Glancing angle deposition (oblique angle deposition)

An array of periodic nanowires can be grown using glancing angle deposition (GLAD) with a highly directional vapor flux produced either from sputtering or e-beam evaporation [184]. When the angle (θ) between the substrate normal and vapor flux nears 70 degrees or higher, the shadowed region of the discrete nuclei leads to the growth of nanorods along the direction of the vapor flux (Fig. 4.17) [185]. Highly porous arrays of nanorods of SnO_2 [186,187], ZnO [185], TiO_2 [187], WO_3 [188,189], Co_3O_4 [190], ITO [187,191], Au-SnO_2 [181,192], Ag-SnO_2 [192], and Pt-WO_3 [193] have been produced using this method. Most researchers [194–197] have reported that sensors using aligned oxide nanorods show significantly higher and faster response as compared to their dense thin film counterparts. The enhanced gas response can be attributed to the high gas accessibility due to the porous film structures. Yao et al. [191] reported that the gas response of aligned ITO nanorods increased as the vapor incident angle (θ) increased from 60 to 86 degrees. This enhancement of gas response also emanated from the increase in porosity of the sensing film. When nanorods become too long, the uniform deposition of catalysts becomes difficult due to the shadow effect in GLAD. Jeon et al. [181] made SnO_2-Pt nano bamboo structures by periodically repeating the growth of SnO_2 nanorods and the deposition of a thin Pt layer, and demonstrated that the resultant sensor could detect trace concentrations of volatile organic compounds.

Hwang et al. [187] prepared various 3D arrays of nanostructures such as helices, rods, zig-zags, and square helices (Fig. 4.18) by the combination of oblique vapor flux and substrate rotation. They fabricated an array of six sensors consisting of TiO_2 helices, TiO_2 vertical posts, SnO_2 helices, a SnO_2 thin film, WO_3 zig-zags, and ITO slanted nanorods, and successfully demonstrated the concept of an

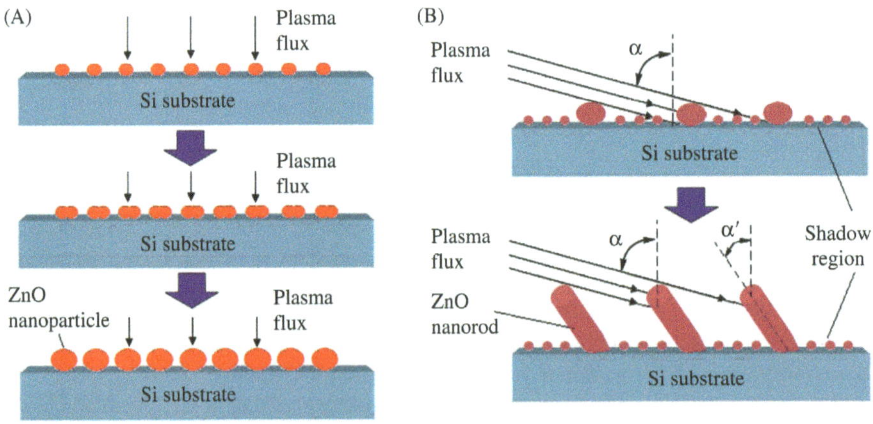

Figure 4.17 Schematic illustration of growth processes of ZnO nanostructures on a Si (100) substrate: (A) nanoparticles deposited along the normal direction; (B) nanorods fabricated by using the oblique angle deposition (OAD) technique.
Source: Reprinted with permission from J. Chu, X. Peng, M. Sajjad, B. Yang, P.X. Feng, Nanostructures and sensing properties of ZnO prepared using normal and oblique angle deposition techniques, Thin Solid Films. 520 (2012) 3493–3498. doi:10.1016/j.tsf.2011.12.066.

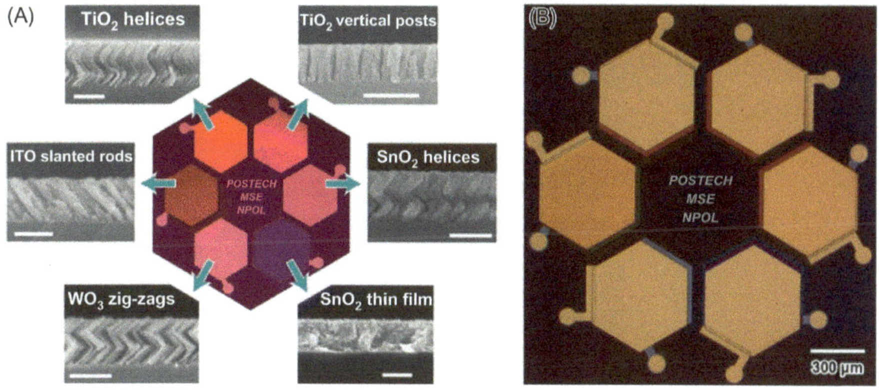

Figure 4.18 (A) Optical microscopy (OM) image of the fabricated prototype e-nose chip after the deposition of six different sensing layers, and corresponding cross-sectional SEM images (scale bars represent 300 nm); (B) OM image of the e-nose after the deposition of top and pad electrodes on the device shown in (A).
Source: Reprinted with permission from S. Hwang, H. Kwon, S. Chhajed, J.W. Byon, J.M. Baik, J. Im, et al., A near single crystalline TiO2 nanohelix array: enhanced gas sensing performance and its application as a monolithically integrated electronic nose, Analyst. 138 (2013) 443–450. doi:10.1039/C2AN35932D.

electronic nose fabricated by GLAD. The control of composition, porosity, catalyst, and morphology of a sensing film is an attractive option for an electronic nose.

The connections between aligned nanowires can be used for the operation of sensors in self-powered mode. Moon et al. [188] prepared transparent gas sensors by the deposition of an array of slanted WO_3 nanorods by GLAD onto a glass substrate with ITO electrodes (Fig. 4.19A). The nanorods were connected with thin necks. When the voltage was applied, the heating occurred primarily and locally at the thin neck regions with high resistance, which increased the temperature of the WO_3 nanorods to the sensing temperature. Thus, the sensor resistance changes upon exposure to analyte gases could be measured as a sensor signal (Fig. 4.19B). For a sensor with a sensing area of $100 \times 170\ \mu m^2$, the power for sensor operation became as low as 21.6 μW in the continuous heating mode and decreased further to 0.18 μW in the pulsed heating mode (Fig. 4.19C). The ultralow power can be explained by the minimal power consumption of the self-heating sensor, preventing any of the heat loss that can occur during heat transfer between the heating and sensing regions.

4.3.6.3 *Atomic layer deposition*

ALD applies precursor materials to the surface of substrates in a pulse mode [198] for the layer-by-layer growth of films. Thus, the film thickness can be controlled to a few angstroms, and the deposition is highly conformal, uniform, and flat. Because the chemiresistive variation in oxide semiconductor gas sensors increases

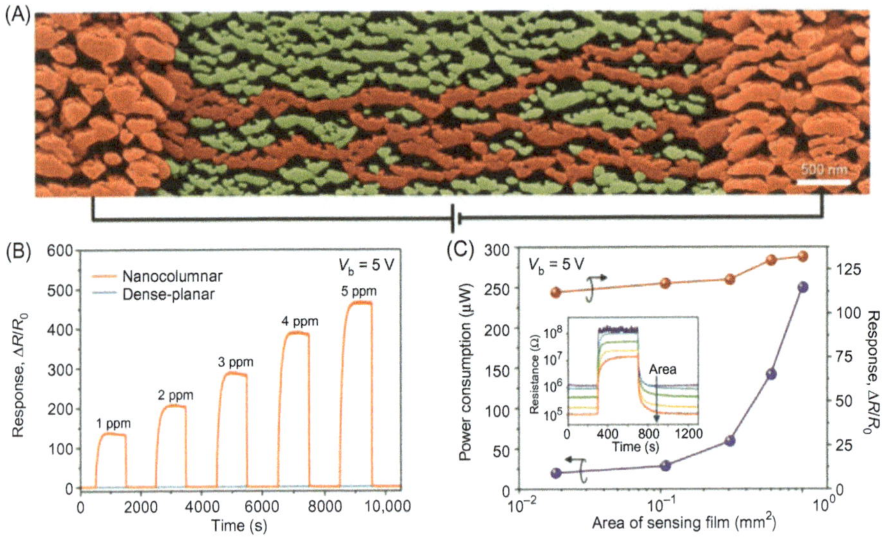

Figure 4.19 (A) 40°-Tilted SEM image of a nanocolumnar WO_3 film between and on ITO IDEs; Parts highlighted in reddish color indicate localized current pathways which meander with narrow necks. (B) Sensing transients of the dense-planar and nanocolumnar WO_3 thin film sensors to 1–5 ppm NO_2 at an applied bias voltage of 5 V. (C) Power consumption and response of nanocolumnar WO_3 thin film sensors to 1 ppm NO_2 as a function of the area of sensing film; inset: response transients to 1 ppm NO_2 of sensors with different areas of sensing film.
Source: Reproduced with permission from H.G. Moon, Y.S. Shim, D.H. Kim, H.Y. Jeong, M. Jeong, J.Y. Jung, et al., Self-activated ultrahigh chemosensitivity of oxide thin film nanostructures for transparent sensors, Sci. Rep. 2 (2012) 1–7. doi:10.1038/srep00588.

dramatically when the particle sizes become comparable to or twice the thickness of the electron depletion layer/hole accumulation layer, ALD is very promising for achieving a high gas response [199]. Moreover, highly conformal deposition on various sacrificial templates or other oxide nanostructures can be used to prepare hollow spheres/tubes with ultrathin shells [200] and core-shell oxide heterostructures [96–99], respectively. Although many source materials for ALD have been suggested [198], the compositional controllability remains low and requires further improvement.

Du et al. [201] prepared SnO_2 thin films of various thicknesses by ALD using *Sn*-chloride precursors and investigated the effect of film thickness on the response to CO. They observed that CO response increased with increasing film thickness, peaked at ~2.5 nm, and then decreased with further thickening to 6 nm. Rosental et al. [202] prepared SnO_2 films by ALD using two different precursors (SnI_4-O_2, and $SnCl_4$-H_2O_2) and investigated their gas-sensing characteristics. They observed a maximum gas response at a film thickness of ~10 nm when SnI_4-O_2 was used as precursor, and the two films prepared from the SnI_4-O_2 and $SnCl_4$-H_2O_2 precursors showed significantly different gas responses to 1% CO, as well as sensor resistances. These results clearly explain that the gas-sensing characteristics depend closely on the ALD source materials, deposition conditions, and microstructures/thicknesses of films.

4.3.6.4 Pulsed layer deposition

Pulsed layer deposition (PLD) allows for the formation of sensing films with a high deposition rate and good controllability of composition [203]. To date, WO_3 [204], SnO_2 [205], NiO [206], MoO_3 [207], and Au-loaded WO_3 [208] thin films have been prepared by PLD for gas sensor applications. The morphology [204] and porosity [205] of sensing films can be controlled by tuning the background gas pressure and reaction temperature. In particular, the change in oxygen partial pressure during PLD can lead to films with different degrees of nonstoichiometry. For example, Stamataki et al. [206] prepared an NiO film by PLD using a Ni target and investigated the effect of oxygen pressure on the electrical properties of NiO thin films. The p-type NiO was prepared at an oxygen partial pressure of 50 Pa at 200°C, while n-type NiO was formed at pressures of 5–30 Pa. This shows that the majority charge carriers of oxide semiconductor gas sensors can be manipulated by controlling the oxygen partial pressure during PLD. However, because the sensors in real applications are operated at 200–400°C in air atmosphere, n-type NiO cannot remain stable during sensor operation and will gradually change to p-type NiO. Accordingly, most oxide semiconductors prepared under low oxygen partial pressures are usually annealed at 400–500°C in air [204,205,207] to allow the stable operation of gas sensors in ambient atmospheres.

4.3.7 Direct growth of nanowires

Oxide semiconductor NWs can be coated on an electrode by slurry deposition with or without employing dielectrophoretic positioning. In both cases, the contact between the NWs and electrode is weak, and the sensor can be vulnerable to mechanical vibration. Oxide semiconductor NWs are usually grown by a vapor-liquid-solid mechanism assisted by Au catalysts. Thus, oxide semiconductor NWs can be grown directly from an Au electrode to establish gas sensors with a network configuration [209,210] when the spacing between the two electrodes is small. Park et al. [211] demonstrated that the number density of NW-to-NW contacts in a sensor can be controlled by controlling the spacing between the two Au electrodes, which would eventually alter the gas response. Direct growth of NWs from an electrode is advantageous for increasing the strength of electric contact between the NWs and the electrode as well as the stability of the sensor against strong vibrations. Moreover, NWs with a highly porous network guarantee a high accessibility of analyte gases to the entire surface of the oxide NWs.

Hsuech et al. [212] fabricated an interdigital pattern of Ga-doped ZnO electrodes on a glass substrate by RF sputtering and lithography-assisted etching to successfully grow ZnO NWs on electrodes for use as gas sensors. Qin et al. [213]

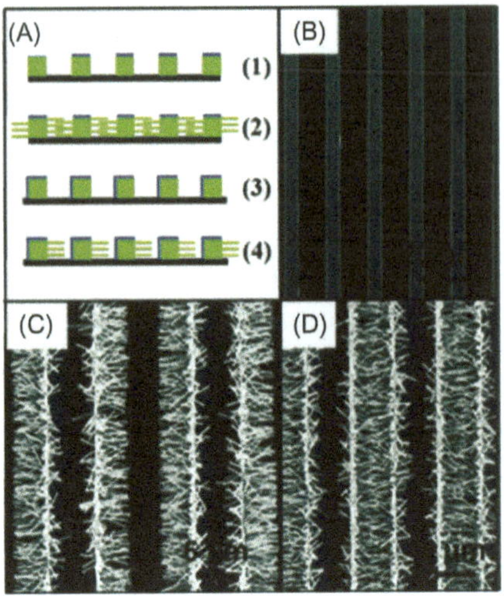

Figure 4.20 (A) Schematic for the growth of patterned and laterally aligned ZnO NW arrays; (B) SEM image of the ZnO pattern covered with a 10 nm Cr shield layer prior to the NW growth; (C, D) SEM images of ZnO NW arrays grown on Si substrates.
Source: Reprinted with permission from Y. Qin, R. Yang, Z.L. Wang, Growth of Horizonatal ZnO Nanowire Arrays on Any Substrate Growth of Horizonatal ZnO Nanowire Arrays on Any Substrate, 112 (2008) 18734–18736. https://doi.org/10.1021/jp808869j.

suggested a method for the horizontal growth of ZnO NWs (Fig. 4.20). For this, they fabricated patterned ZnO electrodes and formed a Cr overlayer to inhibit the growth of ZnO NWs. ZnO NWs were grown from an aqueous zinc nitrate solution containing hexamethylenetetramine. By controlling the coverage of the Cr overlayer at the top and sides of the ZnO electrode, the direction of the lateral NW growth could be controlled.

Direct growth on Au or ZnO electrodes can be employed to fabricate gas sensors on a large scale. For example, Hwang et al. [214] successfully demonstrated that a large number of SnO_2 NW network gas sensors can be fabricated by employing direct and simultaneous growth of NWs on laser-scribed Al_2O_3 substrates containing several sensor elements (Fig. 4.21). This clearly indicates the possibility of mass production of NW-based sensors.

4.3.8 Sensing films on MEMS hotplates

The sensing area is narrow and the membranes are thin in MEMS-based sensors. Thus, reproducible deposition of sensing materials onto MEMS-based transducers is important for real-life applications. To date, various methods have been used to coat sensing materials onto substrates (Table 4.2) [8,10,11,16,17,19,23,65,215–222], which include drop coating [8,17,215], spin coating [216], dielectrophoretic positioning [217], coating of solution-dipped templates [65,218], inkjet printing [223], dip

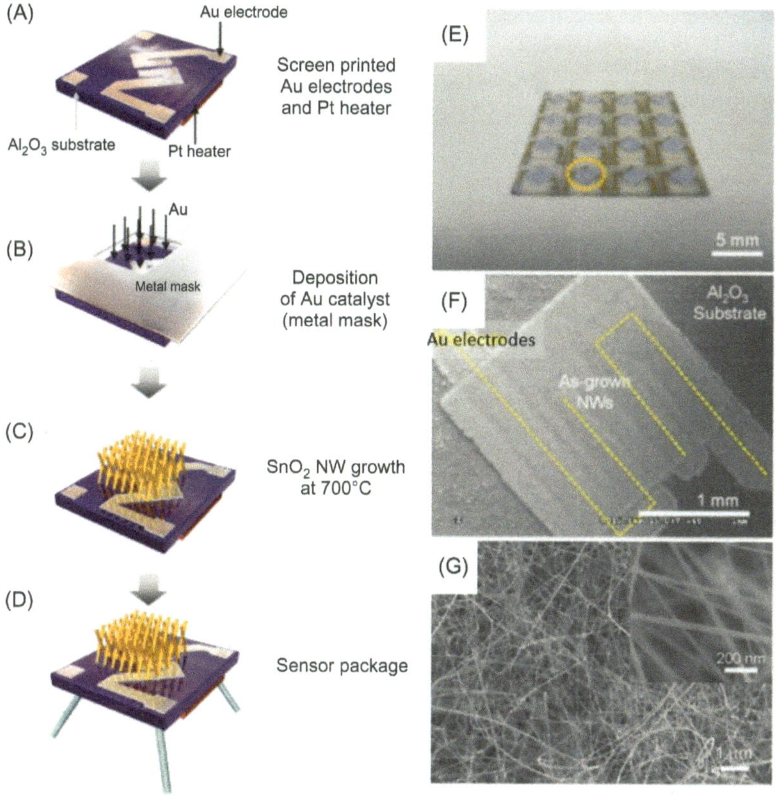

Figure 4.21 Experimental procedures used for the fabrication of SnO_2 NW network gas sensors: (A) Al_2O_3 substrate with two Au electrodes and Pt heater pattern; (B) deposition of Au catalyst using a metal mask onto the defined area; (C) direct growth of SnO_2 NWs via vapor-phase reaction; (D) packaging of sensor elements into a stainless steel holder. (C) and (D) show the schematics of the growth of NW networks on Al_2O_3 substrates; sensor elements: (E) 16 sensor elements prepared by one-pot growth of SnO_2 NW networks on laser-scribed Al_2O_3 substrates; (F) SEM image of a sensor element; (G) SEM image of SnO_2 NW networks.
Source: Reprinted with permission from I.S. Hwang, S.J. Kim, J.K. Choi, J.J. Jung, D.J. Yoo, K.Y. Dong, et al., Large-scale fabrication of highly sensitive SnO2nanowire network gas sensors by single step vapor phase growth, Sensors Actuators, B Chem. 165 (2012) 97–103. doi:10.1016/j.snb.2012.02.022.

pen lithography [215], electrodynamic printing [221], CVD [10,19], and sputtering [11,16,23,222].

Drop coating is one of the most representative methods used to coat sensing materials because the diverse oxide nanostructures can be dispersed in a slurry, and the process is relatively simple. Either a micropipette or microdispenser can be used to generate small droplets, and the precise positioning of the dispenser is

Table 4.2 Methods to deposit sensing materials on MEMS-based transducers [8,10,11,16,17,19,23,65,215–222]

Deposition method	Sensing materials
Drop coating of slurry/precursor	SnO_2 [215], Pd-SnO_2 [17], $BaSnO_3$ NPs [8]
Spin coating of slurry/precursor	ZnO NWs [216]
Dielectrophoretic positioning	ZnO NWs [217]
Coating of solution-dipped colloidal templates and heat treatment	SnO_2 [65,218]
Inkjet printing	In_2O_3 NPs [219]
Dip pen lithography	SnO_2 [220]
Electrohydrodynamic printing	SnO_2 NFs [221], WO_3 NFs [221], In_2O_3 NFs [221], NIO NFs [221]
Chemical vapor deposition	SnO_2 [19], TiO_2 [19], (Ag,Fe,Ni,Co,Cu)-SnO_2 [10]
Sputtering	MoO_3 [16]
Oxidation of sputtered metal	SnO_2 [11,23], ZnO-CuO nanoflakes [222]

possible when a robot-assisted microdispensing system is used. Note that the inner diameter of the microdispenser nozzle should be significantly larger than the oxide nanostructures to prevent clogging, which typically ranges from 50 to 200 μm. After the landing of a slurry droplet on the substrate, the droplet spreads, and the degree of spreading depends on the surface tension of the solvent and the hydrophilic/hydrophobic nature of the substrate. In general, the sensing area after droplet coating, drying, and subsequent heat treatment is larger than the size of the droplets, which typically range from 100 to 400 μm. Accordingly, the deposition of sensing materials onto a narrower area is necessary for further integration of gas sensors. Moreover, it is usually difficult to coat a high viscosity paste by drop coating.

Kang et al. [221] coated slurries containing SnO_2, In_2O_3, WO_3, and NiO nanofibers using an electrohydrodynamic (EHD) printing technique (Fig. 4.22). The diameters of the sensing area (40–70 μm) were substantially smaller than that of the nozzle (90 μm). This is possible because a tiny amount of slurry can be ejected by the application of an intense pulse electric field between the nozzle and substrate. Thus, high viscosity slurries can also be coated by the EHD process. Dip pen nanolithography, the method used to coat a slurry with an AFM-type cantilever with submicrometer precision, was also used to form Au-SnO_2 sensing layer on MEMS transducers [220]. Further investigation is necessary to be able to coat sensing materials on MEMS transducers in a reproducible manner, on a large scale, and with low cost.

4.3.9 Sensing films on flexible substrates

The key advantages of plastic foils as supporting materials are flexibility and transparency, which are compatible with wearable devices. Generally, PI with high thermal stability is used as a flexible supporting material, and the sensing films are

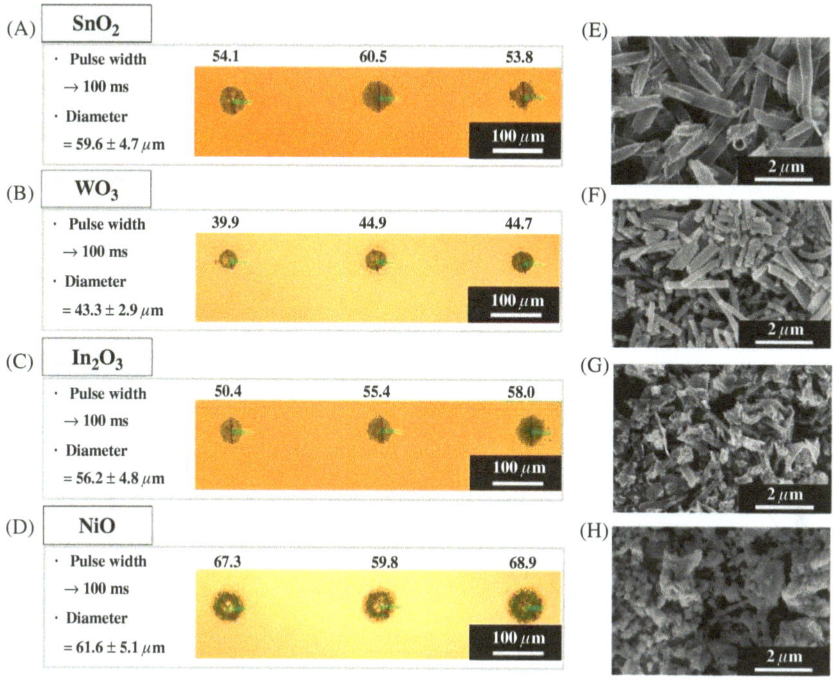

Figure 4.22 EHD printing results of (A) SnO_2, (B) WO_3, (C) In_2O_3, and (D) NiO nanofibers on gold electrodes with 100 ms pulse width. SEM images of EHD printed solution in each pattern: (E) SnO_2, (F) WO_3, (G) In_2O_3 and (H) NiO nanofibers. Single dot pattern consists of nanofiber fragments, and patterns show diameters in the range of 40–70 μm.
Source: Reproduced with permission from K. Kang, D. Yang, J. Park, S. Kim, I. Cho, H.-H. Yang, et al., Micropatterning of metal oxide nanofibers by electrohydrodynamic (EHD) printing towards highly integrated and multiplexed gas sensor applications, Sensors Actuators B Chem. 250 (2017) 574–583. doi:10.1016/j.snb.2017.04.194.

Table 4.3 Methods to deposit sensing materials on flexible substrates [25,26,224–227]

Deposition method	Sensing materials/substrate
Drop coating of slurry	SnO_2/PI [25], ZnO NPs/PET [224], RuO_2 nanosheets/Ag NW-embedded PI [226]
Coating of paste using capillary tip	Pd-SnO_2/PI [26]
Inkjet printing	SnO_2/PI [226]
Aerosol-assisted CVD	WO_3/PI [227]

formed by drop coating [25,224,225], paste coating [26], inkjet printing [226], and CVD [227] (Table 4.3). Good adhesion of the sensing films to the supporting materials despite folding is highly necessary for reliable operation. Also, a sensor or postheat-treatment temperature lower than 350°C is better when PI is used. For this, low-temperature processing of sensing materials, and sensor materials with low operation temperatures are required.

4.4 Conclusions

High-performance oxide semiconductor gas sensors can be designed by the intimate interplay among sensing materials, transducers, and sensing films. Transducers tend to become smaller to achieve low power consumption, rapid warm up, a new mode of sensor operation by temperature-programmed measurement, integration into small devices, realization of small electronic noses, and wireless communication through sensor networks. Highly crystalline metal oxide nanostructures such as nanorods, nanowires, nanosheets, and nanocubes with high surface area-to-volume ratios, and their assembled nanoarchitectures such as hierarchical, hollow, yolk-shell, and mesoporous structures with high gas accessibility, have been suggested as valuable platforms to design high-performance gas sensors with good response, excellent selectivity, and rapid response. For this, diverse synthetic routes have been explored and developed, which include hydrothermal/solvothermal reactions, template-based syntheses, electrospinning, spray pyrolysis, and vapor-phase reactions. Gas transfer through the pores and conduction along the nanostructures in sensing films are other key traits for determining gas-sensing characteristics. For this, reproducible, cost effective, and facile processes for the deposition of sensing films are essential. The principles, advantages, and shortcomings of various film deposition methods for oxide semiconductor gas sensors were explained and compared to find new strategies for designing reliable gas sensors. Progress on transducers, sensing materials, and film deposition methods has shed much light on the technological realization of various new functionalities in oxide semiconductor gas sensors such as enhanced gas-sensing characteristics, small size, ultralow power, integration into other devices, new operation algorithms, and flexible design.

References

[1] N. Barsan, U. Weimar, Conduction model of metal oxide gas sensors, J. Electroceramics 7 (2001) 143–167. Available from: https://doi.org/10.1023/A:1014405811371.

[2] N. Yamazoe, New approaches for improving semiconductor gas sensors, Sensors Actuators B Chem. 5 (1991) 7–19. Available from: https://doi.org/10.1016/0925-4005(91)80213-4.

[3] Y. Shimizu, M. Egashira, Basic aspects and challenges of semiconductor gas sensors, J. MRS Bull. 24 (1999) 18–24. Available from: https://doi.org/10.1557/S0883769400052465.

[4] J.H. Lee, Gas sensors using hierarchical and hollow oxide nanostructures: overview, Sensors Actuators, B Chem. 140 (2009) 319–336. Available from: https://doi.org/10.1016/j.snb.2009.04.026.

[5] T. Wang, Y. Guo, P. Wan, H. Zhang, X. Chen, X. Sun, Flexible transparent electronic gas sensors, Small (2016) 3748–3756. Available from: https://doi.org/10.1002/smll.201601049.

[6] http://www.figarosensor.com/products/822pdf.pdf.

[7] A. Chiba, Development of the TGS gas sensor, in: S. Yamauchi (Ed.), Chemical Sensor Technology, vol. 4, Elsevier, 1992, pp. 1–18.

[8] J.C. Belmonte, J. Puigcorbé, J. Arbiol, A. Vilà, J.R. Morante, N. Sabaté, et al., High-temperature low-power performing micromachined suspended micro-hotplate for gas sensing applications, Sensors Actuators, B Chem. 114 (2006) 826–835. Available from: https://doi.org/10.1016/j.snb.2005.07.057.

[9] K.D. Mitzner, J. Sternhagen, D.W. Galipeau, Development of a micromachined hazardous gas sensor array, Sensors Actuators, B Chem. 93 (2003) 92–99. Available from: https://doi.org/10.1016/S0925-4005(03)00244-2.

[10] S. Semancik, R.E. Cavicchi, M.C. Wheeler, J.E. Tiffany, G.E. Poirier, R.M. Walton, et al., Microhotplate platforms for chemical sensor research, Sensors Actuators, B Chem. 77 (2001) 579–591. Available from: https://doi.org/10.1016/S0925-4005(01)00695-5.

[11] I. Elmi, S. Zampolli, G.C. Cardinali, Optimization of a wafer-level process for the fabrication of highly reproducible thin-film MOX sensors, Sensors Actuators, B Chem. 131 (2008) 548–555. Available from: https://doi.org/10.1016/j.snb.2007.12.033.

[12] A. Tommasi, M. Cocuzza, D. Perrone, C.F. Pirri, R. Mosca, M. Villani, et al., Modeling, fabrication and testing of a customizable micromachined hotplate for sensor applications, Sensors (Switzerland). 17 (2017). Available from: https://doi.org/10.3390/s17010062.

[13] R. Jaaniso, O.K. Tan, Semiconductor Gas Sensors, first ed., Woodhead Publishing Series in Electric and Optical Materials, Cambridge, 2013.

[14] Figaro, TGS8100- Featured Products - FIGARO Engineering inc (2017), pp. 1–4. http://www.figaro.co.jp/en/product/feature/tgs8100.html.

[15] I. Simon, N. Bârsan, M. Bauer, U. Weimar, Micromachined metal oxide gas sensors: opportunities to improve sensor performance, Sensors Actuators B Chem. 73 (2001) 1–26. Available from: https://doi.org/10.1016/S0925-4005(00)00639-0.

[16] V. Guidi, G.C. Cardinali, L. Dori, G. Faglia, M. Ferroni, G. Martinelli, et al., Thin-film gas sensor implemented on a low-power-consumption micromachined silicon structure, Sensors Actuators, B Chem. 49 (1998) 88–92. Available from: https://doi.org/10.1016/S0925-4005(98)00039-2.

[17] D. Briand, A. Krauss, B. Van Der Schoot, U. Weimar, N. Barsan, W. Göpel, et al., Design and fabrication of high-temperature micro-hotplates for drop-coated gas sensors, Sensors Actuators, B Chem. 68 (2000) 223–233. Available from: https://doi.org/10.1016/S0925-4005(00)00433-0.

[18] D. Briand, S. Heimgartner, M.-A. Grétillat, B. van der Schoot, N.F. de Rooij, Thermal optimization of micro-hotplates, J. Micromech. Microeng. 12 (2002) 971–978.

[19] D.C. Meier, J.K. Evju, Z. Boger, B. Raman, K.D. Benkstein, C.J. Martinez, et al., The potential for and challenges of detecting chemical hazards with temperature-programmed microsensors, Sensors Actuators, B Chem. 121 (2007) 282–294. Available from: https://doi.org/10.1016/j.snb.2006.09.050.

[20] S. Semancik, R. Cavicchi, Kinetically controlled chemical sensing using micromachined structures, Acc. Chem. Res. 31 (1998) 279–287. Available from: https://doi.org/10.1021/ar970071b.

[21] S. Chakraborty, A. Sen, H.S. Maiti, Selective detection of methane and butane by temperature modulation in iron doped tin oxide sensors, Sensors Actuators, B Chem. 115 (2006) 610–613. Available from: https://doi.org/10.1016/j.snb.2005.10.046.

[22] S.-J.J. Choi, K.H. Ku, B.J. Kim, I.-D.D. Kim, Novel templating route using Pt infiltrated block copolymer microparticles for catalytic Pt functionalized macroporous WO3 nanofibers and its application in breath pattern recognition, ACS Sensors. 1 (2016) 1124–1131. Available from: https://doi.org/10.1021/acssensors.6b00422.

[23] I. Elmi, S. Zampolli, E. Cozzani, F. Mancarella, G.C. Cardinali, Development of ultra-low-power consumption MOX sensors with ppb-level VOC detection capabilities for emerging applications, Sensors Actuators, B Chem. 135 (2008) 342–351. Available from: https://doi.org/10.1016/j.snb.2008.09.002.

[24] D. Briand, A. Oprea, J. Courbat, N. Bârsan, Making environmental sensors on plastic foil, Mater. Today. 14 (2011) 416–423. Available from: https://doi.org/10.1016/S1369-7021(11)70186-9.

[25] D. Briand, S. Colin, J. Courbat, S. Raible, J. Kappler, N.F. de Rooij, Integration of MOX gas sensors on polyimide hotplates, Sensors Actuators, B Chem. 130 (2008) 430–435. Available from: https://doi.org/10.1016/j.snb.2007.09.013.

[26] J. Courbat, D. Briand, L. Yue, S. Raible, N.F. De Rooij, Drop-coated metal-oxide gas sensor on polyimide foil with reduced power consumption for wireless applications, Sensors Actuators, B Chem. 161 (2012) 862–868. Available from: https://doi.org/10.1016/j.snb.2011.11.050.

[27] M. Aslam, C. Gregory, J.V. Hatfield, Polyimide membrane for micro-heated gas sensor array, Sensors Actuators, B Chem. 103 (2004) 153–157. Available from: https://doi.org/10.1016/j.snb.2004.04.102.

[28] R. Lalauze, C. Pijolat, A New Approach to Selective Detection of Gas By an SnO2 Solid State Sensor, Sensors Actuators B Chem. 5 (1983) 55–63.

[29] J. Tamak, T. Maekawa, N. Miura, N. Yamazoe, CuO-SnO2 element for highly sensitive and selective detection of H2S, J. Electroanal. Chem. 9 (1992) 197. Available from: https://doi.org/10.1016/j.jelechem.2011.01.037.

[30] G. Sberveglieri, P. Benussi, G. Coccoli, Reactively sputtered indium tin oxide polycrystalline thin films as NO and NO 2 gas sensors, Thin Solid Films. 186 (1990). 2013. http://www.sciencedirect.com/science/article/pii/004060909090150C.

[31] A. Ghosh, S. Basu, Spray/CVD deposition and charaterisation of surface modified zinc oxide thick films for gas sensor, Mater. Chem. Phys. 27 (1991) 45–54.

[32] V. Demarne, A. Grisel, A new SnO2 low temperature deposition technique for integrated gas sensors, Sensors Actuators B Chem. 15 (1993) 63–67. Available from: https://doi.org/10.1016/0925-4005(93)85028-9.

[33] C. Xu, J. Tamaki, N. Miura, N. Yamazoe, Grain size effects on gas sensitivity of porous Sn02-based elements, Sensors Actuators B. 3 (1991) 147–155. Available from: https://doi.org/10.1016/0925-4005(91)80207-Z.

[34] N.S. Baik, G. Sakai, N. Miura, N. Yamazoe, Hydrothermally treated sol solution of tin oxide for thin-film gas sensor, Sensors Actuators, B Chem. 63 (2000) 74–79. Available from: https://doi.org/10.1016/S0925-4005(99)00513-4.

[35] Y. Sun, D.J. Riley, M.N.R. Ashfold, Mechanism of ZnO nanotube growth by hydrothermal methods on ZnO film-coated Si substrates, J. Phys. Chem. B. 110 (2006) 15186–15192. Available from: https://doi.org/10.1021/jp062299z.

[36] H.G. Yang, C.H. Sun, S.Z. Qiao, J. Zou, G. Liu, S.C. Smith, et al., Anatase TiO2 single crystals with a large percentage of reactive facets, Nature. 453 (2008) 638–641. Available from: https://doi.org/10.1038/nature06964.

[37] X. Han, M. Jin, S. Xie, Q. Kuang, Z. Jiang, Y. Jiang, et al., Synthesis of tin dioxide octahedral nanoparticles with exposed high-energy {221} facets and enhanced gas-sensing properties, Angew. Chem. Int. Ed. 48 (2009) 9180–9183. Available from: https://doi.org/10.1002/anie.200903926.

[38] H. Wang, K. Dou, W.Y. Teoh, Y. Zhan, T.F. Hung, F. Zhang, et al., Engineering of facets, band structure, and gas-sensing properties of hierarchical Sn2 + -Doped SnO2 nanostructures, Adv. Funct. Mater. 23 (2013) 4847–4853. Available from: https://doi.org/10.1002/adfm.201300303.

[39] S. Tian, F. Yang, D. Zeng, C. Xie, Solution-processed gas sensors based on ZnO nanorods array with an exposed (0001) facet for enhanced gas-sensing properties, J. Phys. Chem. C. 116 (2012) 10586–10591. Available from: https://doi.org/10.1021/jp2123778.

[40] M.M. Titirici, M. Antonietti, A. Thomas, A generalized synthesis of metal oxide hollow spheres using a hydrothermal approach, Chem. Mater. 18 (2006) 3808–3812. Available from: https://doi.org/10.1021/cm052768u.

[41] C.Y. Lee, S.J. Kim, I.S. Hwang, J.H. Lee, Glucose-mediated hydrothermal synthesis and gas sensing characteristics of WO3hollow microspheres, Sensors Actuators, B Chem. 142 (2009) 236–242. Available from: https://doi.org/10.1016/j.snb.2009.08.031.

[42] S.-J. Kim, I.-S. Hwang, C.W. Na, I.-D. Kim, Y.C. Kang, J.-H. Lee, Ultrasensitive and selective C2H5OH sensors using Rh-loaded In2O3 hollow spheres, J. Mater. Chem. 21 (2011) 18560–18567. Available from: https://doi.org/10.1039/C1JM14252F.

[43] F. Caruso, Nanoengineering of inorganic and hybrid hollow spheres by colloidal templating, Science 282 (1998) 1111–1114. Available from: https://doi.org/10.1126/science.282.5391.1111.

[44] J.Y. Lee, J.H. Lee, S.H. Hong, Y.K. Lee, J.Y. Choi, Coating BaTiO3 Nanolayers on spherical Ni powders for multilayer ceramic capacitors, Adv. Mater. 15 (2003) 1655–1658. Available from: https://doi.org/10.1002/adma.200305418.

[45] L. Wang, H. Dou, Z. Lou, T. Zhang, Encapsuled nanoreactors (Au@SnO2): a new sensing material for chemical sensors, Nanoscale 5 (2013) 2686. Available from: https://doi.org/10.1039/c2nr33088a.

[46] K.M. Kim, K. Il Choi, H.M. Jeong, H.J. Kim, H.R. Kim, J.H. Lee, Highly sensitive and selective trimethylamine sensors using Ru-doped SnO2hollow spheres, Sensors Actuators, B Chem. 166–167 (2012) 733–738. Available from: https://doi.org/10.1016/j.snb.2012.03.049.

[47] W.-S. Kim, B.-S. Lee, D.-H. Kim, H.-C. Kim, W.-R. Yu, S.-H. Hong, SnO2 nanotubes fabricated using electrospinning and atomic layer deposition and their gas sensing performance, Nanotechnology. 21 (2010) 245605. Available from: https://doi.org/10.1088/0957-4484/21/24/245605.

[48] N.G. Cho, D.J. Yang, M.J. Jin, H.G. Kim, H.L. Tuller, I.D. Kim, Highly sensitive SnO2 hollow nanofiber-based NO2 gas sensors, Sensors Actuators, B Chem. 160 (2011) 1468–1472. Available from: https://doi.org/10.1016/j.snb.2011.07.035.

[49] S. Cho, D.H. Kim, B.S. Lee, J. Jung, W.R. Yu, S.H. Hong, et al., Ethanol sensors based on ZnO nanotubes with controllable wall thickness via atomic layer deposition, an O2plasma process and an annealing process, Sensors Actuators, B Chem. 162 (2012) 300–306. Available from: https://doi.org/10.1016/j.snb.2011.12.081.

[50] S. Choi, G. Ankonina, D. Youn, S. Oh, J. Hong, Hollow ZnO nnanofibers fabricated using electrospun polymer templates, ACS Nano. 3 (2009) 2623–2631.

[51] S.J. Choi, I. Lee, B.H. Jang, D.Y. Youn, W.H. Ryu, C.O. Park, et al., Selective diagnosis of diabetes using Pt-functionalized WO3 hemitube networks as a sensing layer of

acetone in exhaled breath, Anal. Chem. 85 (2013) 1792–1796. Available from: https://doi.org/10.1021/ac303148a.

[52] N.G. Cho, I.D. Kim, NO2 gas sensing properties of amorphous InGaZnO4 submicron-tubes prepared by polymeric fiber templating route, Sensors Actuators, B Chem. 160 (2011) 499–504. Available from: https://doi.org/10.1016/j.snb.2011.08.017.

[53] N.G. Cho, H.-S. Woo, J.-H. Lee, I.-D. Kim, Thin-walled NiO tubes functionalized with catalytic Pt for highly selective C2H5OH sensors using electrospun fibers as a sacrificial template, Chem. Commun. 47 (2011) 11300. Available from: https://doi.org/10.1039/c1cc13876f.

[54] J. Cao, T. Zhang, F. Li, H. Yang, S. Liu, Enhanced ethanol sensing of SnO2 hollow micro/nanofibers fabricated by coaxial electrospinning, New J. Chem. 37 (2013) 2031. Available from: https://doi.org/10.1039/c3nj00096f.

[55] J. Zhang, S.W. Choi, S.S. Kim, Micro- and nano-scale hollow TiO2fibers by coaxial electrospinning: preparation and gas sensing, J. Solid State Chem. 184 (2011) 3008–3013. Available from: https://doi.org/10.1016/j.jssc.2011.09.014.

[56] J. Cao, H. Dou, H. Zhang, H. Mei, S. Liu, T. Fei, et al., Controllable synthesis and HCHO-sensing properties of In2O3micro/nanotubes with different diameters, Sensors Actuators, B Chem. 198 (2014) 180–187. Available from: https://doi.org/10.1016/j.snb.2014.03.015.

[57] Y.-E. Chang, D.-Y. Youn, G. Ankonina, D.-J. Yang, H.-G. Kim, A. Rothschild, et al., Fabrication and gas sensing properties of hollow SnO2 hemispheres, Chem. Commun. (2009) 4019. Available from: https://doi.org/10.1039/b902542a.

[58] I.D. Kim, A. Rothschild, D.J. Yang, H.L. Tuller, Macroporous TiO2 thin film gas sensors obtained using colloidal templates, Sensors Actuators, B Chem. 130 (2008) 9–13. Available from: https://doi.org/10.1016/j.snb.2007.07.092.

[59] N.G. Cho, I.S. Hwang, H.G. Kim, J.H. Lee, I.D. Kim, Gas sensing properties of p-type hollow NiO hemispheres prepared by polymeric colloidal templating method, Sensors Actuators, B Chem. 155 (2011) 366–371. Available from: https://doi.org/10.1016/j.snb.2010.12.031.

[60] I.D. Kim, A. Rothschild, T. Hyodo, H.L. Tuller, Microsphere templating as means of enhancing surface activity and gas sensitivity of CaCu 3Ti 4O 12 thin films, Nano Lett. 6 (2006) 193–198. Available from: https://doi.org/10.1021/nl051965p.

[61] H.G. Moon, Y.S. Shim, D. Su, H.H. Park, S.J. Yoon, H.W. Jang, Embossed TiO2thin films with tailored links between hollow hemispheres: synthesis and gas-sensing properties, J. Phys. Chem. C. 115 (2011) 9993–9999. Available from: https://doi.org/10.1021/jp2020325.

[62] H.G. Moon, Y.S. Shim, H.W. Jang, J.S. Kim, K.J. Choi, C.Y. Kang, et al., Highly sensitive CO sensors based on cross-linked TiO2hollow hemispheres, Sensors Actuators, B Chem. 149 (2010) 116–121. Available from: https://doi.org/10.1016/j.snb.2010.06.014.

[63] Z. Dai, C.-S. Lee, Y. Tian, I.-D. Kim, J.-H. Lee, Highly reversible switching from P- to N-type NO_2 sensing in a monolayer Fe_2 O_3 inverse opal film and the associated P–N transition phase diagram, J. Mater. Chem. A. 3 (2015) 3372–3381. Available from: https://doi.org/10.1039/C4TA05438E.

[64] Y. Xie, R. Xing, Q. Li, L. Xu, H. Song, Three-dimensional ordered ZnO – CuO inverse opals toward low concentration acetone detection for exhaled breath sensing, Sensors Actuators B. Chem. 211 (2015) 255–262. Available from: https://doi.org/10.1016/j.snb.2015.01.086.

[65] Z. Dai, L. Xu, G. Duan, T. Li, H. Zhang, Y. Li, et al., Fast-response, sensitivitive and low-powered chemosensors by fusing nanostructured porous thin film and IDEs-

microheater chip, Sci. Rep. 3 (2013) 1–7. Available from: https://doi.org/10.1038/srep01669.

[66] C.-S. Lee, Z. Dai, S.-Y. Jeong, C.-H. Kwak, B.-Y. Kim, D.H. Kim, et al., Monolayer Co3O4 inverse opals as multifunctional sensors for volatile organic compounds, Chem. A Eur. J. 22 (2016) 7102–7107. Available from: https://doi.org/10.1002/chem.201505210.

[67] Z. Dai, C.S. Lee, B.Y. Kim, C.H. Kwak, J.W. Yoon, H.M. Jeong, et al., Honeycomb-like periodic porous LaFeO3thin film chemiresistors with enhanced gas-sensing performances, ACS Appl. Mater. Interfaces. 6 (2014) 16217–16226. Available from: https://doi.org/10.1021/am504386q.

[68] R.W.J. Scott, S.M. Yang, G. Chabanis, N. Coombs, D.E. Williams, G.A. Ozin, Tin dioxide opals and inverted opals: near-ideal microstructures for gas sensors, Adv. Mater. 13 (2001) 1468–1472. doi:10.1002/1521-4095(200110)13:19 < 1468::AID-ADMA1468 > 3.0.CO;2-O.

[69] M. D'Arienzo, L. Armelao, A. Cacciamani, C.M. Mari, S. Polizzi, R. Ruffo, et al., One-step preparation of SnO2 and Pt-doped SnO2 as inverse opal thin films for gas sensing, Chem. Mater. 22 (2010) 4083–4089. Available from: https://doi.org/10.1021/cm100866g.

[70] A. Sutti, C. Baratto, G. Calestani, C. Dionigi, M. Ferroni, G. Faglia, et al., Inverse opal gas sensors: Zn(II)-doped tin dioxide systems for low temperature detection of pollutant gases, Sensors Actuators, B Chem. 130 (2008) 567–573. Available from: https://doi.org/10.1016/j.snb.2007.11.048.

[71] F. He, K. Zhao, Z. Huang, X. Li, G. Wei, H. Li, Synthesis of three-dimensionally ordered macroporous LaFeO3perovskites and their performance for chemical-looping reforming of methane, Cuihua Xuebao/Chinese J. Catal. 34 (2013) 1242–1249. Available from: https://doi.org/10.1016/S1872-2067(12)60563-4.

[72] J. Wang, Y. Xu, W. Xu, M. Zhang, X. Chen, Simplified preparation of SnO2inverse opal for Methanol gas sensing performance, Microporous Mesoporous Mater. 208 (2015) 93–97. Available from: https://doi.org/10.1016/j.micromeso.2015.01.038.

[73] R.Q. Xing, L. Xu, Y.S. Zhu, J. Song, W.F. Qin, Q.L. Dai, et al., Three-dimensional ordered SnO2inverse opals for superior formaldehyde gas-sensing performance, Sensors Actuators, B Chem. 188 (2013) 235–241. Available from: https://doi.org/10.1016/j.snb.2013.07.024.

[74] G.L. Messing, S.-C. Zhang, G.V. Jayanthi, Ceramic powder synthesis, J. Am. Ceram. Soc. 76 (1993) 2707–2726. Available from: https://doi.org/10.1111/j.1151-2916.1993.tb04007.x.

[75] Y. Ho Cho, Y. Chan Kang, J.H. Lee, Highly selective and sensitive detection of trimethylamine using wo3 hollow spheres prepared by ultrasonic spray pyrolysis, Sensors Actuators, B Chem. 176 (2013) 971–977. Available from: https://doi.org/10.1016/j.snb.2012.10.044.

[76] J.W. Yoon, J.S. Kim, T.H. Kim, Y.J. Hong, Y.C. Kang, J.H. Lee, A new strategy for humidity independent oxide chemiresistors: dynamic self-refreshing of In2O3 sensing surface assisted by layer-by-layer coated CeO2 nanoclusters, Small 12 (2016) 4229–4240. Available from: https://doi.org/10.1002/smll.201601507.

[77] K. Choi, H. Kim, Y. Chan, J. Lee, Chemical Ultraselective and ultrasensitive detection of H 2 S in highly humid atmosphere using CuO-loaded SnO 2 hollow spheres for real-time diagnosis of halitosis, Sensors Actuators B Chem. 194 (2014) 371–376.

[78] Y.H. Cho, X. Liang, Y.C. Kang, J.H. Lee, Ultrasensitive detection of trimethylamine using Rh-doped SnO2hollow spheres prepared by ultrasonic spray pyrolysis, Sensors

Actuators, B Chem. 207 (2015) 330–337. Available from: https://doi.org/10.1016/j.snb.2014.10.001.

[79] K.-I. Choi, S.-J. Hwang, Z. Dai, Y. Chan Kang, J.-H. Lee, Rh-catalyzed WO3 with anomalous humidity dependence of gas sensing characteristics, RSC Adv. 4 (2014) 53130–53136. Available from: https://doi.org/10.1039/C4RA06654E.

[80] H.J. Kim, H.M. Jeong, T.H. Kim, J.H. Chung, Y.C. Kang, J.H. Lee, Enhanced ethanol sensing characteristics of In2O3-decorated NiO hollow nanostructures via modulation of hole accumulation layers, ACS Appl. Mater. Interfaces. 6 (2014) 18197–18204. Available from: https://doi.org/10.1021/am5051923.

[81] K. Hieda, T. Hyodo, Y. Shimizu, M. Egashira, Preparation of porous tin dioxide powder by ultrasonic spray pyrolysis and their application to sensor materials, Sensors Actuators, B Chem. 133 (2008) 144–150. Available from: https://doi.org/10.1016/j.snb.2008.02.002.

[82] T. Hyodo, H. Inoue, H. Motomura, K. Matsuo, T. Hashishin, J. Tamaki, et al., NO2 sensing properties of macroporous In2O 3-based powders fabricated by utilizing ultrasonic spray pyrolysis employing polymethylmethacrylate microspheres as a template, Sensors Actuators, B Chem. 151 (2010) 265–273. Available from: https://doi.org/10.1016/j.snb.2010.09.002.

[83] A.A. Firooz, T. Hyodo, A.R. Mahjoub, A.A. Khodadadi, Y. Shimizu, Synthesis and gas-sensing properties of nano- and meso-porous MoO3-doped SnO2, Sensors Actuators, B Chem. 147 (2010) 554–560. Available from: https://doi.org/10.1016/j.snb.2010.03.021.

[84] J.W. Yoon, S.H. Choi, J.S. Kim, H.W. Jang, Y.C. Kang, J.H. Lee, Trimodally porous SnO2nanospheres with threedimensional interconnectivity and size tunability: a one-pot synthetic route and potential application as an extremely sensitive ethanol detector, NPG Asia Mater. 8 (2016) 1–8. Available from: https://doi.org/10.1038/am.2016.16.

[85] G. Sakai, N. Matsunaga, K. Shimanoe, N. Yamazoe, Theory of gas-diffusion controlled sensitivity for thin ® lm semiconductor gas sensor, Sensors Actuators B Chem. 80 (2001) 125–131. Available from: https://doi.org/10.1016/S0925-4005(01)00890-5.

[86] L. Wang, Z. Lou, R. Zhang, T. Zhou, J. Deng, T. Zhang, Hybrid Co3O4/SnO2 core-shell nanospheres as real-time rapid-response sensors for ammonia gas, ACS Appl. Mater. Interfaces. 8 (2016) 6539–6545. Available from: https://doi.org/10.1021/acsami.6b00305.

[87] P. Sun, Y. Sun, J. Ma, L. You, G. Lu, W. Fu, et al., Synthesis of novel SnO2/ZnSnO3core-shell microspheres and their gas sensing properities, Sensors Actuators, B Chem. 155 (2011) 606–611. Available from: https://doi.org/10.1016/j.snb.2011.01.017.

[88] H.M. Jeong, J.H. Kim, S.Y. Jeong, C.H. Kwak, J.H. Lee, Co3O4-SnO2 hollow hetero-nanostructures: facile control of gas selectivity by compositional tuning of sensing materials via galvanic replacement, ACS Appl. Mater. Interfaces. 8 (2016) 7877–7883. Available from: https://doi.org/10.1021/acsami.6b00216.

[89] Y.-J. Chen, C.-L. Zhu, L.-J. Wang, P. Gao, M.-S. Cao, X.-L. Shi, Synthesis and enhanced ethanol sensing characteristics of α-Fe_2 O_3/SnO_2 core–shell nanorods, Nanotechnology 20 (2009) 45502. Available from: https://doi.org/10.1088/0957-4484/20/4/045502.

[90] Y.J. Chen, G. Xiao, T.S. Wang, F. Zhang, Y. Ma, P. Gao, et al., α-MoO3/TiO2core/shell nanorods: controlled-synthesis and low-temperature gas sensing properties, Sensors Actuators, B Chem. 155 (2011) 270–277. Available from: https://doi.org/10.1016/j.snb.2010.12.034.

[91] S. Si, C. Li, X. Wang, Q. Peng, Y. Li, Fe2O3/ZnO core-shell nanorods for gas sensors, Sensors Actuators, B Chem. 119 (2006) 52–56. Available from: https://doi.org/10.1016/j.snb.2005.11.050.

[92] J. Zhang, X. Liu, L. Wang, T. Yang, X. Guo, S. Wu, et al., Synthesis and gas sensing properties of α-Fe2O3@ZnO core-shell nanospindles, Nanotechnology 22 (2011). Available from: https://doi.org/10.1088/0957-4484/22/18/185501.

[93] J. Cao, Z. Wang, R. Wang, S. Liu, T. Fei, L. Wang, et al., Synthesis of core–shell α-Fe_2 O_3 @NiO nanofibers with hollow structures and their enhanced HCHO sensing properties, J. Mater. Chem. A. 3 (2015) 5635–5641. Available from: https://doi.org/10.1039/C4TA06892K.

[94] J. Cao, Z. Wang, R. Wang, S. Liu, T. Fei, L. Wang, et al., Core–shell Co_3 O_4/α-Fe_2 O_3 heterostructure nanofibers with enhanced gas sensing properties, RSC Adv. 5 (2015) 36340–36346. Available from: https://doi.org/10.1039/C5RA03675E.

[95] L. Xu, R. Zheng, S. Liu, J. Song, J. Chen, B. Dong, et al., NiO@ZnO heterostructured nanotubes: coelectrospinning fabrication, characterization, and highly enhanced gas sensing properties, Inorg. Chem. 51 (2012) 7733–7740. Available from: https://doi.org/10.1021/ic300749a.

[96] S.W. Choi, J.Y. Park, S.S. Kim, Synthesis of SnO2-ZnO core-shell nanofibers via a novel two-step process and their gas sensing properties, Nanotechnology 20 (2009). Available from: https://doi.org/10.1088/0957-4484/20/46/465603.

[97] A. Katoch, S.-W. Choi, G.-J. Sun, S.S. Kim, An approach to detecting a reducing gas by radial modulation of electron-depleted shells in core–shell nanofibers, J. Mater. Chem. A. 1 (2013) 13588. Available from: https://doi.org/10.1039/c3ta13087h.

[98] S.W. Choi, A. Katoch, G.J. Sun, J.H. Kim, S.H. Kim, S.S. Kim, Dual functional sensing mechanism in SnO2-ZnO core-shell nanowires, ACS Appl. Mater. Interfaces 6 (2014) 8281–8287. Available from: https://doi.org/10.1021/am501107c.

[99] J.Y. Park, S.W. Choi, J.W. Lee, C. Lee, S.S. Kim, Synthesis and gas sensing properties of TiO2-ZnO core-shell nanofibers, J. Am. Ceram. Soc. 92 (2009) 2551–2554. Available from: https://doi.org/10.1111/j.1551-2916.2009.03270.x.

[100] I.S. Hwang, S.J. Kim, J.K. Choi, J. Choi, H. Ji, G.T. Kim, et al., Synthesis and gas sensing characteristics of highly crystalline ZnO-SnO2core-shell nanowires, Sensors Actuators, B Chem. 148 (2010) 595–600. Available from: https://doi.org/10.1016/j.snb.2010.05.052.

[101] N. Singh, A. Ponzoni, R.K. Gupta, P.S. Lee, E. Comini, Synthesis of In2O3-ZnO core-shell nanowires and their application in gas sensing, Sensors Actuators, B Chem. 160 (2011) 1346–1351. Available from: https://doi.org/10.1016/j.snb.2011.09.073.

[102] H.S. Woo, C.W. Na, I.D. Kim, J.H. Lee, Highly sensitive and selective trimethylamine sensor using one-dimensional ZnO-Cr 2O 3 hetero-nanostructures, Nanotechnology 23 (2012). Available from: https://doi.org/10.1088/0957-4484/23/24/245501.

[103] C.W. Na, S.Y. Park, J.H. Chung, J.H. Lee, Transformation of ZnO nanobelts into single-crystalline Mn 3O4 nanowires, ACS Appl. Mater. Interfaces. 4 (2012) 6565–6572. Available from: https://doi.org/10.1021/am301670x.

[104] C.W. Na, H.-S. Woo, H.-J. Kim, U. Jeong, J.-H. Chung, J.-H. Lee, Controlled transformation of ZnO nanobelts into CoO/Co3O4 nanowires, CrystEngComm. 14 (2012) 3737. Available from: https://doi.org/10.1039/c2ce06681e.

[105] H.-J. Kim, J.-H. Lee, Highly sensitive and selective gas sensors using p-type oxide semiconductors: overview, Sensors Actuators B Chem. 192 (2014) 607–627. Available from: https://doi.org/10.1016/j.snb.2013.11.005.

[106] L. Wang, Y. Kang, Y. Wang, B. Zhu, S. Zhang, W. Huang, et al., CuO nanoparticle decorated ZnO nanorod sensor for low-temperature H2S detection, Mater. Sci. Eng. C. 32 (2012) 2079–2085. Available from: https://doi.org/10.1016/j.msec.2012.05.042.
[107] S.J. Kim, C.W. Na, I.S. Hwang, J.H. Lee, One-pot hydrothermal synthesis of CuO-ZnO composite hollow spheres for selective H2S detection, Sensors Actuators, B Chem. 168 (2012) 83–89. Available from: https://doi.org/10.1016/j.snb.2012.01.045.
[108] X. Liang, T.H. Kim, J.W. Yoon, C.H. Kwak, J.H. Lee, Ultrasensitive and ultraselective detection of H2S using electrospun CuO-loaded In2O3nanofiber sensors assisted by pulse heating, Sensors Actuators, B Chem. 209 (2015) 934–942. Available from: https://doi.org/10.1016/j.snb.2014.11.130.
[109] N.S. Ramgir, C.P. Goyal, P.K. Sharma, U.K. Goutam, S. Bhattacharya, N. Datta, et al., Selective H2S sensing characteristics of CuO modified WO3thin films, Sensors Actuators, B Chem. 188 (2013) 525–532. Available from: https://doi.org/10.1016/j.snb.2013.07.052.
[110] T.S. Wang, Q.S. Wang, C.L. Zhu, Q.Y. Ouyang, L.H. Qi, C.Y. Li, et al., Synthesis and enhanced H2S gas sensing properties of α-MoO3/CuO p-n junction nanocomposite, Sensors Actuators, B Chem. 171–172 (2012) 256–262. Available from: https://doi.org/10.1016/j.snb.2012.03.058.
[111] C.W. Na, H.-S. Woo, I.-D. Kim, J.-H. Lee, Selective detection of NO2 and C2H5OH using a Co3O4-decorated ZnO nanowire network sensor, Chem. Commun. 47 (2011) 5148. Available from: https://doi.org/10.1039/c0cc05256f.
[112] S.W. Choi, A. Katoch, J.H. Kim, S.S. Kim, Prominent reducing gas-sensing performances of n-SnO2nanowires by local creation of p-n heterojunctions by functionalization with p-Cr2O3nanoparticles, ACS Appl. Mater. Interfaces. 6 (2014) 17723–17729. Available from: https://doi.org/10.1021/am504164j.
[113] C.H. Kwak, H.S. Woo, J.H. Lee, Selective trimethylamine sensors using Cr2O3- decorated SnO2 nanowires, Sensors Actuators, B Chem. 204 (2014) 231–238. Available from: https://doi.org/10.1016/j.snb.2014.07.084.
[114] D.R. Miller, S.A. Akbar, P.A. Morris, Nanoscale metal oxide-based heterojunctions for gas sensing: a review, Sensors Actuators, B Chem. 204 (2014) 250–272. Available from: https://doi.org/10.1016/j.snb.2014.07.074.
[115] T. Li, W. Zeng, Z. Wang, Quasi-one-dimensional metal-oxide-based heterostructural gas-sensing materials: a review, Sensors Actuators B Chem. 221 (2015) 1570–1585.
[116] M. Gillet, C. Lemire, E. Gillet, K. Aguir, The role of surface oxygen vacancies upon WO3 conductivity, Surf. Sci. 532–535 (2003) 519–525. Available from: https://doi.org/10.1016/S0039-6028(03)00477-1.
[117] C. Liu, B. Wang, T. Liu, P. Sun, Y. Gao, F. Liu, et al., Facile synthesis and gas sensing properties of the flower-like NiO-decorated ZnO microstructures, Sensors Actuators, B Chem. 235 (2016) 294–301. Available from: https://doi.org/10.1016/j.snb.2016.05.064.
[118] L. Zhang, Z. Gao, C. Liu, Y. Zhang, Z. Tu, X. Yang, et al., Synthesis of TiO_2 decorated Co_3 O_4 acicular nanowire arrays and their application as an ethanol sensor, J. Mater. Chem. A. 3 (2015) 2794–2801. Available from: https://doi.org/10.1039/C4TA06440B.
[119] J. Liu, S.Z. Qiao, J.S. Chen, X.W. (David) Lou, X. Xing, G.Q. (Max) Lu, Yolk/shell nanoparticles: new platforms for nanoreactors, drug delivery and lithium-ion batteries, Chem. Commun. 47 (2011) 12578. Available from: https://doi.org/10.1039/c1cc13658e.

[120] P. Rai, S.M. Majhi, Y.-T. Yu, J.-H. Lee, Noble metal@metal oxide semiconductor core@shell nano-architectures as a new platform for gas sensor applications, RSC Adv. 5 (2015) 76229–76248. Available from: https://doi.org/10.1039/C5RA14322E.
[121] F. Gyger, A. Sackmann, M. Hübner, P. Bockstaller, D. Gerthsen, H. Lichtenberg, et al., Pd@SnO2 and SnO2 @Pd Core@Shell nanocomposite sensors, Part. Part. Syst. Charact. 31 (2014) 591–596. Available from: https://doi.org/10.1002/ppsc.201300241.
[122] P. Rai, J.W. Yoon, C.H. Kwak, J.H. Lee, Role of Pd nanoparticles in gas sensing behaviour of Pd@In2O3 yolk–shell nanoreactors, J. Mater. Chem. A. 4 (2016) 264–269. Available from: https://doi.org/10.1039/C5TA08873A.
[123] P. Rai, J.-W. Yoon, H.-M. Jeong, S.-J. Hwang, C.-H. Kwak, J.-H. Lee, Design of highly sensitive and selective Au@NiO yolk–shell nanoreactors for gas sensor applications, Nanoscale 6 (2014) 8292–8299. Available from: https://doi.org/10.1039/C4NR01906G.
[124] E. Ogel, S.A. Müller, A. Sackmann, F. Gyger, P. Bockstaller, E. Brose, et al., Comparison of the catalytic performance and carbon monoxide sensing behavior of Pd-SnO2 core@shell nanocomposites, ChemCatChem. 9 (2017) 407–413. Available from: https://doi.org/10.1002/cctc.201601132.
[125] Y.J. Hong, M.Y. Son, Y.C. Kang, One-pot facile synthesis of double-shelled SnO2 yolk-shell-structured powders by continuous process as anode materials for li-ion batteries, Adv. Mater. 25 (2013) 2279–2283. Available from: https://doi.org/10.1002/adma.201204506.
[126] Y.J. Hong, M.Y. Son, B.K. Park, Y.C. Kang, One-pot synthesis of yolk-shell materials with single, binary, ternary, quaternary, and quinary systems, Small 9 (2013) 2224–2227. Available from: https://doi.org/10.1002/smll.201202840.
[127] Y.J. Hong, J.W. Yoon, J.H. Lee, Y.C. Kang, One-pot synthesis of Pd-loaded SnO2 yolk-shell nanostructures for ultraselective methyl benzene sensors, Chem. - A Eur. J. 20 (2014) 2737–2741. Available from: https://doi.org/10.1002/chem.201304502.
[128] J.W. Yoon, Y.J. Hong, G.D. Park, S.J. Hwang, F. Abdel-Hady, A.A.A. Wazzan, et al., Kilogram-scale synthesis of Pd-Loaded quintuple-shelled Co3O4 microreactors and their application to ultrasensitive and ultraselective detection of methylbenzenes, ACS Appl. Mater. Interfaces. 7 (2015) 7717–7723. Available from: https://doi.org/10.1021/acsami.5b00706.
[129] J.S. Kim, J.W. Yoon, Y.J. Hong, Y.C. Kang, F. Abdel-Hady, A.A. Wazzan, et al., Highly sensitive and selective detection of ppb-level NO2using multi-shelled WO3yolk-shell spheres, Sensors Actuators, B Chem. 229 (2016) 561–569. Available from: https://doi.org/10.1016/j.snb.2016.02.003.
[130] J.-W. Yoon, Y.J. Hong, Y. Chan Kang, J.-H. Lee, High performance chemiresistive H_2 S sensors using Ag-loaded SnO_2 yolk–shell nanostructures, RSC Adv. 4 (2014) 16067–16074. Available from: https://doi.org/10.1039/C4RA01364F.
[131] T.H. Kim, J.W. Yoon, Y.C. Kang, F. Abdel-Hady, A.A. Wazzan, J.H. Lee, A strategy for ultrasensitive and selective detection of methylamine using p-type Cr2O3: morphological design of sensing materials, control of charge carrier concentrations, and configurational tuning of Au catalysts, Sensors Actuators, B Chem. 240 (2017) 1049–1057. Available from: https://doi.org/10.1016/j.snb.2016.09.098.
[132] D. Zhang, Z. Liu, C. Li, T. Tang, X. Liu, S. Han, et al., Detection of NO2 down to ppb levels using individual and multiple In2O3 nanowire devices, Nano Lett. 4 (2004) 1919–1924. Available from: https://doi.org/10.1021/nl0489283.

[133] K.M. Kim, H.M. Jeong, H.R. Kim, K. Il Choi, H.J. Kim, J.H. Lee, Selective detection of NO2 using Cr-doped CuO nanorods, Sensors (Switzerland) 12 (2012) 8013–8025. Available from: https://doi.org/10.3390/s120608013.

[134] M.H. Seo, M. Yuasa, T. Kida, J.S. Huh, K. Shimanoe, N. Yamazoe, Gas sensing characteristics and porosity control of nanostructured films composed of TiO2 nanotubes, Sensors Actuators, B Chem. 137 (2009) 513–520. Available from: https://doi.org/10.1016/j.snb.2009.01.057.

[135] K. Suematsu, Y. Shin, Z. Hua, K. Yoshida, M. Yuasa, T. Kida, et al., Nanoparticle cluster gas sensor: controlled clustering of SnO2nanoparticles for highly sensitive toluene detection, ACS Appl. Mater. Interfaces. 6 (2014) 5319–5326. Available from: https://doi.org/10.1021/am500944a.

[136] T. Kida, S. Fujiyama, K. Suematsu, M. Yuasa, K. Shimanoe, Pore and particle size control of gas sensing films using SnO2 nanoparticles synthesized by seed-mediated growth: design of highly sensitive gas sensors, J. Phys. Chem. C. 117 (2013) 17574–17582. Available from: https://doi.org/10.1021/jp4045226.

[137] W.-Y. Chung, G. Sakai, K. Shimanoe, N. Miura, D.-D. Lee, N. Yamazoe, Preparation of indium oxide thin film by spin-coating method and its gas-sensing properties, Sensors Actuators B Chem. 46 (1998) 139–145. Available from: https://doi.org/10.1016/S0925-4005(98)00100-2.

[138] N.S. Baik, G. Sakai, K. Shimanoe, N. Miura, N. Yamazoe, Hydrothermal treatment of tin oxide sol solution for preparation of thin-film sensor with enhanced thermal stability and gas sensitivity, Sensors Actuators, B Chem. 65 (2000) 97–100. Available from: https://doi.org/10.1016/S0925-4005(99)00403-7.

[139] G. Sakai, N.S. Baik, N. Miura, N. Yamazoe, Gas sensing properties of tin oxide thin films fabricated from hydrothermally treated nanoparticles, Sensors Actuators B Chem. 77 (2001) 116–121. Available from: https://doi.org/10.1016/S0925-4005(01)00682-7.

[140] D.D. Vuong, G. Sakai, K. Shimanoe, N. Yamazoe, Hydrogen sulfide gas sensing properties of thin films derived from SnO 2 sols different in grain size, Sensors Actuators, B Chem. 105 (2005) 437–442. Available from: https://doi.org/10.1016/j.snb.2004.06.034.

[141] S. Capone, M.G. Manera, A. Taurino, P. Siciliano, R. Rella, S. Luby, et al., Fe3O4/γ-Fe2O3nanoparticle multilayers deposited by the langmuir-blodgett technique for gas sensors application, Langmuir. 30 (2014) 1190–1197. Available from: https://doi.org/10.1021/la404542u.

[142] W. Cheng, Y. Ju, P. Payamyar, D. Primc, J. Rao, C. Willa, et al., Large-area alignment of Tungsten oxide nanowires over flat and patterned substrates for room-temperature gas sensing, Angew. Chemie - Int. Ed. 54 (2015) 340–344. Available from: https://doi.org/10.1002/anie.201408617.

[143] S. Choudhury, C.A. Betty, K.G. Girija, S.K. Kulshreshtha, Room temperature gas sensitivity of ultrathin SnO[sub 2] films prepared from Langmuir-Blodgett film precursors, Appl. Phys. Lett. 89 (2006) 71914. Available from: https://doi.org/10.1063/1.2336725.

[144] S. Kumar, Z. Peng, H. Shin, Z.L. Wang, P.J. Hesketh, Ac dielectrophoresis of tin oxide nanobelts suspended in ethanol: manipulation and visualization, Anal. Chem. 82 (2010) 2204–2212. Available from: https://doi.org/10.1021/ac901287z.

[145] J. Guilera, C. Fàbrega, O. Casals, F. Hernández-Ramírez, S. Wang, S. Mathur, et al., Facile integration of ordered nanowires in functional devices, Sensors Actuators, B Chem. 221 (2015) 104–112. Available from: https://doi.org/10.1016/j.snb.2015.06.069.

[146] X. Li, E. Chin, H. Sun, P. Kurup, Z. Gu, Fabrication and integration of metal oxide nanowire sensors using dielectrophoretic assembly and improved post-assembly processing, Sensors Actuators, B Chem. 148 (2010) 404–412. Available from: https://doi.org/10.1016/j.snb.2010.05.062.
[147] X. Li, Y. Wang, Y. Lei, Z. Gu, Highly sensitive H2S sensor based on template-synthesized CuO nanowires, RSC Adv. 2 (2012) 2302. Available from: https://doi.org/10.1039/c2ra00718e.
[148] X. Li, Z. Gu, J. Cho, H. Sun, P. Kurup, Tin-copper mixed metal oxide nanowires: synthesis and sensor response to chemical vapors, Sensors Actuators, B Chem. 158 (2011) 199–207. Available from: https://doi.org/10.1016/j.snb.2011.06.004.
[149] S. Ghashghaie, E. Marzbanrad, B. Raissi, C. Zamani, R. Riahifar, Effect of low frequency electric field parameters on chain formation of ZnO nanoparticles for gas sensing applications, J. Am. Ceram. Soc. 95 (2012) 1843–1850. Available from: https://doi.org/10.1111/j.1551-2916.2012.05133.x.
[150] J. Esmaeilzadeh, S. Ghashghaie, P.S. Khiabani, A. Hosseinmardi, E. Marzbanrad, B. Raissi, et al., Effect of dispersant on chain formation capability of TiO2 nanoparticles under low frequency electric fields for NO2 gas sensing applications, J. Eur. Ceram. Soc. 34 (2014) 1201–1208. Available from: https://doi.org/10.1016/j.jeurceramsoc.2013.11.022.
[151] E.K. Heidari, C. Zamani, E. Marzbanrad, B. Raissi, S. Nazarpour, WO3-based NO2sensors fabricated through low frequency AC electrophoretic deposition, Sensors Actuators, B Chem. 146 (2010) 165–170. Available from: https://doi.org/10.1016/j.snb.2010.01.073.
[152] X. Liu, T.-J. Tarn, F. Huang, J. Fan, Recent advances in inkjet printing synthesis of functional metal oxides, Particuology 19 (2015) 1–13. Available from: https://doi.org/10.1016/j.partic.2014.05.001.
[153] M. Pashchanka, R.M. Prasad, R.C. Hoffmann, A. Gurlo, J.J. Schneider, Inkjet-printed nanoscaled CuO for miniaturized gas-sensing devices, Eur. J. Inorg. Chem. (2013) 1481–1487. Available from: https://doi.org/10.1002/ejic.201201152.
[154] X. Liu, Y. Shen, R. Yang, S. Zou, X. Ji, L. Shi, et al., Inkjet printing assisted synthesis of multicomponent mesoporous metal oxides for ultrafast catalyst exploration, Nano Lett. 12 (2012) 5733–5739. Available from: https://doi.org/10.1021/nl302992q.
[155] M.A. Andio, P.N. Browning, P.A. Morris, S.A. Akbar, Comparison of gas sensor performance of SnO2nano-structures on microhotplate platforms, Sensors Actuators, B Chem. 165 (2012) 13–18. Available from: https://doi.org/10.1016/j.snb.2011.12.045.
[156] C.J. Chang, S.T. Hung, C.K. Lin, C.Y. Chen, E.H. Kuo, Selective growth of ZnO nanorods for gas sensors using ink-jet printing and hydrothermal processes, Thin Solid Films. 519 (2010) 1693–1698. Available from: https://doi.org/10.1016/j.tsf.2010.08.153.
[157] J. Kukkola, M. Mohl, A.-R. Leino, G. Tóth, M.-C. Wu, A. Shchukarev, et al., Inkjet-printed gas sensors: metal decorated WO3 nanoparticles and their gas sensing properties, J. Mater. Chem. 22 (2012) 17878. Available from: https://doi.org/10.1039/c2jm32499g.
[158] B. Li, S. Santhanam, L. Schultz, M. Jeffries-EL, M.C. Iovu, G. Sauvé, et al., Inkjet printer chemical sensor array based on polythiophene conductive polymers, Sensors Actuators, B Chem. 123 (2007) 651–660. Available from: https://doi.org/10.1016/j.snb.2006.09.064.
[159] T.Y. Hill, T.L. Reitz, H. Huang, Controlling micro ceramic patterns via multiple/sequential drop-on-demand inkjet printing of dilute colloidal suspensions, Mater. Res. Bull. 84 (2016) 437–444. Available from: https://doi.org/10.1016/j.materresbull.2016.08.035.

[160] D. Sun, C. Chang, S. Li, L. Lin, Near-field electrospinning, Nano Lett. 6 (2006) 839–842. Available from: https://doi.org/10.1021/nl0602701.
[161] G. Zheng, W. Li, X. Wang, D. Wu, D. Sun, L. Lin, Precision deposition of a nanofibre by near-field electrospinning, J. Phys. D. Appl. Phys. 43 (2010) 415501. Available from: https://doi.org/10.1088/0022-3727/43/41/415501.
[162] D. Chen, S. Lei, Y. Chen, A single polyaniline nanofiber field effect transistor and its gas sensing mechanisms, Sensors. 11 (2011) 6509–6516. Available from: https://doi.org/10.3390/s110706509.
[163] F. Ruggieri, D. Di Camillo, L. Lozzi, S. Santucci, A. De Marcellis, G. Ferri, et al., Preparation of nitrogen doped TiO2 nanofibers by near field electrospinning (NFES) technique for NO2 sensing, Sensors Actuators, B Chem. 179 (2013) 107–113. Available from: https://doi.org/10.1016/j.snb.2012.10.094.
[164] Y. Wang, I. Ramos, Detection of moisture and methanol gas using a single electrospun tin oxide nanofiber, IEEE Sens. J. 7 (2007) 1347–1348. Available from: https://doi.org/10.1109/JSEN.2007.905045.
[165] T. Sahm, Flame spray synthesis of tin dioxide nanoparticles for gas sensing, Sensors Actuators B Chem. 98 (2004) 148–153. Available from: https://doi.org/10.1016/j.snb.2003.10.003.
[166] L. Mädler, A. Roessler, S.E. Pratsinis, T. Sahm, A. Gurlo, N. Barsan, et al., Direct formation of highly porous gas-sensing films by in situ thermophoretic deposition of flame-made Pt/SnO2 nanoparticles, Sensors Actuators B Chem. 114 (2006) 283–295. Available from: https://doi.org/10.1016/j.snb.2005.05.014.
[167] C. Liewhiran, N. Tamaekong, A. Wisitsoraat, S. Phanichphant, Highly selective environmental sensors based on flame-spray-made SnO2nanoparticles, Sensors Actuators, B Chem. 163 (2012) 51–60. Available from: https://doi.org/10.1016/j.snb.2011.12.097.
[168] T. Samerjai, N. Tamaekong, C. Liewhiran, A. Wisitsoraat, A. Tuantranont, S. Phanichphant, Selectivity towards H2 gas by flame-made Pt-loaded WO3 sensing films, Sensors Actuators, B Chem. 157 (2011) 290–297. Available from: https://doi.org/10.1016/j.snb.2011.03.065.
[169] T. Sahm, W. Rong, N. Bârsan, L. Mädler, U. Weimar, Sensing of CH4, CO and ethanol with in situ nanoparticle aerosol-fabricated multilayer sensors, Sensors Actuators, B Chem. 127 (2007) 63–68. Available from: https://doi.org/10.1016/j.snb.2007.07.001.
[170] W.D. Sproul, D.J. Christie, D.C. Carter, Control of reactive sputtering processes, Thin Solid Films. 491 (2005) 1–17. Available from: https://doi.org/10.1016/j.tsf.2005.05.022.
[171] M. Stankova, X. Vilanova, J. Calderer, E. Llobet, J. Brezmes, I. Gràcia, et al., Sensitivity and selectivity improvement of rf sputtered WO3microhotplate gas sensors, Sensors Actuators, B Chem. 113 (2006) 241–248. Available from: https://doi.org/10.1016/j.snb.2005.02.056.
[172] J.F. Chang, H.H. Kuo, I.C. Leu, M.H. Hon, The effects of thickness and operation temperature on ZnO: Al thin film CO gas sensor, Sensors Actuators, B Chem. 84 (2002) 258–264. Available from: https://doi.org/10.1016/S0925-4005(02)00034-5.
[173] H. Gong, J.Q. Hu, J.H. Wang, C.H. Ong, F.R. Zhu, Nano-crystalline Cu-doped ZnO thin film gas sensor for CO, Sensors Actuators, B Chem. 115 (2006) 247–251. Available from: https://doi.org/10.1016/j.snb.2005.09.008.
[174] M. Penza, C. Martucci, G. Cassano, NOx gas sensing characteristics of WO3 thin films activated by noble metals (Pd, Pt, Au) layers, Sensors Actuators B Chem. 50 (1998) 52–59. Available from: https://doi.org/10.1016/S0925-4005(98)00156-7.

[175] S.J. Ippolito, S. Kandasamy, K. Kalantar-zadeh, W. Wlodarski, Hydrogen sensing characteristics of WO3 thin film conductometric sensors activated by Pt and Au catalysts, Sensors Actuators B Chem. 108 (2005) 154–158. Available from: https://doi.org/10.1016/j.snb.2004.11.092.
[176] Z. Liu, T. Yamazaki, Y. Shen, T. Kikuta, N. Nakatani, Influence of annealing on microstructure and NO2-sensing properties of sputtered WO3thin films, Sensors Actuators, B Chem. 128 (2007) 173–178. Available from: https://doi.org/10.1016/j.snb.2007.06.001.
[177] J. Singh, D.E. Wolfe, Nano and macro-structured component fabrication by electron beam-physical vapor deposition (EB-PVD), J. Mater. Sci. 40 (2005) 1–27.
[178] M.H. Madhusudhana, A N. Chandorkar, Response study of electron-beam evaporated thin-film tin oxide gas sensors, Sensors Actuators B Chem. 9 (1992) 1–8.
[179] F. Teimoori, K. Khojier, N.Z. Dehnavi, Investigation on the electrical and methane gas-sensing properties of ZnO thin films produced by different methods, J. Electron. Mater. 45 (2016) 4881–4889. Available from: https://doi.org/10.1007/s11664-016-4709-9.
[180] J. Tamaki, C. Naruo, Y. Yamamoto, M. Matsuoka, Sensing properties to dilute chlorine gas of indium oxide based thin ® lm sensors prepared by electron beam evaporation, Sensors And Actuators. 83 (2002) 190–194.
[181] J.-M. Jeon, Y.-S. Shim, S.D. Han, D.H. Kim, Y.H. Kim, C.-Y. Kang, et al., Vertically ordered SnO_2 nanobamboos for substantially improved detection of volatile reducing gases, J. Mater. Chem. A. 3 (2015) 17939–17945. Available from: https://doi.org/10.1039/C5TA03293H.
[182] I.S. Hwang, J.K. Choi, H.S. Woo, S.J. Kim, S.Y. Jung, T.Y. Seong, et al., Facile control of C2H5OH sensing characteristics by decorating discrete Ag nanoclusters on SnO2 nanowire networks, ACS Appl. Mater. Interfaces. 3 (2011) 3140–3145. Available from: https://doi.org/10.1021/am200647f.
[183] S.-Y. Jeong, J.-W. Yoon, T.-H. Kim, H.-M. Jeong, C.-S. Lee, Y. Chan Kang, et al., Ultra-selective detection of sub-ppm-level benzene using Pd–SnO_2 yolk–shell micro-reactors with a catalytic Co_3 O_4 overlayer for monitoring air quality, J. Mater. Chem. A. 5 (2017) 1446–1454. Available from: https://doi.org/10.1039/C6TA09397C.
[184] H. Kwon, S.H. Lee, J.K. Kim, Three-dimensional metal-oxide nanohelix arrays fabricated by oblique angle deposition: fabrication, properties, and applications, Nanoscale Res. Lett. 10 (2015). Available from: https://doi.org/10.1186/s11671-015-1057-2.
[185] J. Chu, X. Peng, M. Sajjad, B. Yang, P.X. Feng, Nanostructures and sensing properties of ZnO prepared using normal and oblique angle deposition techniques, Thin Solid Films. 520 (2012) 3493–3498. Available from: https://doi.org/10.1016/j.tsf.2011.12.066.
[186] C. Oros, M. Horprathum, A. Wisitsoraat, T. Srichaiyaperk, B. Samransuksamer, S. Limwichean, et al., Ultra-sensitive NO2sensor based on vertically aligned SnO2nanorods deposited by DC reactive magnetron sputtering with glancing angle deposition technique, Sensors Actuators, B Chem. 223 (2016) 936–945. Available from: https://doi.org/10.1016/j.snb.2015.09.104.
[187] S. Hwang, H. Kwon, S. Chhajed, J.W. Byon, J.M. Baik, J. Im, et al., A near single crystalline TiO2 nanohelix array: enhanced gas sensing performance and its application as a monolithically integrated electronic nose, Analyst. 138 (2013) 443–450. Available from: https://doi.org/10.1039/C2AN35932D.

[188] H.G. Moon, Y.S. Shim, D.H. Kim, H.Y. Jeong, M. Jeong, J.Y. Jung, et al., Self-activated ultrahigh chemosensitivity of oxide thin film nanostructures for transparent sensors, Sci. Rep. 2 (2012) 1–7. Available from: https://doi.org/10.1038/srep00588.

[189] H.G. Moon, Y.R. Choi, Y.S. Shim, K. Il Choi, J.H. Lee, J.S. Kim, et al., Extremely sensitive and selective NO probe based on villi-like WO3nanostructures for application to exhaled breath analyzers, ACS Appl. Mater. Interfaces. 5 (2013) 10591–10596. Available from: https://doi.org/10.1021/am402456s.

[190] L. Li, T. Sasaki, Y. Shimizu, N. Koshizaki, Controlled cobalt oxide from two-dimensional films to one-dimensional nanorods and zero-dimensional nanoparticles: morphology-dependent optical carbon monoxide gas-sensing properties, J. Phys. Chem. C. 113 (2009) 15948–15954. Available from: https://doi.org/10.1021/jp905715g.

[191] K. Yao, R. Toole, P. Basnet, Y. Zhao, Highly sensitive double-layered nanorod array gas sensors prepared by oblique angle deposition, Appl. Phys. Lett. 104 (2014). Available from: https://doi.org/10.1063/1.4866047.

[192] K. Yoo, S. Han, H. Moon, S.-J. Yoon, C.-Y. Kang, Highly sensitive H2S sensor based on the metal-catalyzed SnO2 nanocolumns fabricated by glancing angle deposition, Sensors. 15 (2015) 15468–15477. Available from: https://doi.org/10.3390/s150715468.

[193] M. Horprathum, T. Srichaiyaperk, B. Samransuksamer, A. Wisitsoraat, P. Eiamchai, S. Limwichean, et al., Ultrasensitive hydrogen sensor based on Pt-decorated WO3 nanorods prepared by glancing-angle dc magnetron sputtering, ACS Appl. Mater. Interfaces. 6 (2014) 22051–22060. Available from: https://doi.org/10.1021/am505127g.

[194] T.S. Bjørheim, S. Stølen, T. Norby, Ab initio studies of hydrogen and acceptor defects in rutile TiO(2), Phys. Chem. Chem. Phys. 12 (2010) 6817–6825. Available from: https://doi.org/10.1039/b925823j.

[195] J.B. Varley, A. Janotti, A.K. Singh, C.G. Van de Walle, Hydrogen interactions with acceptor impurities in SnO2: first-principles calculations, Phys. Rev. B. 79 (2009) 245206. Available from: https://doi.org/10.1103/PhysRevB.79.245206.

[196] J.B. Varley, H. Peelaers, A. Janotti, C.G. Van de Walle, Hydrogenated cation vacancies in semiconducting oxides, J. Phys. Condens. Matter. 23 (2011) 334212. Available from: https://doi.org/10.1088/0953-8984/23/33/334212.

[197] A. Janotti, C.G. Van de Walle, Hydrogen multicentre bonds, Nat. Mater. 6 (2007) 44–47. Available from: https://doi.org/10.1038/nmat1795.

[198] S.M. George, Atomic layer deposition: an overview, Chem. Rev. 110 (2010) 111. Available from: https://doi.org/10.1021/cr900056b.

[199] A.J. Niskanen, A. Varpula, M. Utriainen, G. Natarajan, D.C. Cameron, S. Novikov, et al., Atomic layer deposition of tin dioxide sensing film in microhotplate gas sensors, Sensors Actuators, B Chem. 148 (2010) 227–232. Available from: https://doi.org/10.1016/j.snb.2010.05.018.

[200] C. Marichy, N. Donato, M.G. Willinger, M. Latino, D. Karpinsky, S.H. Yu, et al., Tin dioxide sensing layer grown on tubular nanostructures by a non-aqueous atomic layer deposition process, Adv. Funct. Mater. 21 (2011) 658–666. Available from: https://doi.org/10.1002/adfm.201001572.

[201] X. Du, S.M. George, Thickness dependence of sensor response for CO gas sensing by tin oxide films grown using atomic layer deposition, Sensors Actuators, B Chem. 135 (2008) 152–160. Available from: https://doi.org/10.1016/j.snb.2008.08.015.

[202] A. Rosental, A. Tarre, A. Gerst, J. Sundqvist, A. Hårsta, A. Aidla, et al., Gas sensing properties of epitaxial SnO2thin films prepared by atomic layer deposition, Sensors Actuators, B Chem. 93 (2003) 552–555. Available from: https://doi.org/10.1016/S0925-4005(03)00236-3.
[203] T.K. M. Jelínek, V. Myslík, M. Vrňata, R. Fryĉek, P. Fitl, F. Vysoužil, Conductive gas sensors prepared by PLD, in: NAPSB, 2011: pp. 1–5.
[204] A. Ponzoni, V. Russo, A. Bailini, C.S. Casari, M. Ferroni, A. Li Bassi, et al., Structural and gas-sensing characterization of tungsten oxide nanorods and nanoparticles, Sensors Actuators, B Chem. 153 (2011) 340–346. Available from: https://doi.org/10.1016/j.snb.2010.10.045.
[205] R. Dolbec, M.A. El Khakani, A.M. Serventi, R.G. Saint-Jacques, Influence of the nanostructural characteristics on the gas sensing properties of pulsed laser deposited tin oxide thin films, Sensors Actuators, B Chem. 93 (2003) 566–571. Available from: https://doi.org/10.1016/S0925-4005(03)00229-6.
[206] M. Stamataki, D. Tsamakis, N. Brilis, I. Fasaki, A. Giannoudakos, M. Kompitsas, Hydrogen gas sensors based on PLD grown NiO thin film structures, Phys. Status Solidi Appl. Mater. Sci. 205 (2008) 2064–2068. Available from: https://doi.org/10.1002/pssa.200778914.
[207] S.S. Sunu, E. Prabhu, V. Jayaraman, K.I. Gnanasekar, T. Gnanasekaran, Gas sensing properties of PLD made MoO3films, Sensors Actuators, B Chem. 94 (2003) 189–196. Available from: https://doi.org/10.1016/S0925-4005(03)00342-3.
[208] H. Kawasaki, T. Ueda, Y. Suda, T. Ohshima, Properties of metal doped tungsten oxide thin films for NOxgas sensors grown by PLD method combined with sputtering process, Sensors Actuators, B Chem. 100 (2004) 266–269. Available from: https://doi.org/10.1016/j.snb.2003.12.052.
[209] M.W. Ahn, K.S. Park, J.H. Heo, J.G. Park, D.W. Kim, K.J. Choi, et al., Gas sensing properties of defect-controlled ZnO-nanowire gas sensor, Appl. Phys. Lett. 93 (2008) 2006–2009. Available from: https://doi.org/10.1063/1.3046726.
[210] S.W. Choi, S.H. Jung, S.S. Kim, Functionalization of selectively grown networked SnO2nanowires with Pd nanodots by γ-ray radiolysis, Nanotechnology 22 (2011). Available from: https://doi.org/10.1088/0957-4484/22/22/225501.
[211] J.Y. Park, S.W. Choi, S.S. Kim, Junction-tuned SnO2 nanowires and their sensing properties, J. Phys. Chem. C. 115 (2011) 12774–12781. Available from: https://doi.org/10.1021/jp202113x.
[212] T.J. Hsueh, C.L. Hsu, S.J. Chang, I.C. Chen, Laterally grown ZnO nanowire ethanol gas sensors, Sensors Actuators, B Chem. 126 (2007) 473–477. Available from: https://doi.org/10.1016/j.snb.2007.03.034.
[213] Y. Qin, R. Yang, Z.L. Wang, Growth of Horizonatal ZnO Nanowire Arrays on Any Substrate, J. Phys. Chem. C 112 (2008) 18734–18736. https://doi.org/10.1021/jp808869j.
[214] I.S. Hwang, S.J. Kim, J.K. Choi, J.J. Jung, D.J. Yoo, K.Y. Dong, et al., Large-scale fabrication of highly sensitive SnO2nanowire network gas sensors by single step vapor phase growth, Sensors Actuators, B Chem. 165 (2012) 97–103. Available from: https://doi.org/10.1016/j.snb.2012.02.022.
[215] H. Long, A. Harley-Trochimczyk, T. He, T. Pham, Z. Tang, T. Shi, et al., In situ localized growth of porous tin oxide films on low power microheater platform for low temperature CO detection, ACS Sensors. 1 (2016) 339–343. Available from: https://doi.org/10.1021/acssensors.5b00302.

[216] Q. Wan, Q.H. Li, Y.J. Chen, T.H. Wang, X.L. He, J.P. Li, et al., Fabrication and ethanol sensing characteristics of ZnO nanowire gas sensors, Appl. Phys. Lett. 84 (2004) 3654–3656. Available from: https://doi.org/10.1063/1.1738932.

[217] F. Shao, J.D. Fan, F. Hernández-Ramírez, C. Fàbrega, T. Andreu, A. Cabot, et al., NH3 sensing with self-assembled ZnO-nanowire μhP sensors in isothermal and temperature-pulsed mode, Sensors Actuators, B Chem. 226 (2016) 110–117. Available from: https://doi.org/10.1016/j.snb.2015.11.109.

[218] C.J. Martinez, B. Hockey, C.B. Montgomery, S. Semancik, Porous tin oxide nanostructured microspheres for sensor applications, Langmuir. 21 (2005) 7937–7944. Available from: https://doi.org/10.1021/la050118z.

[219] S.E. Moon, H.K. Lee, N.J. Choi, H.T. Kang, J. Lee, S.D. Ahn, et al., Low power consumption micro C2H5OH gas sensor based on micro-heater and ink jetting technique, Sensors Actuators, B Chem. 217 (2015) 146–150. Available from: https://doi.org/10.1016/j.snb.2014.10.034.

[220] S. Santra, A.K. Sinha, A. De Luca, S.Z. Ali, F. Udrea, P.K. Guha, et al., Mask-less deposition of Au–SnO2 nanocomposites on CMOS MEMS platform for ethanol detection, Nanotechnology 27 (2016) 125502. Available from: https://doi.org/10.1088/0957-4484/27/12/125502.

[221] K. Kang, D. Yang, J. Park, S. Kim, I. Cho, H.-H. Yang, et al., Micropatterning of metal oxide nanofibers by electrohydrodynamic (EHD) printing towards highly integrated and multiplexed gas sensor applications, Sensors Actuators B Chem. 250 (2017) 574–583. Available from: https://doi.org/10.1016/j.snb.2017.04.194.

[222] B. Behera, S. Chandra, An innovative gas sensor incorporating ZnO – CuO nanoflakes in planar MEMS technology, Sensors Actuators B. Chem. 229 (2016) 414–424. Available from: https://doi.org/10.1016/j.snb.2016.01.079.

[223] A. Gurlo, R. Riedel, In situ and operando spectroscopy for assessing mechanisms of gas sensing, Angew. Chem. Int. Ed. 46 (2007) 3826–3848. Available from: https://doi.org/10.1002/anie.200602597.

[224] Z.Q. Zheng, J.D. Yao, B. Wang, G.W. Yang, Light-controlling, flexible and transparent ethanol gas sensor based on ZnO nanoparticles for wearable devices, Sci. Rep. 5 (2015) 1–8. Available from: https://doi.org/10.1038/srep11070.

[225] S.-J. Choi, J.-S. Jang, H.J. Park, I.-D. Kim, Optically sintered 2D RuO_2 nanosheets: temperature-controlled NO_2 reaction, Adv. Funct. Mater. 27 (2017) 1606026. Available from: https://doi.org/10.1002/adfm.201606026.

[226] M. Rieu, M. Camara, G. Tournier, J.P. Viricelle, C. Pijolat, N.F. de Rooij, et al., Fully inkjet printed SnO2gas sensor on plastic substrate, Sensors Actuators, B Chem. 236 (2016) 1091–1097. Available from: https://doi.org/10.1016/j.snb.2016.06.042.

[227] S. Vallejos, I. Gràcia, E. Figueras, J. Sánchez, R. Mas, O. Beldarrain, et al., Microfabrication of flexible gas sensing devices based on nanostructured semiconducting metal oxides, Sensors Actuators, A Phys. 219 (2014) 88–93. Available from: https://doi.org/10.1016/j.sna.2014.09.001.

Application of SMOX-based sensors

5

Kuniyuki Izawa[1], Heiko Ulmer[2], Anna Staerz[3], Udo Weimar[3] and Nicolae Barsan[4]
[1]Figaro Engineering Inc., Mino, Japan, [2]Sensirion AG, Stäfa, Switzerland, [3]University of Tübingen, Tübingen, Germany, [4]Institute for Theoretical and Physical Chemistry, University of Tübingen, Tübingen, Germany

Today, SMOX-based sensors are used in a wide array of applications. This chapter will begin with a review on the historical development of gas-based sensors and the emergence of SMOX-based sensors. The current and possible future applications will then be reviewed. The requirements which commercial SMOX-based sensors must meet will be discussed. The chapter offers a comprehensive overview of the development and application of SMOX sensors.

5.1 Application of SMOX-based sensors

5.1.1 The history of gas and the ensuing need for detection

For centuries, the ideas of Greek philosophers dominated chemistry. Alchemists attempted to explain chemical phenomena using modifications of Aristotle's four basic elements: earth, wind, air, and fire [1].

It was not until the early 17th century, that Jan Baptista van Helmont first identified substances which are "more subtle or fine ... than a vapor, mist, or distilled oiliness, although... many times thicker than air" as gases (the Greek word for chaos) [2]. Only after the scientific revolution and the work of Lavoisier, in the late 18th century, did "gas" become a standard chemical term [2]. Lavoisier noted that burning phosphor and sulfur in air resulted in a weight increase, indicating that air could also be "captured" during combustion, leading him to believe that air reacted with the burning material. Later, Lavoisier named the part of air which reacts with the metallic or organic substance during combustion, oxygen (Greek for "acid-generator" because he incorrectly thought oxygen was necessary for all acids) [1]. A contemporary of Lavoisier, Joseph Priestley, identified that oxygen was necessary for respiration. In the 1780s, based on experiments with a guinea pig, Lavoisier found a quantitative association between inhaled oxygen and exhaled CO_2 [3]. In the 1850s, Pettenkoffer, the first professor of Hygiene at the Ludwig-Maximilians-University of Munich, suggested that the indoor CO_2 concentration should, thus, be monitored (every 15 minutes) [4]. He suggested that a CO_2 concentration under 1000 ppm (700 in bedrooms) indoors indicated adequate

Gas Sensors Based on Conducting Metal Oxides. DOI: https://doi.org/10.1016/B978-0-12-811224-3.00005-6

ventilation [3]. This marked the beginning of indoor air-quality monitoring. The indoor CO_2 concentration is still used today as a quick (albeit indirect) indicator of air quality.

The work of Lavoisier and his contemporaries on combustion also laid the necessary groundwork for the second industrial revolution. At the turn of the 19th century, Philippe Lebon first developed and patented a lamp powdered by gas extracted from wood [5]. Around the same period, William Murdoch an employee of Boulton and Watt, the Birmingham steam engine manufacturing firm, built an illumination device powered by inflammable gas from coal [5]. The use of coal lighting prospered and by 1820, the London and Westminster Gas Light and Coke Company provided over 30,000 customers with gas [5]. As a result, the need for coal in Britain skyrocketed and mining rose from 2.43 million tons in 1700–224 million tons in 1900 [6]. Before the first industrial revolution, coal mines were near the surface, but in order to meet the increased needs, mines became deeper and more dangerous [6]. Deeper coal seams have a higher gas content than surface coal [7], in particular, heightened concentrations of methane (firedamp) and H_2S (stinkdamp) [8]. Methane is a colorless, odorless and explosive gas. H_2S forms flammable mixtures with air in the range of 4.5%–45%. It is highly toxic and can cause death at exposures above 500 ppm [9]. Mining disasters became more common. In 1907, there were 18 mine disasters in the United States. The Monograph coal mine explosion alone claimed the lives of 326 miners [10]. A detection method was necessary to lower the fatality rate. In the early 1900s, John Scott Haldane, a physiologist, suggested that miners take small animals, such as canaries, into mines in order to detect dangerous levels of gas in their working environment [11]. Canaries have a loud chirp. If the birds stopped chirping and began to rattle their cage then there was a gas leak, and the miners were warned [11]. In the 1920s significant advances were made in the field of gas detection [12], and in 1925, Dr. Jiro Tsuji, founder of Reiken Keiki, developed a method for detecting combustible gases based on light wave interference [12]. Other technologies rapidly followed. In 1957, Oliver Johnson patented a gas detection device based on the combustion of gases on a hot platinum wire (pellistor) [13]. In 1968, Naoyoshi Taguchi brought the first SMOX-based gas sensors on the market for the detection of explosive gases.

In addition to increased mining, the widespread use of domestic gas for lighting also presented risks for people. If the appliances were not properly maintained, incomplete combustion resulted in the formation of CO, which is toxic and binds 20 times more strongly to hemoglobin than oxygen [14]. The ability to sense CO became relevant since the discovery of fire, and the human olfactory system has not evolved to sense it. As a result, in the 19th century many people unknowingly suffered from chronic carbon monoxide poisoning. One famous example is the American author Edgar Allan Poe, who possibly suffered from multiple chemical sensitivity as a result of a leaky gas lamp fixture in his home [15]. In a portrait at the Maryland historical society, Poe is depicted with a drooping eye and mouth which could be attributed to neurological damage as a result of prolonged exposure to the toxic gas [16]. In fact, the dark imagery for which Poe's later works are famous may be the result of delusion and paranoia, two symptoms commonly attributed to chronic CO poisoning [15]. Additional symptoms include fatigue, nausea, memory impairment and even hearing loss [17]. After being found delirious on the streets of Baltimore, Poe died in 1849. Some experts believe that Poe's death was the direct result of his chronic exposure to CO [16]. Despite the widespread poisoning, it was not until 1925 that the first carbon monoxide detector was patented by Chester Gordon and James Lowe, from the American Telephone and Telegraph Company [18]. The detector was based on a color change as a result of the reaction of palladium chloride with CO and only offered a single-use solution [18]. A slew of "reagent cards" for the detection of CO followed, the reaction did not, however, necessarily indicate the CO concentration. Exposure to low concentrations over an extended period of time resulted in the same result as short exposures to high concentrations [19]. The number of CO-related deaths continued to increase until the 1980s [20]. As a result, audible detectors were needed. In 1980, the marketing of TGS203 by Figaro Engineering Inc. began. The carbon monoxide sensor was based on tin dioxide [21]. As of 1999, it was one Figaro's bestselling tin dioxide-based sensors [21]. In the 1990s, Motorola developed a SMOX-based sensor for the detection of CO, the MGS1100. The mass-produced sensor was based on silicon micromachining, a technology patented by Motorola. By 1998, however, Motorola ceased work on the chemical sensor [22]. Despite initially seeming promising, due to lack of selectivity SMOX-based sensors are not widely used for CO detection today. Even companies like Figaro began to research more selective options for CO detection. As a result, Figaro developed and launched electrochemical CO sensors in the middle of 2000s. In 2016, Figaro announced that it was taking TGS203 and TGS2442 (both SMOX-based CO detectors) off the market [23]. The only SMOX-based sensor now available from Figaro for CO is the TGS3870 SMOX, which is advertised as a sensor for CO and methane [24]. In 2014, at the 15th international Meeting on Chemical Sensors, a Figaro representative held a talk about a newly developed electrochemical CO sensor, characterized by ultra-compactness, resulting from a water-reservoir-less structure. In April 2016, Figaro brought the ultra-compact electrochemical CO sensor, TGS5141, onto the market [25].

By the turn of the 19th century, domestic cooking and heating also became important uses for coal gas [26]. Although natural gas has a higher energy content than manufactured gas, and the first natural gas company, the Fredonia Gas Light Company, was founded in 1858, coal gas remained popular because the necessary infrastructure was already in place [26]. In order to effectively use natural gas, transportation pipelines were required, and it was not until after World War II that a large interstate pipeline system for natural gas was brought online in the United States [26]. The last nonindustrial manufactured gas works in the United States closed in the 1960s [26]. At the turn of the 20th century, research had begun on another fuel, liquid propane gas (LPG), and in 1913 an American chemist, Walter O. Snelling, patented its production on an industrial scale [27]. The commercial production began in the 1920s, and in 1922 the US sales topped 223 thousand gallons [27]. By 1929, the sales had skyrocketed to almost 10 million gallons in the United States alone [27]. By the 1960s, the use of LPG and natural gas to power lighting, heating, and for cooking was on the rise in developed nations [28].

5.1.1.1 *The beginning of SMOX-based sensors used for domestic safety*

As the domestic use of LPG and natural gas continued to increase, so did the number of explosions as a result of leaks. A gas sensor was needed to protect people from gas leaks and explosion accidents [29]. It is known that the dawn of SMOX-based gas sensors was in 1962, when Seiyama et al. at Kyushu University published that the resistance change of ZnO thin films, as a result of gas exposure, could be used to detect gaseous components [30]. This publication was the first report of gas sensors based on the resistance changes of metal oxides from adsorption and reactions of gases on the metal-oxide surface. It marks the beginning of R&D on SMOX-based gas sensors, a field which even today, remains active internationally. In the same year, Taguchi, an independent inventor, filed a patent for a SMOX-based gas sensor [31]. His invention was largely driven by the increased number of domestic gas explosions in Japan as a result of LPG use for cooking. In 1968, after a long period of trial and error, Taguchi finally completed the development of the world's first commercial SMOX-based gas sensors, the TGS (Taguchi Gas Sensor) 109. It was based on tin dioxide sensitized by supporting palladium oxide. TGS109 was manufactured for residential gas alarms. At that time, existing residential gas alarms used catalytic-type gas sensors which were originally developed for mine safety and had been commercialized in Japan since 1963. Despite the high number of domestic accidents, these gas alarms were not widely affordable [32]. The gas alarms using TGS109, however, were inexpensive, costing roughly about one-third as the previous gas alarms [32]. Unlike catalytic-type gas sensors, SMOX-based gas sensors do not require a circuit to amplify the gas sensor output signal, because the sensor signal as a response to target gases is so substantial. Furthermore, TGS109 was designed to enable a gas alarm to operate without a high-cost circuit for voltage stabilization. The more widespread use of domestic gas sensors lead to a drastic decrease in gas accidents (Fig. 5.1).

5.1.1.2 *In-cabin air quality monitoring*

After their discovery, the next big developmental push for SMOX-based gas sensors came in the 1980s. A sensor was needed to regulate the air intake into the car passenger cabin. Car exhaust fumes contain toxic components like CO, NO_x, and many different hydrocarbons [33]. In order to automatically prevent these toxins from entering the passenger cabin, a sensor-controlled ventilation system was required. The sensing environment for automotive sensors is extremely harsh, due to the presence of not only the exhaust gases, but also water, salt, dust/sand, and extreme temperatures (−40°C−125°C) in addition to vibration and shock [34]. In order to be used in automobiles, the packaging must be extremely robust and the sensor easily integrated. In addition, to minimize cost, only sensors which result from standardized high-volume manufacturing processes can be successfully introduced into the automotive industry [34].

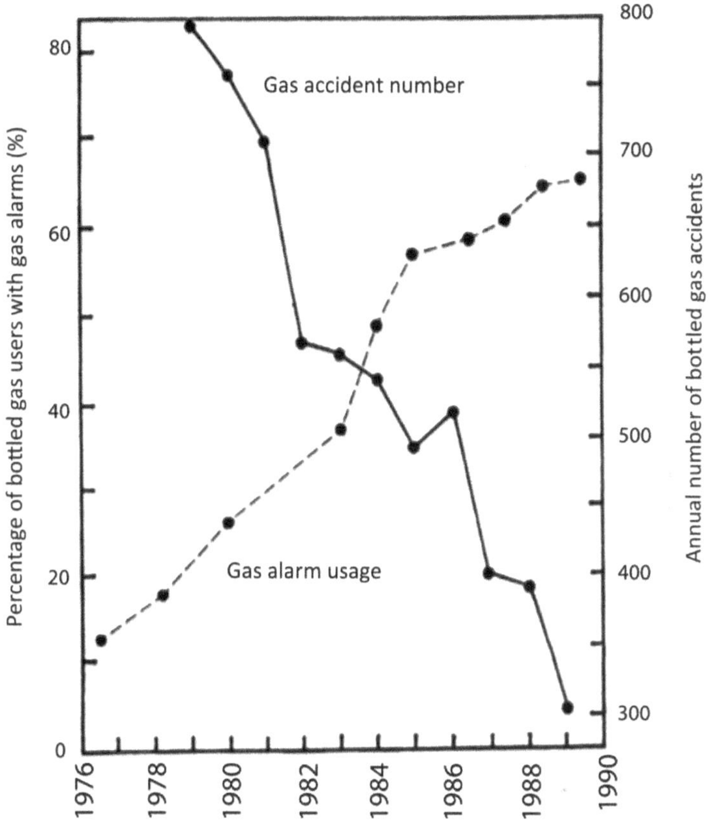

Figure 5.1 Shows the correlation between the decline in gas explosions and the use of gas alarms in Japan. Rights still must be obtained (see [29], Figure 1, obtained from publicity materials by the gas alarm industries association of Japan, May 1990).

The first control unit was developed in 1986 for gasoline-powered BMWs [35]. The unit contained a commercially available sensor from Figaro (TGS822) which detects volatile organic compounds and CO. The system was followed by a two-sensor system from Bosch that could also detect nitrous oxides, making it useful for diesel cars [35]. In the 1990s, the one-sensor system from Paragon Electronic GmbH, now Paragon AG, entered the market [36]. As use in the automotive industry increased, the pressure for reduced costs became higher, driving the development of smaller and more easily packaged sensors. This resulted in the combination of gas-sensing thick films with microelectromechanical system (MEMS) substrates [35]. Research began in 1997 on this approach and today this type of sensor production is standard [35,37]. The sensing area can be heated in continuous mode, using only 40 mW. By thermally decoupling the sensing area, the elemental frame stays cool, enabling a cost-effective direct mounting on the printed circuit [38]. Another noteworthy technology that contributed to the development of automotive cabin air

Figure 5.2 In this schematic, the location of the air cabin flap module is shown.

quality control systems was the sensor output signal processing for detecting sensor response signals to relatively low concentration of gases under noisy environments, where variations in ambient temperature and humidity cause fluctuations of sensor signal. Dr. Hanns Rump was one of those who developed fundamental sensor signal processing technology in the early stages of this automotive application [39]. From this important application of SMOX-based sensors, one can learn that a successful gas sensor application needs both good gas sensor technology and excellent signal processing technology. This is true and just as valid for today's various other applications of gas sensors (Fig. 5.2).

5.1.2 Applications in development and future applications

5.1.2.1 Advancement in in-cabin monitoring and inclusion on airplanes and space stations

Today, the automotive HVAC market is mature and the percentage of cars with semi- or fully-automatic HVAC systems is forecasted to increase each year [40]. In terms of volume, it is still the most important application for SMOX-based sensors. Ideally, in the future, the electronic control systems (ECU) would not only prevent combustion gases from entering the cabin, but would also monitor gases originating in the car cabin, for example, bioeffluents, smoke, etc. In 2006, Ulmer et al. examined the applicability of a SMOX-based sensor module [41] (Fig. 5.3).

The situation in the car is more complicated than outside the car. Gases originating in the car cabin are more varied and are usually present in low concentrations. To address these difficulties, Ulmer et al. suggested a two-step approach. In the first step, the total VOC concentration is determined. In the next step, the response of an array containing three SMOX-based sensors to the odor is evaluated, using multivariate data analysis. The result is compared to stored patterns of typical odor events. Based on the similarity of the results to the saved pattern, a weight factor is then applied [38]. Overall, the results of the SMOX-based array were in good accordance with the human-sensory impression. Using the two-step approach, it

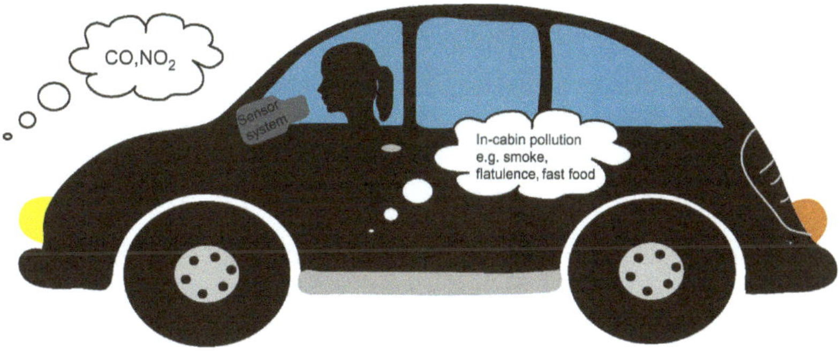

Figure 5.3 This schematic shows where, in the future, the two gas sensor modules should be used in the automobile.

was possible to differentiate between different in-cabin odor origins, such as fast food, smoking, and bioeffluents [38].

ECUs similar to those already implemented in automobiles may be useful for other applications, for example, in-cabin air control of passenger airplanes. To ensure a comfortable journey for all on board, there must be a continuous supply of fresh air in the passenger cabin during flight. Good air quality is ensured by replacing the complete cabin volume with outside air in less than three minutes [42]. Air quality measurements are currently not performed [42]. The ventilation system in a passenger airplane is very complex. The atmosphere on earth is comprised mainly of oxygen (21%) and nitrogen (78%) [43]. Most passenger airplanes fly between 24,000 and 40,000 ft. At this altitude, the atmospheric oxygen pressure is so low that after 5 minutes a person's blood oxygen saturation would be reduced to around 50% [43]. In order to ensure that passengers' blood remains saturated with oxygen during the flight, passenger planes have air-pressurized cabins. During flight, the compressed air is usually taken from the engine (bleed air) [44]. The concentration of ozone is much higher at flight altitude than at sea level, making it necessary to catalytically convert ozone into oxygen. Engine defects or saturation of the catalytic converter could result in contamination of in-cabin air; for example, through the presence of lubrication oil vapor or ozone [42]. System failures could be discovered by routinely monitoring air for traces of ozone and lubrication oil. Additionally, the collection of the bleed air through the monitors creates friction which must be compensated by additional thrust, leading to increased fuel consumption [42]. By regularly monitoring the air quality within the passenger cabin, ventilation could be adjusted as needed, reducing fuel consumption and ensuring good air quality. Within the realm of the European research program Clean Sky 2, the Fraunhofer IBP is working on a sensor system that enables the fresh air supply to be controlled, based on the existing air quality. The system uses an optimized metal oxide–based sensor [45]. The same qualities that make SMOX-based sensors attractive for application in the ECUs of cars, small size, robustness, and fast response time, make them a practical option for application in airplanes. Large aeronautic companies, such as Boeing, are interested in integrating air quality detectors into airplanes, if the sensor systems are proven reliable and accurate [46].

The issue of providing good passenger air is exacerbated in space. The international space station flies approximately 248 miles over earth and has been continually occupied since 2000. Over time, microorganisms are known to grow on space stations. After 15 years of operation, 108 bacterial and 126 fungal species were identified on the Russian space station MIR [47]. The growth of microorganisms can cause health problems for crew members and can lead to corrosion of the station [48]. In order to prevent the widespread contamination of microorganisms aboard the ISS, it is important to monitor the progression of microbial growth while the station is active and in flight. Since 2009, the DLR (German Aerospace Center) has been working on a portable detection system to be used on board the ISS to monitor microbial contamination [49]. Microbial volatile organic compounds (mVOCs) are produced as a result of a microbes' metabolism [48]. Different microbes produce varying quantities and combinations of mVOCS. The commercially available portable e-nose from AIRSENSE Analytics GmbH, Schwerin, Germany contains 10 metal oxide–based sensors that respond to hydrogen, sulfur, natural gas, alcohols, chlorine, aromatic, and aliphatic gases [48]. An odor fingerprint attributed to a certain microbe can be recognized through a complex evaluation of the sensor responses [48]. A similar system, the Russian module Zvesda, has already been tested for several months aboard the ISS. The first results are very promising.

Because they are small, robust and inexpensive, SMOX modules are already widely used for automobile passenger comfort. They have also proven to be helpful on the ISS to protect the health of astronauts and extend the life of the space station, by monitoring microbial contamination. In the future, potentially SMOX-based sensor arrays to could be used to increase passenger comfort and reduce fuel consumption on airplanes.

5.1.2.2 Indoor air quality monitoring

Today the average American spends 69% of their time indoors [50]. Indoor air is often more polluted than the outdoors [3], and as modern homes are made more ecologically friendly they are also more airtight. Thus, the need for active ventilation to prevent the build-up of indoor pollution is of utmost importance. Already in the 1850s, Pettenkoffer, the first professor for Hygiene at the Ludwig-Maximilians-University of Munich, suggested that the indoor CO_2 concentration could be used as a marker of indoor air quality [4]. Today, there are numerous different domestic CO_2 sensors on the market. Most domestic CO_2 sensors currently available are based on nondispersive infrared (NDIR) sensors. Optical detection methods, such as NDIR, are expensive, making them less suited for widespread domestic use. A cheap alternative is needed. CO_2 does not oxidize or reduce the metal oxide surface, making detection difficult with SMOX-based sensors [51]. By adding materials, such as barium or lanthanum, that have a high affinity for CO_2, the sensing properties of SnO_2 could, however, be drastically increased [52,53]. This could be used to create SMOX-based gas sensors for CO_2 in the future.

Although CO_2 is still a commonly used indicator of indoor air pollution, it is a nontoxic gas and only an indirect indicator, as its main source is the building occupants. In 2004, the European Union released a guideline for indoor air quality identifying CO, the VOCs formaldehyde, benzene and naphthalene, as well as the oxidizing gas nitrogen dioxide, as first-priority indoor pollutants [54]. Domestic VOCs can originate from paints, varnishes, cleaning products, etc. [55]. The concentration of VOCs has not only been found to correlate to the CO_2 concentration indoors, but also allows for the detection of odorous events unrelated to level of human occupation, making it a more accurate indicator of indoor air quality [56]. VOCs are potential contributors to sick building syndrome [57]. The term is applied when multiple tenants of a building have nonspecific health issues, such as nausea, irritated eyes, nose, skin, and throat [57], and the syndrome first appeared in the 1970s. As a result of the Arab Oil Embargo, there was an attempt to reduce energy consumption, and indoor ventilation was reduced to 5 cfm per occupant (today, a minimum of 15 cfm per person is recommended). It is believed that this strong reduction in ventilation led to the appearance of sick building syndrome [58].

To prevent a widespread reoccurrence of sick building syndrome, good ventilation is more important today than ever before. Some governments, such as France, have passed stringent requirements for the energy consumption of a new building, but have neglected the wellbeing of its occupants [59]. Airtight homes are, overall, more energy-efficient than conventional housing. It is important, however, that mechanical ventilation systems work constantly in airtight homes; if not, poor air quality can negatively affect the wellbeing of the occupants [59]. The individual and varying habits of people make fixed programs for ventilation unpractical, and require inhabitants to interact with the control system to optimize air quality [60]. A study done on newly built energy-efficient homes in France found that mechanical ventilation systems were not user-friendly and most inhabitants used the manual mode instead of the programmable mode [59]. Overall, smart home systems remain complex to operate and the integration of elements from different systems can be very difficult [61]. One solution would be fully automated systems based on widely distributed sensors that could react locally to events. In fact, Sir James Dyson, a billionaire inventor, predicts that within the next 10 years, the "crude controls" in place now will no longer be necessary. Instead, homes will automatically adjust to our momentary needs [62].

For domestic application it is only necessary to detect the presence of VOCs in general, and not to differentiate between the different substances. The high and rapid sensor response of some metal oxides to VOCs make SMOX-based sensors ideal. In the middle of the 1980s, Matsushita Seiko, now Panasonic Ecology Systems [63], and other Japanese home appliance manufacturers began to use SMOX sensors for automatic control of residential air purifiers, these were based on indoor air quality measurements by SMOX sensors for cigarette smoke and VOCs in combination with signal processing software. They were followed by many other manufacturers of air purifiers in Korea and China. Today, indoor air quality monitoring for air purifiers is one of the most important applications of SMOX-based sensors, comparable to the automotive application in terms of volume. In the future, indoor air quality could be inexpensively monitored at numerous points within the homes using SMOX-based VOCs sensors. Domestic VOC monitors based on SMOX are already commercially available. At the Sensor + Test 2015, AMS released its iAM USB [64]. The compact VOCs monitor plugs into your USB port and costs under US$30 (AMS homepage 15.11.2016). Dyson already offers innovative air purifiers that can be connected with a smartphone, from which indoor air quality measured by a particle sensor and a SMOX sensor for VOCs both built in the air purifier can be monitored remotely, and whose airflow can be adjusted accordingly.

In 2015, 70% of questioned consumers were excited about the potential cost savings and energy efficiency of smart homes [65]. Through the wide distribution of SMOX-based VOC sensors in the home, the necessary monitoring to enable fully automated ventilation could be realized relatively inexpensively.

5.1.2.3 *Industrial safety*

Coal mines are often cited as the origin of gas sensors, see Section 5.1.1. Despite the numerous different sensing methods available today, gas monitoring, however, within mines remains challenging. Due to the vast lengths and depths of coal mine tunnels, a large sampling density is necessary to ensure the safety of miners. In the past, sensors needed to be directly connected to the processing server, making a large amount of cable and coordination necessary. Today, research exists on wireless monitoring systems [8]. These would be much more flexible and would allow for widespread detection within the mine. For this application, SMOX-based sensors are ideally suited because they are small and inexpensive. They are robust and can handle the harsh environment within mines. The first entirely wireless prototype detection system for mines contains SMOX sensors for CO, methane, propane, and butane. Researchers are optimistic that the entire sensing system (electronics and a sensor node) will cost under $400, with each additional sensor node only costing $250 [8]. By using MEMS technology and pulsed heating, SMOX-based sensors can be powered for several years using a standard battery. Battery-operated detectors not only lead to an increase in stationary devices, but also enabled the development of small portable personal safety devices.

In addition to miners, garbage collectors, and firefighters are in the top ten dangerous jobs in many developed countries. Exposure to toxic gases plays a role in the high mortality rates of all three jobs [66]. Personal safety apparel is already commonplace in these dangerous professions, and textiles are positioned close to the body. This combination makes clothing the ideal place to integrate and couple safety sensors with biometric measurements. Small battery-operated and wireless SMOX-based sensors are well suited, and there are already several prototypes for different smart safety clothing articles. The European project Proetex was aimed at developing smart gear for firefighters. A prototype safety boot with an integrated gas sensor was developed [67].

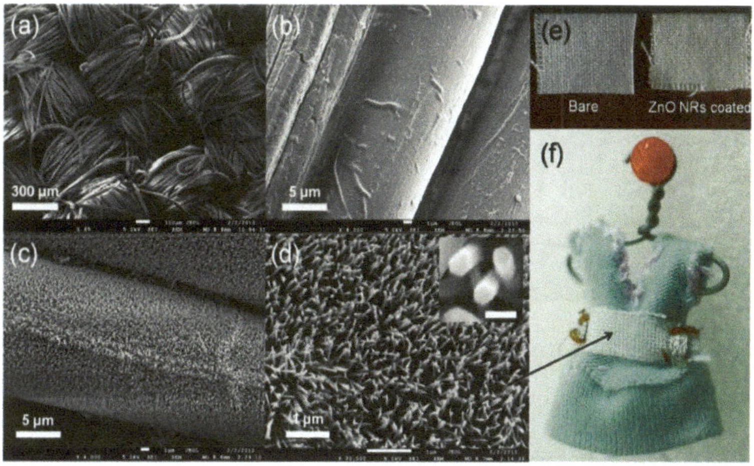

Figure 5.4 Here the integration of SMOX fibers in fabric is shown. SEM images of (A) cotton fabric substrate; (B) bare cotton fibers; (C) ZnO NR-coated cotton fibers; (D) high magnification images of ZnO NRs on a cotton fiber; (E) photograph of bare and ZnO NR-coated fabric; and (F) ZnO NRs-on-fabric multifunctional sensing device [70,71].
Source: Reprinted with permission from Z.H. Lim, Z.X. Chia, M. Kevin, S.W. Wong, G.W. Ho, A facile approach towards ZnO nanorods conductive textile for room temperature multifunctional sensors, Sensors and Actuators B: Chemical 151, 1 (2010), https://doi.org/10.1016/j.snb.2010.09.037.

Research is currently being done on flexible sensors which could be directly integrated into clothing. In the past, ZnO nanorods and nanowires have successfully been synthesized onto fabric materials for several different purposes [68,69]. Lim et al. created a conductive fabric by growing ZnO nanorods onto cotton. Copper wires were then attached to the ZnO as electrodes. Using this setup, it was possible to detect H_2 at room temperature. The ZnO nanofiber fabric was stable during mechanical stretching [70]. In the next step, flexible hotplates will need to be coupled with fabric substrates for better sensing results. The field of wearable gas sensing is still very much in its infancy, but the future seems highly promising and would revolutionize personal safety in dangerous jobs (Fig. 5.4).

5.1.2.4 Antiterrorism detection

Recently, in an attempt to protect citizens, governments have begun to invest in chemical, biological, radioactive, nuclear, and explosive preparedness technology. The detection of explosives, illicit drugs, and chemical warfare agents (CWAs) using SMOX-based sensors has been examined.

Most explosives and illicit drugs are low-vapor-pressure molecular solids under normal ambient conditions [72]. Typically, to detect residue of explosives or drugs invisible to the naked eye, the item is swabbed. The substances on the swab are vaporized, using a preheated metal rod [72]. The current state-of-the-art for security

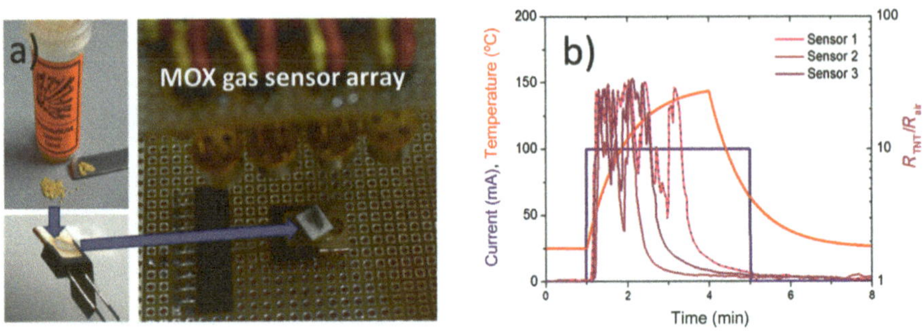

Figure 5.5 The result of an array containing three SMOX-based sensors to the vaporization of TNT. (A) Here the sensor array and test setup is shown. (B) the result of an array based on three SMOX sensors to the vaporized TNT is shown.
Source: Reprinted with permission from G. Mueller, A. Hackner, S. Beer, J. Goebel, Solid-state gas sensors: sensor system challenges in the civil security domain, Materials (Basel) 9 (2016), https://doi.org/10.3390/ma9010065.

screening is ion mobility spectroscopy, but SMOX-based detectors would be a low-cost alternative. Most explosives can be divided into two subgroups: military explosives and improvised explosives [72]. Once vaporized, explosives produce a cloud of reaction products which include NO_2, NO, NH_3, H_2, CO, CO_2, N_2, O_2, and CH_3 [73]. NO_2 can easily be detected using SMOX-based sensors [74]. Fig. 5.5 shows the results of a sensor array to the vaporization of TNT.

The sensors respond with a series of oxidizing signals, the individual grains deflagrate at random times due to the differing thermal contacts of the TNT grains to the heater surface [72]. Although SMOX-based sensors typically show higher responses to oxidizing gases over reducing gases, in a factor 10–100, it is possible that if a swab contains very little explosive residue, but a large quantity of reducing substances, the signal maybe hidden [72]. Better selectivity is required. Maurer et al. were able to differentiate between explosives and nonexplosives by interpreting the results of a sensor array using principal component analysis [75].

CWAs are easily produced and often result in mass casualty terrorist attacks [76]. The first record of chemical warfare was in 1915, when the Germans used chlorine gas for the first time during World War I. By the end of the war, 3000 different chemicals had been examined for possible use as weapons and around 50 had been tested on the battlefield. In addition to chlorine, tear gases, mustard gas, phosgene and diphosgene were used during World War I. The largest single chemical warfare attack on civilians was in the Halabja chemical attack in 1988, where around 5000 people were killed in an Iraqi nerve attack. In the past, the simplistic production of some CWAs have made them an attractive option for terrorist groups, for example, the Sarin attacks by a Japanese cult in 1994/1995 [76]. CWAs are a threat to not only soldiers, but also civilians. Robust, portable, rapidly-responding, and inexpensive detection systems are necessary for the battlefield and civilian protection [77]. Based on research funded by the United States' Navy, arrays containing SMOX-based sensors are one of the most promising detection methods for CWAs [77]. SMOX-based sensors are small, fast responding, inexpensive, sensitive, and robust. SnO_2 is the most commonly used n-type metal oxide for SMOX-based sensors, and shows promise for use in CWAs detection [78,79]. By adding different metal oxides and noble metals, SnO_2-based sensors show high responses to several CWAs [22]. Dimethyl methylphosphonate (DMMP) is commonly used as a flame retardant and is one ingredient needed to create Sarin, the CWA used in the Japanese attacks in the 1990s. Adding 3% Al_2O_3, In_2O_3, ZnO or ZrO_2, and SnO_2 showed higher sensor responses to ppb concentrations of DMMP. These additives also increased the sensor response to dipropylene glycol methyl ether, a vesicant stimulant [22]. Sensors based on SnO_2 with the additive ZnO and ZrO_2 also showed high responses to acetonitrile, a simulant blood agent. By using an array of two or more sensors, highly-selective devices can be realized [22]. In the future, SMOX-based sensor devices could be used to detect explosives and CWAs. Today, terrorist attacks pose a very real threat for civilians all over the world. SMOX-based detectors could be used for the prevention and quick detection of threats to civilians.

5.1.2.5 Food freshness

As soon as fruits/vegetables are picked or animals are slaughtered, the spoiling process begins. With the development of mechanical refrigeration in the 1800s, it became possible to transport perishable foods over great distances, and in 1882 meat was successfully shipped from New Zealand to England for the first time [80]. Today, before food is consumed, it is produced, processed, distributed, sold, and purchased. In the United States, various studies found that vegetables or fruits travel over 1000 food miles (the distance traveled between production and consumption) [81]. To ensure freshness, companies issue expiration dates. If, however, storage conditions are optimal, food can still be enjoyable even after this date. On the other hand, if conditions are not optimal, food may prematurely spoil. Many people are overly cautious and excessively throw away food. The average German throws away 80 kg of food per year, and over half the food was still edible [82]. A more accurate determination of freshness may lead to a reduction in the amount of waste. The first intelligent chemical sensor array system for odor classification was developed by Dodd and Persaud in 1982 [83]. The later-coined electronic nose (e-nose) most commonly contains an array of nonselective (traditionally SMOX-based) sensors whose signal is evaluated using a pattern recognition routine [84]. E-noses have been used to determine the storage history of eggs [85], bacteria growth in milk [86], meat spoilage [87,88], fish freshness [89] and fruit ripeness [90]. E-noses have been commercially available since 1995. Today, numerous different generic e-nose systems are commercially available, for example, The Portable Electronic Nose device from Airsense Analytics (Schwerin, Germany) or Aeonose (Zutphen, the Netherlands). In addition, some companies have brought e-noses onto the market used specifically to identify spoiled food, FOODsniffer (California, United States). But despite promising research and the numerous commercially available options, e-noses have not been widely accepted in the food

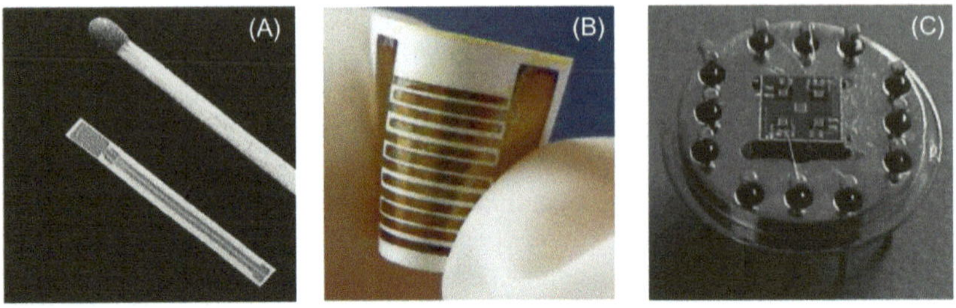

Planar-type gas sensor Flexible gas sensor Micromachined gas sensor

Figure 5.6 A variety of SMOX-based sensor types.
Source: Reprinted with permission from G. Neri, First fifty years of chemoresistive gas sensors, Chemosensors 3 (2015) 1–20, doi:10.3390/chemosensors3010001 [96].

industry. More specialized e-nose systems are needed that allow for the continual monitoring during the food production chain.

In order to monitor food from the initial production to the final consumption, flexible inexpensive sensors with low power consumption are needed. One potential solution would be sensor-enabled RFID tags. RFID tags are widely used wireless transponders with integrated memory chips [91]. The ability to inkjet print SMOX-based sensors would allow the sensors to be directly integrated onto RFID chips. Today, there are already reports of inkjet-printed metal oxide–based sensors [92,93]. Rieu et al. report the successful printing of SnO_2 sensors onto polyimide foil. The sensor was fully inkjet-printed and flexible [93]. In addition to being flexible, printing reduces the production costs and allows for extremely fast deposition times (a fraction of a second) [92]. In order to be used for smart packaging, sensors must be mass producible and inexpensive, and the power consumption problem must be addressed.

Smart packaging is already a multi-billion dollar industry [94]. By more accurately monitoring food freshness, food could be used more efficiently and waste reduced. A great deal of research on the applicability of e-noses for food freshness monitoring exists. By adding gas sensors to RFID tags, more accurate expiration dates could be determined for food based on factors encountered during distribution [95]. This would result in reduced food waste (Fig. 5.6).

5.1.2.6 Breath analysis

Alcohol has been consumed by humans for centuries, but it was not until the large-scale manufacturing of cars in the early 20th century that the need to measure intoxication levels became necessary [97]. The amount of alcohol in the breath correlates to one's blood alcohol concentration. In 1938, Rolla Harger invented the Drunk-o-meter. A person had to breathe into a balloon. The air was then released into a chemical solution; if alcohol was present the solution would change color [97]. In 1954, Robert Borkenstein improved Harger's invention, creating a portable tool called the breathalyzer. Today, most modern police-grade breath-alcohol detectors use fuel-cell technology. SMOX-based detectors for personal use also exist. SMOX-based detectors are substantially cheaper than most fuel-cell-based devices [98]. This difference in price will become substantial when considering the mass production of sensors and in particular for their integration into cellphones. Around 68% of Americans have a smartphone [99]. In the future, gas sensors would ideally be integrated directly into the phone. The total cost of sensors in the iPhone 5 was approximately $6.50 [100]. Based on this number, it is apparent that in order to be viable for integration into a smartphone, any sensor can cost only fractions of a dollar. SMOX-based sensors are inexpensive, and by using inkjet printing technology, it is possible that the price will be even lower in the future (see the subsection on Food Freshness). In addition to cost, small size and energy efficiency are important criteria. At the Sensor + Test in 2015, Cambridge CMOS Sensors (now AMS), demonstrated a SMOX-based sensor for ethanol (CCS803) that was cheap, small, and energy-efficient enough to be accommodated on the PCBs of a smartphone [101].

Breath analysis can, however, be used for more than alcohol intoxication detection. It was already known in ancient Greece that a patient's breath contains information about their health. In 1784, Lavoisier discovered that CO_2 was the product of respiration (see the subsection on indoor air quality). It was not, however, until the 1970s that Linus Pauling was able to separate and identify 250 other gaseous constituents of breath [102]. Using highly sensitive analytical devices, the detection of over 1000 substances in breath became possible. As the detection devices became more sophisticated, the need for systematic and exact sampling became evident. Exhaled breath consists mainly of oxygen, CO_2, water, and nitrogen [103]. The concentration of biomarkers in breath is typically low, within the ppt or ppb range, see Table 5.1 for examples.

In addition to the low concentration of biomarkers, the complexity of respiration makes accurate sampling very difficult. Simply put, exhaled air contains different fractions of breath. Alveolar air is the part of breath that has participated in the oxygen-carbon dioxide exchange. The rest of exhaled breath comes from the physiologic dead space. For a more detailed description on respiration and breath sampling, see [98]. Depending on the origin of the targeted biomarker, breath must be sampled differently. The origin of the biomarker in the body and possible sources of contamination must be considered. Take, for example, the biomarker NO. Nitrogen oxide is a biomarker for asthma. In 2003, the FDA approved a device from Aerocrine Inc. Sweden for medical diagnostics that monitors fractional exhaled nitric oxide [105]. One of the keys to gaining FDA approval was the standardization of sampling [105]. Once sampling methods have been optimized for a

Table 5.1 Several diseases and the resulting gas concentrations in human breath are listed [104]

Biomarker	Disease	Diagnostic concentration range
Acetone	Diabetes	1.8 ppm
Toluene	Lung cancer	10 ppb
CO	Bronchitis	8.6 ppm
H_2S	Halitosis	0.5 ppm
NH_3	Renal failure	14.7 ppm

certain breath biomarker and it is possible to monitor a patient's health using expensive sophisticated analytic methods like mass spectrometry, then the process of miniaturization can begin. By replacing sophisticated analytical detectors, for example, mass spectrometers, with SMOX-based sensors, breath analyzers could be made smaller, cheaper, and easy to operate. For quality control, the results can then be compared to those attained using laboratory equipment. A great deal of research has been done on the suitability of different metal oxides for breath analysis: SnO, ZnO, WO_3 [106], and In_2O_3. In 2012, Pratsinis et al. first reported the development of a prototype breath detector for acetone, using Si:WO_3-based sensors [107]. In 2015, Righetonni et al. then reported the successful detection of acetone in breath, using a commercially available breath sampler and Si:WO_3-based sensors [108]. The breath sampler is modular; meaning that, in the future, other SMOX-based sensors can easily be tested for acetone. In addition, the sampler can be used to test if SMOX-based sensors can detect other biomarkers found in alveolar air, such as toluene.

Volatile sulfuric compounds (VSCs) produced intra-orally by, for example, bacteria, are responsible for 90% of halitosis (colloquially "bad breath") cases [109]. The other 10% have extra-oral origins, for example, liver disease [109]. VSCs have a highly pungent smell, even in very low concentrations. Heightened concentrations of hydrogen sulfide (H_2S), and methyl mercaptan (CH_3SH) were found in the mouth breath of patients suffering from intra-oral halitosis [110]. In the case of patients with extra-oral halitosis, heightened concentrations of dimethyl sulfide (CH_3SCH_3) were found in mouth breath [110]. In order to determine if halitosis is the result of a serious disease, VSCs must be selectively detected. Using gas chromatography, different VSCs can be identified and the cause of halitosis recognized [111]. Through coupling with a gas chromatogram (GC), gases can be measured selectively, using SMOX-based sensors. Commercial options already exist. The OralChroma device from Fis Inc. (Hyogo Japan) is small, portable, and relatively inexpensive ($6000) [112]. The device uses a highly sensitive In_2O_3-based sensor for detection [112].

Sampling was one of the largest issues preventing breath analysis for medical diagnostics. Today, through a thorough scientific understanding of human respiration and standardization of measurements, accurate breath sampling is possible. Today, FDA-approved breath analyzers for biomarkers are on the market that couple these samplers with sophisticated analytical detection methods. SMOX-based detectors already exist for the detection of ethanol in breath, showing that breath analysis is possible, using these small inexpensive sensors. In the future, the miniaturization of breath analyzers may be possible by coupling sophisticated samplers with SMOX-based detectors.

5.1.3 The requirements of commercial SMOX-based gas sensors – example domestic safety

Detailed technical requirements for this application will be introduced as one can learn the general technical requirements for SMOX-based sensor from them. Requirements for performance of residential gas alarms specified in the internationally acknowledged standards are selectively summarized in Table 5.2. In the

Table 5.2 Excerption of requirements from the technical standards for residential gas alarms

Title of standard	**JIA E 001–15 standard for certification for town gas alarms**	**Standard for certification for liquefied petroleum gas alarms and bulk reservoir gas detectors**	**EN50194-1:2009 electrical apparatus for the detection of combustible gases in domestic premises Part 1: Test methods and performance requirements**	**ANSI/UL 1484–2008 (R2013) standard for safety for residential gas detectors**
Publisher	Japan Gas Appliances Inspection Association, JIA	The High Pressure Gas Safety Institute of Japan, KHK	European Committee for Electrotechnical Standardization, CENELEC	Underwriters Laboratories Inc., UL
Date of the latest revision at the time of writing	February 2015	April 2013	August 2010	December 2000
Target gas	Natural gas and other gases	LP gas	Town gas, natural gas, and LP gas	Natural gas and LP gas
Test gas	Methane for natural gas	ISO-butane	Methane for natural gas, butane for LP gas	Methane for Natural gas, propane for LP gas
Range of alarm set point	1000–12,500 ppm methane for natural gas alarm	500–3000 ppm of test gas	3%–20%LEL of test gas	25%LEL or less of test gas
Accuracy	No alarm in 1000 ppm and alarm in 12,500 ppm methane for natural gas alarm	(A) No alarm in 500 ppm and alarm in 3500 ppm of test gas(B) No alarm in 500 ppm and alarm in 3000 ppm of test gas	(A) Declared alarm set point ± 3%LEL of test gas(B) Declared alarm set point ± 6%LEL of test gas	Alarm at U or less of the target gas, where $U = (K + I)/2$ K: 25%LEL of test gas I: Initial detection threshold
Selectivity	• No alarm in 1000 ppm ethanol in 38–40°C, >85% RH • No alarm in 500 ppm hydrogen	NA	No alarm in 2000 ppm ethanol for 30 min for natural gas apparatus/in 500 ppm ethanol for 30 min for LP gas apparatus	NA

(*Continued*)

Table 5.2 (Continued)

Title of standard	JIA E 001–15 standard for certification for town gas alarms	Standard for certification for liquefied petroleum gas alarms and bulk reservoir gas detectors	EN50194-1:2009 electrical apparatus for the detection of combustible gases in domestic premises Part 1: Test methods and performance requirements	ANSI/UL 1484–2008 (R2013) standard for safety for residential gas detectors
Fluctuation of alarm conc. by atmospheric temp. and humidity	• To satisfy accuracy in <−10°C • To satisfy accuracy and selectivity in >50°C, 35%–45%RH • To satisfy accuracy in 35–40°C, >85%RH	• To satisfy accuracy of (A) in −10°C • To satisfy accuracy of (B) in >40°C, <40%RH and 35–40°C, >85%RH or >40°C, >85%RH • To satisfy accuracy (B) in 40–45°C, >95%RH	• To satisfy accuracy (B) in −10 ± 1°C and 40 ± 1°C • To satisfy accuracy (B) in 15 ± 2°C, 30 ± 5%RH and 40 ± 2°C, 90 ± 5%RH	• To satisfy the accuracy in 0°C and 49°C, 30%–50%RH • To satisfy the accuracy in 30°C, 0%–5%RH and 30°C, 85 ± 5%RH • To satisfy the accuracy in 40 ± 2°C, 93 ± 2% RH for 168 h
Response time to alarm	Alarm within 60 s following exposure to 12,500 ppm methane	Alarm within 30 s following exposure to 4500 ppm of the test gas	Alarm within 30 s following exposure to 25%LEL of the test gas	NA
Corrosion test	To satisfy accuracy and selectivity following exposure to 0.4 ppm sulfur dioxide in 40–50°C for 10 days, then to clean air for 24 h	To satisfy accuracy following exposure to 0.4 ppm sulfur dioxide in 40–50°C for 10 days, then to clean air for 24 h	To satisfy accuracy (B) following consecutive exposure to 5 ppm nitrogen monoxide for 1 h, 2 ppm sulfur dioxide for 1 h, the same conc. of ethanol for 30 min as selectivity test, 10 ppm hexamethyldisiloxane for 40 min	• To satisfy accuracy after exposure to moist 0.1% hydrogen sulfide for 10 days • To satisfy accuracy after exposure to moist 1% carbon dioxide + 0.5% sulfur dioxide for 10 days

Long-term stability	To satisfy the accuracy and selectivity following 10 days with repeated exposure to 500 ppm hydrogen at 35°C, 60%RH for 30 min twice a day, then to clean air	To satisfy accuracy of (B) following energizing for 2 months	To satisfy accuracy of (B) during the test in clean air for 3 months	No false alarms for 90 days in a typical residential atmosphere with wind of 300 ft/min, or without wind
Lifetime	NA	NA	NA	At least 3 years from date of manufacture or from date of service installation
Impact test	Gas alarm is dropped twice in different directions from a height of 30 cm on a 3 cm thick wood board placed on concrete floor.	Gas alarm is dropped twice from a height of 30 cm on a 3 cm thick wood board placed on concrete floor. A 50 g steel sphere is dropped from a height of 1 m on a gas alarm placed on a 1 cm thick wooden platform.	EN60335-1:2002 Clause 21 The spring hammer test with (1 ± 0.2)J impact energy	A 4.08 J impact is applied to the reverse side of a plywood board on which a gas detector is mounted.
Vibration test	Vibration at 10 hertz and peak to peak amplitude of 5 mm is applied to a gas alarm packaged for transportation in three different directions for 20 min each.	Vibration at 10 hertz and peak to peak amplitude of 5 mm is applied to a packaged gas alarm in three different directions for 20 min each.	NA	Vibration at the max. resonant frequency and amplitude of 0.25 mm is applied to gas alarm for 1/4 h, or at 35 hertz for 4 h.

subsequent paragraphs, detailed requirements relevant to SMOX-based gas sensors will be discussed with reference to these mandatory requirements for the performance of residential gas alarms.

5.1.3.1 Mechanical strength

Commercial gas sensors must be mechanically stable during manufacturing processes, transportation, and handling of gas alarms. Depending on the application, different stability requirements exist. For example, the specific requirements regarding mechanical strength of residential gas alarms are shown in Table 5.2, including the vibration test and impact test. In addition, if a cyclic heating condition with periodical temperature change is applied to a sensing element, resistance to thermal shock caused by such cyclic heating should also be verified.

In the case of intermittent heating applied to a sensing element, a conducting test of SMOX-based gas sensors energized in high relative background humidity is an effective way of verification. Mechanical stress can be caused by a volume change of capillary-condensed water, which may be present in a porous gas-sensing element during the heater power-off period. This occurs when condensed water changes from a liquid to a vapor at the instant of heater power-on during the intermittent heating.

Certain steps made in an attempt to enhance the mechanical strength of SMOX-based gas sensors may also alter their sensing properties. An example is the use of ethyl silicate, a typical binder, to increase the sintering strength of the gas-sensing element. However, it is reported that the use of such a binder can change the sensor properties, for example, decrease the sensor resistance by an order of magnitude and increase the sensor signals to gases [29]. Another example is the change of secondary particle size in the sensing element in order to mitigate contraction stresses. Contraction which arises during the drying and sintering process of the sensing element can cause cracks or residual stress and ultimately results in a decreased mechanical strength of the gas-sensing element. Variation in the secondary particle size, however, may also affect the sensor signals to gases. This can change the gas concentration profile of the sensing element in the steady state. The balance between gas diffusion and consumption through the sensing layer can be altered. It can even result in a change of the film thickness. In conclusion, designs for enhancement of mechanical strength and controlling sensor properties should be considered in sensor development simultaneously, rather than separately.

5.1.3.2 Setting of alarm point

Gas detectors and alarms are normally calibrated to a predefined alarm set point through exposure to a calibration gas with an appropriate concentration, prior to factory shipment. In this calibration process, the sensor resistance to the calibration gas is memorized in an electric circuit as an alarm set point. While gas alarm manufacturers can define alarm set points arbitrarily in the range required by the standards as given in Table 5.2. Around one-tenth of lower explosion limit (LEL) of target gas is typically used as an alarm set point; for instance, 5000 ppm methane, of which LEL is 5%, for town gas alarms.

During storage periods, without energizing, the sensor resistance normally drifts from the original value, which is defined at the stable state reached after sufficient energizing before the storage period. Usually, the sensor resistance will recover to the original level after being energized for a certain long period. The length of energizing required for the resistance recovery depends on the length of and the ambient conditions during storage [29]. Due to this drift, a pre-heating process is required in the manufacturing process of gas alarms using SMOX-based gas sensors prior to an alarm point setting process in order to ensure accurate alarm threshold, by removing influence of this drift. For example, when a commercially available LP gas sensor TGS 2610 is used for manufacturing gas alarms, more than 48 hours of pre-heating is required. This pre-heating process results in an increased cost. Therefore, efforts to shorten this pre-heating period would be beneficial.

5.1.3.3 Sensor signal

A typical relation between the gas concentration and sensor resistance of commercial sensors (methane sensor TGS2611), is shown in Fig. 5.7A. The solid line is the sensor resistances (R_s) normalized by the alarm set point (R_{std}), sensor resistance in 5000 ppm methane. The sensor resistance in air is expressed without relation to the x-axis for the sake of expediency. The sensor signal of SMOX-based gas sensors are generally defined as the ratio of the sensor resistance in air (R_a) to that in gases (R_g); R_a/R_g. In particular, the sensor signal at the alarm set point (R_a/R_{std}) is important for gas alarms. Sufficiently high initial R_a/R_{std} will ensure that R_a will never reach the R_{std} level during the product lifetime, even if R_a changes due to the long-term humidity effect and poisoning caused by particular environmental gaseous substances. These issues are described in greater detail later.

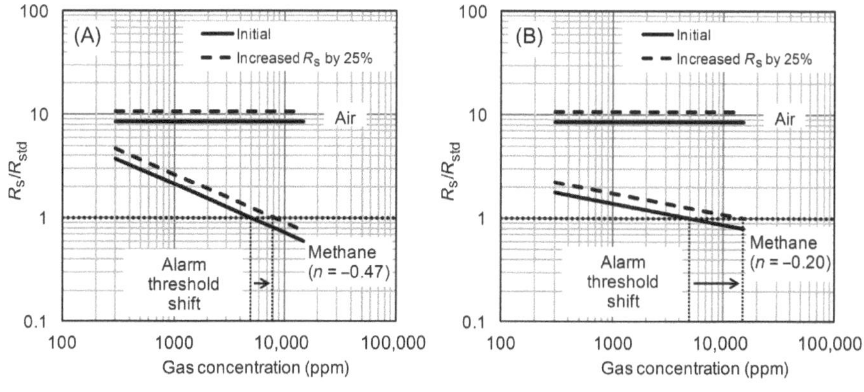

Figure 5.7 Sensitivity characteristics of methane sensor and alarm threshold shifts: (A) TGS2611 and (B) hypothetical sensor.

It is well known that R_s is proportional to power of gas concentration (C), as can be seen in Fig. 5.7 with the following equations, where a, a' and n are constants.

$$R_s = aC^n \ or \ \log(R_s) = a' + n \cdot \log(C) \quad (5.1)$$

In addition to ensuring sufficient sensor signal, large sensitivity (absolute value of n in the equations above) is also highly important. This importance becomes clear when examining the dependence of a hypothetical sensor's resistance to a methane concentration, where $n = -0.20$ is shown in Fig. 5.7B as a hypothetical example for the sake of discussion, as compared with the Fig. 5.7A ($n = -0.47$). Suppose sensor resistance in methane in both figures increases by 25%, keeping their sensitivities, the alarm threshold for a sensor with $n = -0.20$ in the Fig. 5.7B would shift from 5000 to 15,000 ppm, exceeding the requirement by the standards (must alarm below 12,500 ppm), as shown in the Table 5.1, in contrast with that, for a sensor with $n = -0.47$ in the Fig. 5.7A, the alarm threshold would shift to 8000 ppm. To be more tolerant to the sensor resistances drifts, making the sensitivity steeper is strongly required for SMOX-based gas sensors. In research and development of SMOX-based gas sensors, discussing more about the sensitivity would be invaluable; to allow not only gas alarms, but also other applications using SMOX-based gas sensors, to be more precise.

5.1.3.4 Selectivity

The infancy of the residential gas alarm industry in Japan was plagued by a large number of false alarm problems. Causes of false alarms can be roughly categorized into two sections. One is that the gas sensors inherently have cross sensitivity to high concentrations of interference gases which can abruptly appear in daily life: alcohol vapor during cooking for town gas alarms installed in proximity to the ceiling, and reducing gases contained in insecticide sprays for LP gas alarms installed in proximity to the floor. Some of the standards for gas alarms include a selectivity test using ethanol as given in Table 5.2. For instance, no alarm shall be generated when gas alarms are subjected to 1000 ppm ethanol vapor in 38°C–40°C, 85%RH in the JIA standard, and to 2000 ppm ethanol for natural gas apparatus in the European standard. In order to meet these requirements, adsorbent materials are commonly used in the filter cap of a sensor, such as activated carbon for methane detection, which can adsorb interference gases with only minimal adsorption of target gases. For prevention of false alarms caused by insecticide sprays, a solution using signal processing to ignore temporary sensor signal changes is commonly utilized. Another cause of false alarms is that the sensor resistance in air has decreased and/or sensor signals to interference gases have increased over a period of time. This issue will be described in more detail in the section 5.1.3.7 on long-term stability.

According to the report regarding extension of the validity period of residential town gas alarms issued in Japan in 1985, gas alarms were collected from the end users' houses after 4 years of usage, and alarm thresholds of the gas alarms were measured. As a result, a good correlation was found between the incidence rate of the false alarms and the alarm threshold to hydrogen at the time of the collection [113]. A certain concentration of hydrogen is generated due to an incomplete combustion of a gas cooking stove in some cases, and in the JIA standard, no alarm shall be generated in 500 ppm hydrogen for town gas alarms, as shown in Table 5.2. Since filtration of hydrogen by using adsorbents is theoretically difficult, the selectivity is achieved by a sensing element itself through sensing material technology and film thickness optimization, etc.

SMOX-based gas sensors respond to water vapor, which is omnipresent in real world applications and is usually the interference gas found in the highest concentration. Generally, sensor resistances decrease as a result of increased humidity. This influence can be lowered by incorporating a temperature sensor and a relative humidity sensor, and compensating the SMOX-based gas sensors signal based on the absolute humidity information calculated from these two sensors' output signals. For residential gas alarms, the effect of absolute humidity is generally compensated using only a temperature sensor, under the assumption that typical seasonal changes in humidity should be similar to those of the ambient temperature in a country or an area where the gas alarms are used. This is possible because a relatively wide tolerance of required accuracy exists. In addition, the use of only one component reduces cost.

5.1.3.5 Gas response speed

Depending on the application, the gas sensor response time can be very important. Most standard residential gas alarms are required to issue an alarm within the specified time limit when subjected to a specified concentration of test gases, as shown in Table 5.2. Hence, the gas response time of SMOX-based gas sensors for residential gas alarms can be defined as the time required for sensor resistance to reach the value at the alarm set point when the sensor is subjected to the specified concentration of test gases. For example, to meet the European standard, gas alarms need to alarm within 30 seconds when exposed to 1/4LEL of test gases. Typical response curves of the two commercial LP gas sensors with or without a filter cap incorporating zeolite-based adsorbent material, to 4500 ppm iso-butane, are shown in Fig. 5.8A (TGS2610-C00 intended for gas leak detectors) and Fig. 5.8B (TGS2610-D00), respectively. These sensors' configurations are the same, except for filter and cap size. The gas response times of the two sensors are 6 and 24 seconds, respectively. This implies that generally SMOX-based gas sensors without a filter can easily satisfy the requirements of the standards, while adsorbent materials incorporated in a filter cap can seriously interfere with the diffusion of target gases.

A filter may, however, be necessary for certain applications. There are two required functions of a filter. The originally required function is to adsorb interference gases, such as ethanol. The other is to prevent a gas-sensing element from being poisoned by gaseous substances which can cause irreversible changes in the sensing properties. After the infancy of residential gas alarms, the latter has become more important, due to decreased resistance of miniaturized sensing elements to such substances. Indeed, if required only to improve selectivity, the amount of the adsorbent materials can be boldly reduced to quicken the response speed in modern SMOX-based gas sensors.

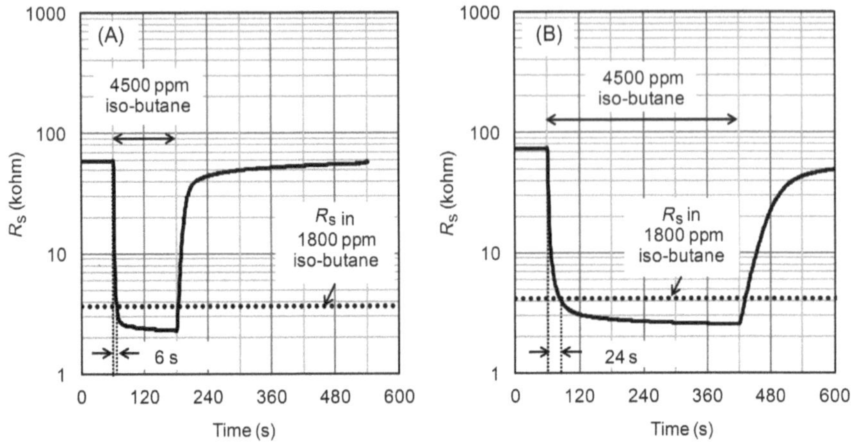

Figure 5.8 Gas response curves to iso-butane and response time to alarm: (A) TGS2610-C00 (without filter) and (B) TGS2610-D00 (with filter).

5.1.3.6 Warm-up behavior

SMOX-based gas sensors show a characteristic change in sensor resistance that appears immediately after power-on, called the initial action transient. During this transient period, the sensor resistance sharply decreases initially, and subsequently rises to a stable state, according to the ambient conditions. Shape and time length of the transient response may vary, depending upon storage history. In one instance, an initial action transient of a commercial LP gas sensor, TGS2610, that had been stored in normal air for 60 days, is shown in Fig. 5.9. The sensor resistance drops for the first few seconds, then reaches the stable state about a minute after power-on. This transient will make the sensor resistance reach the alarming point, which may cause a false alarm at the beginning of the use of gas alarms in actual sites. In order to avoid this nuisance alarm, an initial delay circuit is commonly incorporated into the gas alarm design. Gas alarms are allowed to have a start-up period of 5 minutes in the standards. Hence, this initial action transient is not a serious problem for the real world application of gas sensors. However, in the case of applying intermittent heating to SMOX-based gas sensors with the intention of trying to save power consumption, which is a relevant issue for enabling gas alarms to be battery operated, this transient takes on great significance. Discussions about the mechanism of this transient and efforts to shorten this transient period would be strongly needed when intermittent heating is applied to SMOX-based gas sensors, and would also be useful for applications using SMOX-based gas sensors for on-demand usage, such as portable gas detectors.

5.1.3.7 Long-term stability

Unlike industrial gas detectors, residential gas alarms are often expected to have a product lifetime of more than 5 years without maintenance in the marketplace.

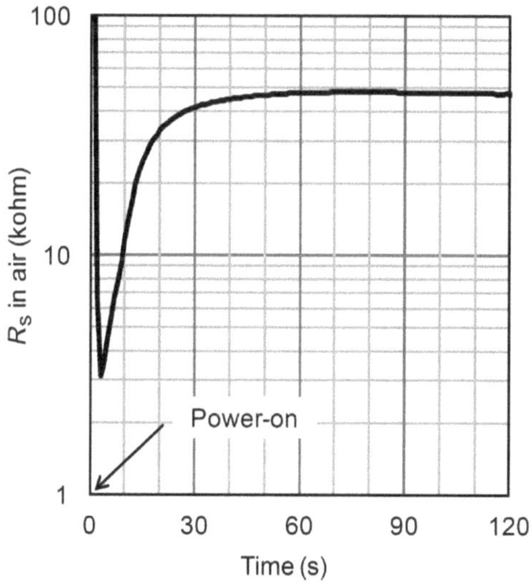

Figure 5.9 Sensor resistance change following heater power-on (initial action transient) of TGS2610 stored in normal air for 6 months.

Hence, long-term stability of SMOX-based gas sensors is of vital importance for residential gas alarms. Some of the technical standards include requirements regarding long-term stability. For instance, gas alarms need to maintain the accuracy defined in the KHK standard and the European standard after energizing in normal air for 2 and 3 months respectively, as shown in Table 5.2. Satisfaction of these requirements is essential, but is not sufficient to guarantee a long product lifetime. In this context, gas alarm manufacture and SMOX-based gas sensors manufacture have been tackling this issue with their own unique ways, including acceleration test and field test. In this paragraph, three key points for ensuring the long-term stability of SMOX-based gas sensors are discussed, based on past experiences and efforts made in this industry in conjunction with the related information.

5.1.3.8 Thermal effects

By analyzing the SMOX-based gas sensors (TGS109) incorporated in the town gas alarms which were used in a residential environment for 4 years (introduced in the former paragraph in terms of selectivity), it was discovered that the alarms installed in close proximity to cooking stoves had a tendency to become more sensitive to hydrogen, which may cause a false alarm. It was inferred that the temperature of the sensor elements increased due to the unique design of TGS109, which causes the sensor element to generate Joule heating as the sensor resistance decreases when more reducing gases and water vapor are present in such ambient environment. In the TGS109 sensors, which had become more sensitive to hydrogen in the analysis, a growth of the constituent tin dioxide grains and a change of the distributed state of the sensitizer of palladium were observed [113]. Matsuura et al.

reported that even at the temperature where changes of crystallite size and specific surface area of neat tin dioxide are negligible, decreases in sensor resistance in air and in some gases occur over time, and certain additives effectively mitigate those changes [114]. Because long-period energizing potentially causes sensor characteristics changes as described above, thermal stability of new sensing materials need to be verified by tests, such as the thermal acceleration test and long-term energizing test, when they will be applied to sensor development.

5.1.3.9 Moisture effect

A seasonal change in alarm threshold, reversible increase in summer and decrease in winter, is often observed in field tests of residential gas alarms. The main reason for this change is an increase in sensor resistance under high absolute humidity and decrease under low absolute humidity conditions during a long energizing period as implied by Fig. 5.10. TGS2610 sensor was used in this field test, and the sensor resistances were obtained in a laboratory under a controlled temperature and humidity standard conditions. This moisture effect is well-known in the industry as the long-term moisture effect, and was studied on neat tin dioxide by academics, as well [115]. Another moisture effect, called the short-term moisture effect, shows a quick decrease in sensor resistance, with an increase in absolute humidity.

In order to verify reliability of residential gas alarms in relation to long-term accuracy, a test should be performed under conditions which simulate actual humidity changes in the real environment. For example, continuously energized samples are subjected cyclically to high temperature and high humidity conditions (summer), and low temperature and low humidity conditions (winter) for 1 or 2 months, respectively. Reduction in the long-term moisture effect is also beneficial for gas

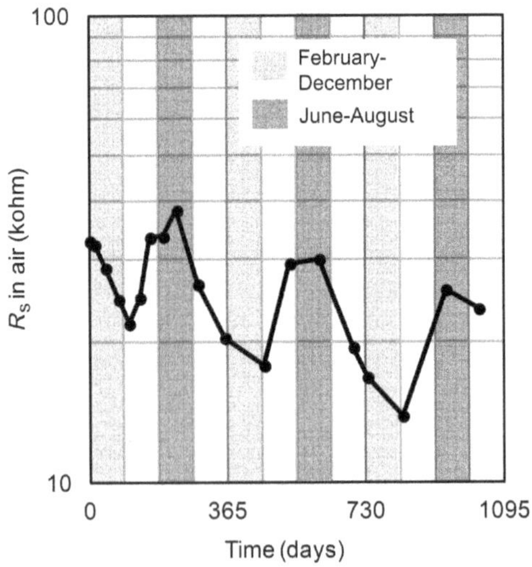

Figure 5.10 Seasonal resistance change of TGS2610 tested in domestic kitchen.

detectors having a gas concentration indication and gas monitors with a low gas concentration range which require a high signal-to-noise ratio. Therefore, basic research and effort for mitigating the long-term moisture effect would be required, not only for improvement of the long-term accuracy of residential gas alarms, but also for the expansion of the contributive range of SMOX-based gas sensors.

5.1.3.10 *Poisoning by siloxanes*

In a field test of residential gas alarms, though only in rare cases, a sudden decrease in alarm threshold to hydrogen and some other gases may occur irreversibly, resulting from a decrease in sensor resistance in air and in those gases. As a typical example, the sensor resistance changes in air, hydrogen, iso-butane, and methane of SMOX-based gas sensors utilizing a zeolite-based filter during the field test of over 4 years and the alarm thresholds converted from the sensor resistances, are shown in Fig. 5.11A and B, respectively. The sensor resistances plotted in Fig. 5.11 were measured in a laboratory with a gas chamber in which temperature and humidity were controlled at standard conditions. Fig. 5.11 shows a significant drop of the sensor resistances and the alarm thresholds after 3.5 years except for those in methane, due to breakdown of the capability of its filter. From the adsorbents inside of the filter utilized in the sensor which has exhibited sensor resistance changes similar to the changes in the above sensor used in the field test, relatively large amounts of siloxanes, such as octamethylcyclotetrasiloxane (D4) and decamethylcyclopentasiloxane (D5) are always detected. Therefore, it implies that this phenomenon may be caused by siloxanes which may be generated from products for daily living containing silicone, such as caulking agent and hair dressing. The sensor having the same configuration as the sensor used in Fig. 5.11 was tested for durability to siloxane exposure, as well. In this test, the sensor had been exposed to mixed siloxane

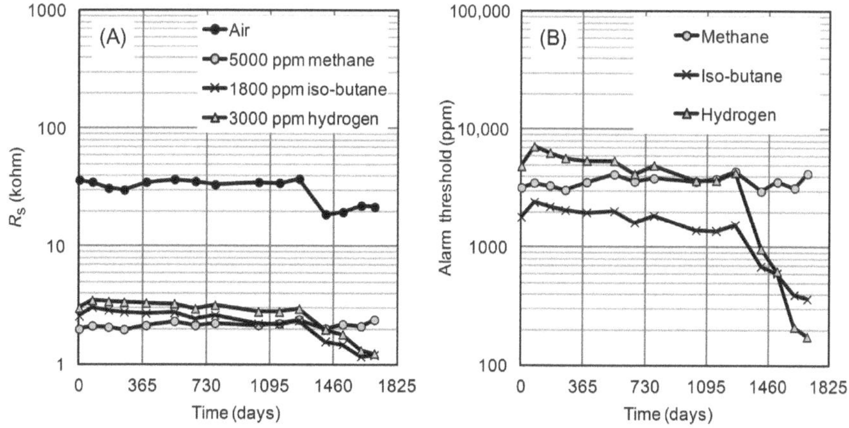

Figure 5.11 Long-term stability of combustible gas sensor with zeolite-based filter tested in domestic kitchen: (A) sensor resistance changes and (B) alarm threshold changes.

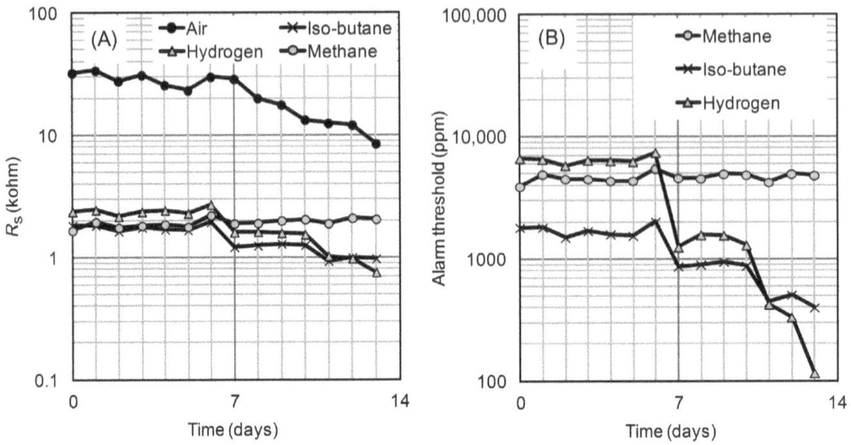

Figure 5.12 Effect of siloxane exposure on combustible gas sensor with zeolite-based filter: (A) sensor resistance changes and (B) alarm threshold changes.

of octamethyltrisiloxane, D4, and D5 of 10 ppm each for 2 weeks, and the sensor resistances and alarm thresholds were obtained in the same manner as the field test. As shown by the test results in Fig. 5.12A and B, the sensor resistances and alarm thresholds in test gases, except methane, suddenly decreased following exposure to siloxane for 6 days, similar to the results of the field test in Figs. 5.11A and 5.12B, respectively. The European standard includes the siloxane exposure test, which requires exposure to 10 ppm hexamethyldisiloxane for 40 minutes, as shown in Table 5.2. But the results of the above described field test and the siloxane exposure test imply that more severe siloxane contamination could be present in real indoor environment. Performing validation tests for siloxane durability under adequate test conditions is an important issue for gas sensors in a developmental stage.

Paradoxically, in comparison with gas alarms using catalytic gas sensors, ones using SMOX-based gas sensors are often preferred choices, due to a difference in siloxane effect. When being subjected to siloxane exposure strongly, gas alarms using catalytic gas sensors become less sensitive, while gas alarms using SMOX-based gas sensors become more sensitive, finally reaching a state of continuous alarming even in clean air, making it a fail-safe design. However, false alarm causes a nuisance to end users, and a cost is incurred for unnecessary calls for service personnel by gas suppliers, which is still an issue. In addition, there is some concern that downsizing of volume of a gas sensing element, that is, MEMS-type SMOX-based gas sensors, for reducing power consumption would result in a decrease in durability against siloxane exposure in general, and that increasing air-tightness of dwellings for heating energy conservation would cause an increase in siloxane concentration in the indoor environment. In this context, constant efforts to improve resistance to siloxane exposure for SMOX-based gas sensors would be indispensable. Enhancement of a filter capability to filter out siloxane would be one of the promising means for improvement. When developing such a filter, it should be taken into consideration that there may be a serious trade-off between improvement of filtration capability and gas response speed (response time to alarm in Table 5.2), because adsorbents in a filter not only adsorb siloxane vapors, but also may interfere with diffusion of target gases.

5.1.3.11 Battery operation and combination of methane and carbon monoxide detections

Battery-operable residential gas alarms are awaited in Japan in response to a recent decreasing trend in the number of wall sockets near the installation sites of gas alarms in this country [116]. Installing gas alarms without the constraint of a commercial power source and uninterrupted operation in the event of a power outage may be a common need in any country. In the Japanese gas alarm market, power consumption of a gas sensor required for battery operation is estimated at less than 0.1 mW, considering a product life of 5 years, current battery technology, and allowable cost and size of a gas alarm [10]. This requirement for power consumption could become a reasonable benchmark value for SMOX-based gas sensors for battery-operable residential gas alarms in general. In order to satisfy this requirement for SMOX-based gas sensors, reduction in thermal capacity of a sensing element and introduction of intermittent heating with an extremely short heating period are essential. MEMS technology will be able to realize an extremely miniaturized and thermally insulated sensing element of SMOX-based gas sensors, providing quick thermal response, which seems to be the most promising solution.

Intermittent heating may raise a new issue, depending on heating conditions and sensing materials. As an example, the effects of exposure to 1% ethanol vapor on two MEMS-type SMOX-based gas sensors are shown in Fig. 5.13A and B. In order

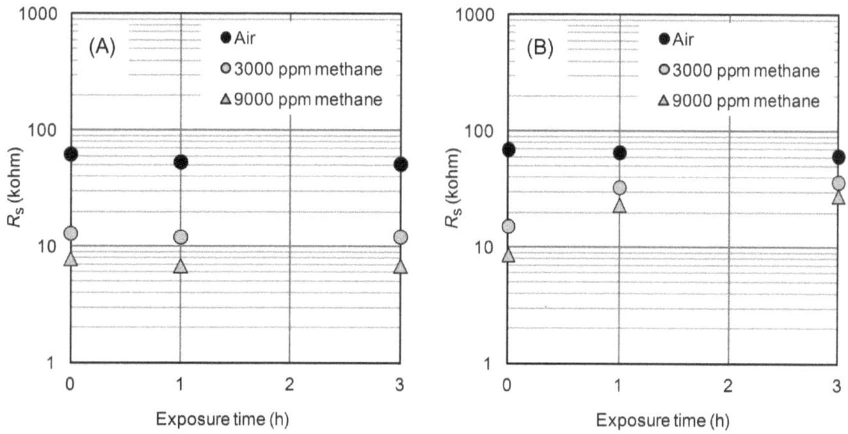

Figure 5.13 Effect of ethanol exposure on MEMS-type sensors without filter: (A) with continuous heating and (B) with intermittent heating.

to make phenomena accelerated in this test, filters were removed, and the sensors were exposed to a relatively high concentration of ethanol vapor. The sensor resistances plotted in Fig. 5.13A and B were measured with exposure to the ethanol vapor. The configuration of the two sensors was identical, except for the heating conditions, of which one was continuous heating, and the other was intermittent heating. The power-on period was for 100 ms during a 30 seconds cycle, and the sensor output was obtained at the end of each power-on period. They had a double layer sensing element made of two different materials; Pt-SnO_2 for the lower layer and Pd-SnO_2 for the upper layer. Fig. 5.13A and B indicate that the sensor signal to methane of the sensor driven under intermittent heating decreases gradually, while the other sensor driven under continuous heating maintains the sensor signal to methane. This implies that the accumulation of polymerized substances of ethanol may have been caused on the sensing element by the intermittent heating, although such substances have not been observed directly yet. In this way, intermittent heating could cause phenomena that have not been seen under continuous heating, and should be taken into account.

The number of accidents in connection with gas appliances in Japan decreased significantly in the mid-1980s, as the use of gas alarms and other safety equipment for households grew. On the other hand, the number of accidents of carbon monoxide poisoning did not decrease significantly. As a countermeasure against the continuing accidents, a carbon monoxide alarm utilizing SMOX-based gas sensors was developed and launched in 1986 [116]. Methane, mainly used for town gas, and carbon monoxide, generated by incomplete combustion, are both lighter than air, unlike LP gas, and both gases diffuse into a space near the ceiling in the event of gas leakage and imperfect combustion. Hence, functions to detect methane and carbon monoxide can be combined into one enclosure for town gas alarms, so that end users do not have to purchase two separate gas alarms. Indeed, such combination town gas alarms were launched in 1994, and accounted for around 80% of the residential town gas alarms sold in Japan in 2012 [116].

A gas sensor capable of detecting both methane and carbon monoxide with one sensor is desirable for use in gas alarms in order to reduce the cost of gas alarms. Some capable SMOX-based gas sensors exist, for example, the commercially available sensor TGS3870. The sensing element is a bead of a few hundred micrometers in diameter. This micro-bead element allows the sensor to work at two specific temperatures periodically with the higher heater voltage being applied for 5 seconds to detect methane, and the lower heater voltage for 15 seconds to detect carbon monoxide. This heating profile meets the requirement of gas response time to alarm within 60 seconds, in the standard of JIA. The power consumption of TGS3870 is 38 mW, which is one of the lowest among SMOX-based gas sensors for gas alarms, except MEMS-type SMOX-based gas sensors. But this sensor still cannot satisfy the requirement of power consumption within 0.1 mW for battery-operable gas alarms. If a MEMS-type SMOX-based gas sensor capable of both methane and carbon monoxide detection is realized, it would be welcomed by the industry. It would realize battery-operated combination town gas alarms with the lower cost, compared to a combination gas alarm utilizing a MEMS-type SMOX-based gas sensor for methane detection and an electrochemical carbon monoxide sensor, which consumes no power.

The goal of implementing residential gas detectors is to save life and property by preventing gas accidents. This is already apparent from the story of the birth of SMOX-based gas sensors. In this context, R&D for MEMS-type SMOX-based gas sensors will lay the groundwork for realizing battery-operated gas alarms that can be used even when the power source infrastructure is poor. In spite of technical challenges, MEMS-type SMOX-based gas sensors capable of the detection of both methane and carbon monoxide will be beneficial, expanding a degree of safety and contributing to a cost reduction of town gas alarms. This will lead to further growth in their domestic use.

5.2 Conclusion

In this chapter, the historical development of SMOX-based sensors was examined. Over the years, SMOX-based sensors have offered an inexpensive, fast-responding, robust, and compact option to other detection methods. SMOX sensors were initially developed in the 1960s as domestic gas alarms in Japan. The next big step in the development of SMOX-based sensors came in the 1980s, with the invention of ECU for automobiles. Today, these applications remain very relevant. In addition, a wide array of other applications has been researched. Although the use of SMOX-based sensors is not a universal solution for gas detection, for example, different sensing methods are better suited for the detection of certain gases, such as CO and CO_2, they are one of the most robust, compact, and inexpensive options. Through the years, standards and requirements have been developed for commercial gas sensors and numerous SMOX-based gas sensors are now commercially available from a wide array of vendors all over the world, see Table 5.3. By surveying history and evaluating recent developments of SMOX-based sensors for residential gas alarms, several requirements for commercial gas sensors can be identified: mechanical strength, properly defined alarm points, high sensor signals, selectivity, fast response time, minimal power consumption, low cost, and long-term stability (Table 5.2).

With proper consideration and by using the insights gained from the historical development, in the future SMOX-based sensors can be successfully used in many different fields.

Table 5.3 A list of commercially available SMOX-based sensors. Sensors beginning with TGS are manufactured by Figaro Engineering. Sensors beginning with AS-M are manufactured by AMS. Sensors beginning with CCS were originally manufactured by Cambridge CMOS (now AMS). Sensors beginning with MICS are manufactured by SGX Sensortech. Sensors beginning with MQ are manufactured by Hanwei Electronics CO. LTD.

Sensor	Metal oxide	Gas	Application
TGS2610-C00/D00	Unknown	Butane and propane	Domestic/industrial Safety, automotive
TGS2611-C00/E00	Unknown	Methane	Domestic/industrial safety
TGS2612	Unknown	Methane, propane, butane	Domestic/industrial safety
TGS3870	Unknown	Methane, co	Domestic/industrial safety
TGS832-A00	SnO_2 based	Halocarbon gas	Industrial safety
TGS832-F01	SnO_2 based	Halocarbon gas	Industrial safety
TGS2444	SnO_2 based	NH_3	Industrial safety
TGS3830	SnO_2 based	Halocarbon gas	Industrial safety
TGS6812-D00	Unknown	Hydrogen, methane, butane, propane	Industrial safety
TGS2600	Unknown	Air contaminants (hydrogen, ethanol, etc.)	Air quality, appliance Control
TGS2602	Unknown	Air contaminants (VOCs, NH_3, H_2S, etc.)	Air quality, appliance Control
TGS2603	Unknown	Air contaminants (trimethylamine, methyl mercaptan, etc.)	Air quality, automative, Appliance control
TGS8100 (MEMS)	Unknown	Air contaminants (hydrogen, ethanol, etc.)	Air quality, appliance Control
AS-MLV	Unknown	VOCS	Air quality
AS-MLV-P2 (MEMS)	Unknown	VOCS	Air quality
CCS801	Unknown	Air contaminants (CO and VOCS)	Air quality
CCS803	Unknown	Ethanol	Ethanol detection
CCS811	Unknown	VOCs	Air quality
MICS-5524	Unknown	CO, Ethanol, H_2, NH_3, CH_4	Domestic safety/breath Analysis

MICS-2714	Unknown	NO_2, H_2	Nonspecified
MICS-5914	Unknown	NH_3, C_2H_5OH, H_2, C_3H_8, C_4H_{10} (iso-butane)	Domestic safety/air quality
MICS-4514	Unknown	CO, NO_2, C_2H_5OH, H_2, NH_3, CH_4	Automotive
MICS-6814	Triple Sensor	CO, CH_4, NH_3, C_2H_5OH, H_2, C_3H_8, C_4H_{10} (iso-butane)	Industrial safety, Automotive
MQ-2	SnO_2 based	C_3H_8, C_4H_{10}, CH_4, C_2H_5OH, H_2	Domestic/industrial safety
MQ-3	SnO_2 based	C_2H_5OH	Breath analysis
MQ-4	SnO_2 based	CH_4, C_3H_8, C_4H_{10}	Domestic/industrial safety
MQ-5	SnO_2 based	C_3H_8, C_4H_{10} (iso-butane)	Domestic/industrial safety
MQ-6	SnO_2 based	C_3H_8, C_4H_{10} (iso-butane)	Domestic/industrial safety
MQ-7	SnO_2 based	CO	Domestic safety
MQ-8	SnO_2 based	H_2	Domestic/industrial safety
MQ-9	SnO_2 based	CO, CH_4, C_3H_5, C_4H_{10}	Domestic/industrial safety
MQ-131	SnO_2 based	O_3. Cl_2, NO_2	Domestic/industrial safety
MQ-135	Unknown	NH_3, H_2, C_6H_6	Domestic/industrial safety
MQ-136	SnO_2 based	SO_2	Domestic/industrial safety
MQ-309A	SnO_2 based	CO, CH_4	Nonspecified
MQ-306A	SnO_2 based	CH_2, C_3H_8, C_4H_{10}, H_2	Nonspecified
MQ-303A	SnO_2 based	C_2H_5OH	Breath analysis
MQ-216	Unknown	C_3H_8, C_4H_{10}, CH_4, C_2H_5OH,	Domestic/industrial safety
MQ-214	Unknown	C_3H_8, C_4H_{10}, CH_4	Domestic/industrial safety
SB-11A-00	SnO_2 based	CH_4, C_3H_8, C_xH_y, H_2	Domestic safety
SB-12A-00	SnO_2 based	CH_4, C_xH_y, H_2	Domestic safety
SB-12C-00	SnO_2 based	CH_4	Automobile
SB-15-00	SnO_2 based	C_3H_8, C_xH_y	Industrial safety
SB-15B-00	SnO_2 based	C_3H_8, C_xH_y	Domestic safety
SB-500-12	SnO_2 based	CO	Domestic safety
SB-96-00	SnO_2 based	C_3H_8, CO, C_xH_y	Domestic safety
SB-95-12	SnO_2 based	CH_4, C_xH_y, CO	Domestic safety
SB-19-00	SnO_2 based	H_2	Industrial safety

(*Continued*)

Table 5.3 (Continued)

Sensor	Metal oxide	Gas	Application
SB-42A-11	Unknown	H_2	Industrial safety
SB-31-00	SnO_2 based	VOCs, C_2H_5OH	Industrial safety
SB-51-00	SnO_2 based	Sulfides	Industrial safety
SB-41A-00	SnO_2 based	Freon	Industrial safety
SB-42A-00	SnO_2 based	Freon	Industrial safety
SB-43-00	SnO_2 based	Freon	Industrial safety
SB-53-00	Unknown	Ammonia	Industrial safety
SB-AQ1–06	SnO_2 based	C_xh_y, H_2, Vocs, alcohol	Air quality/automobile
SB-AQ1-00	SnO_2 based	VOCs, alcohol	Air quality
SB-AQ9-00	SnO_2 based	CO	Air quality
SB-30-04	SnO_2 based	VOCs, alcohol	Breath analysis
SM-30-00	SnO_2 based	Alcohol	Breath analysis
SP-11-00	SnO_2 based	CH_4, C_3H_8, C_xH_y, H_2	Domestic safety
SP-12A-00	SnO_2 based	CH_4	Domestic safety
SP-15A-00	SnO_2 based	C_3H_8	Domestic safety
SP-19-01	SnO_2 based	H_2	Industrial safety
SP-31-00	SnO_2 based	VOCs, alcohol	Industrial safety
SP-41-00	SnO_2 based	Freon	Industrial safety
SP-42A-00	SnO_2 based	Freon	Industrial safety
SP-42AF-00	SnO_2 based	Freon	Industrial safety
SP-53B-00	SnO_2 based	Ammonia	Industrial safety
SP3S-AQ2-01	SnO_2 based	CO, VOCs, alcohol	Air quality/automobile
SP3–61-00	In/SnO_2 based	Ozone	Air quality

References

[1] C.E. Perrin, The Chemical Revolution, in: R.C. Olby, G.N. Cantor, J.R.R. Christie, M. J.S. Hodge (Eds.), Companion to Hist. Mod. Sci, Routledge, 1990, pp. 264–277.
[2] B.B. Chastian, J.B. van Helmont, Online Encycl. Br. (1998). https://doi.org/10.1002/ejoc.201200111.
[3] J. Sundell, On the history of indoor air quality and health, Indoor Air. 14 (2004) 51–58. Available from: https://doi.org/10.1111/j.1600-0668.2004.00273.x.
[4] M. von Pettenkofer, Über den Luftwechsel in Wohngebäuden, Literarisch-Artistische Anstalt Der J. G. Cotta'schen Buchhandlung, Munich, 1858.
[5] L. Tomory, Progressive Enlightenment: The origin of the Gaslight Industry, MIT Press, 2012.
[6] Industrial Revolution, (2016) 1–5. http://industrialrevolution.org.uk/coal-mines-industrial-revolution/ (accessed 22.12.16).
[7] W. Irving, O. Tailakov, D. Kruger, D. Williams, Ch 4 emissions : coal mining and handling, IPCC Good Pract. Uncertain. Manag. Natl. Greenh. Gas Invent (2001) 129–144.
[8] S.K. Srivastava, Real Time Monitoring System for Mine Safety using Wireless Sensor Network (Multi-Gas Detector), National Institute of Technology Rourkela, 2015, http://ethesis.nitrkl.ac.in/6942/1/Real_Shrivastav_2015.pdf.
[9] M. Ryan, T. Harvey, J. Bride, M. Kizil, Report on hydrogen sulphide experience at Southern Colliery, in: Coal Oper. Conf., Wollongong Canada, 1998.
[10] United States Department of Labor Mine Safety and Health Administration, Historical Data on Mine Disasters in the United States, (2017) 7–9. https://arlweb.msha.gov/MSHAINFO/FactSheets/MSHAFCT8.htm. (accessed 06.06.17).
[11] K.C. Sekhar, S.S.C. Chakra Rao, J.S. Haldane, The father of oxygen therapy, Indian J. Anaesth. 58 (2014) 350–352. Available from: https://doi.org/10.4103/0019-5049.135087.
[12] B. Kleven, A Summary of Gas Detection, Advanced Calibration Designs, Inc., 2016
[13] K. Johnson, Gas Detection Apparatus, US33698664A, 1964.
[14] L. Karalliedde, C. Keshishian, Carbon monoxide poisoning, in: D. Baker, L. Karalliedde, V. Murray, R. Maynard, N.H. Parkinson (Eds.), Essentials Toxicol. Heal. Prot. a Handb. F. Prof., 2nd ed., 2008: pp. 1082–1083.
[15] L.E. Otterbein, Quoth the Raven: carbon monoxide and nothing more, Med. Gas Res. 3 (2013) 7. Available from: https://doi.org/10.1186/2045-9912-3-7.
[16] L. Lippman, Poe-Purri: Add carbon monoxide poisoning to the many theories about the death of Edgar Allen-Poe, Balt. Sun. (2000).
[17] WHO Regional Office for Europe, Guidelines for indoor air quality, 2010, https://doi.org/10.1186/2041-1480-2-S2-I1
[18] C.S. Gordon, J.T. Lowe, Carbon-Monoxide detector, US1644014A, 1925.
[19] K. Grosskopf, Carbon monoxide detecting device, US 3112999A, 1961.
[20] M. Janik, M. Ublova, S. Kucerova, P. Hejna, Carbon monoxide-related fatalities: a 60-year single institution experience, J. Forensic Leg. Med. 48 (2017) 23–29. Available from: https://doi.org/10.1016/j.jflm.2017.04.002.
[21] Data Sheet: TGS 203 - Carbon Monoxide Sensor, (n.d.). http://pdf.dzsc.com/88888/2008617161056961.pdf.
[22] R. Jaaniso, O.K. Tan, eds., Semiconductor Gas Sensors, Woodhead Publishing Series in Electric and Optical Materials, n.d.

[23] Figaro Announcement of Final Orders for TGS 203 and TGS2442, 2016. (n.d.). http://www.figaro.co.jp/en/search/index.cgi?query = TGS + 203 + &x = 0&y = 0&index = figaro &config = &set = 1.
[24] Figaro Sensor, Datasheet: TGS 3870-B00, (2013) 2. http://www.figarosensor.com/products/docs/TGS3870-B00(1213).pdf.
[25] Figaro Announcement: Ultra-compact Electrochemical CO Sensor TGS5141 Now Available, (2017). http://www.figaro.co.jp/en/topic/2016/04/co-sensor-tgs5141.html.
[26] J. Tabak, Natural Gas and Hydrogen, Infobase Publishing, New York, 2009.
[27] P. Zloty, History of LPG- the first 100 years, (2013). https://gazeo.com/up-to-date/news/2013/History-of-LPG-the-first-100-years,news,6662.htm.
[28] Annual Energy Review, U.S. Energy Information Administration, 2015. https://www.eia.gov/totalenergy/data/annual/index.php#summary
[29] K. Ihokura, J. Watson, The Stannic Oxide Gas Sensor - Principles and Application, CRC Press, 1994. pp. 208.
[30] T. Seiyama, A. Kato, K. Fujishi, M. Nagatani, A. New, Detector for gaseous components using semiconductive thin films, Ananl Chem (1962) 2–3.
[31] N. Taguchi, Gas-Detecting Device, US3695848, 1971.
[32] Editing Committee The Company History, The 50th anniversary book of New Cosmos Electric, Osaka, 2010.
[33] H. Nakagawa, S. Okazaki, S. Asakura, K. Fukuda, H. Akimoto, H. Takahashi, S. Shigemori, An automated car ventilation system, Sensors Actuators B (2000) 133–137.
[34] J. Marek, H.-P. Trah, Y. Suzuki, I. Yokomori, Sensor Applications: Sensors for Automotive Applications, 4th ed., WILEY-VCH Verlag, 2003.
[35] N. Barsan, G. Gauglitz, A. Oprea, E. Osterberg, G. Proll, K. Rebner, K. Schierbaum, F. Schleifenbaum, U. Weimar, Ullmann's Encyclopedia of Industrial Chemistry, Wiley VCH GmbH & Co KGaA., 2016.
[36] E-Newsletter: Sensor Business Digest - Sensor Industry Developments and Trends, (2001) 2001–2003. http://www.sensorsmag.com.
[37] J.W. Gardner, A. Pike, N.F. de Rooij, M. Koudelka-Hep, P.A. Clerc, A. Hierlemann, et al., Integrated array sensor for detecting organic solvents, Sens. Actuat. B Chem. 26 (1995) 135–139. Available from: https://doi.org/10.1016/0925-4005(94)01573-Z.
[38] M. Blaschke, T. Tille, P. Robertson, S. Mair, U. Weimar, H. Ulmer, MEMS gas-sensor array for monitoring the perceives car-cabin air quality, IEEE Sens. J. 6 (2006) 1–12.
[39] H. Hölter, H. Rump, Sensing System for Controlling Air Circulation Valves in Motor Vehicles, US4930407A, 1987.
[40] Automatic HVAC Air Recirculation Made Easy with Stepper Motors, (2014) 9. http://www.onsemi.com/pub_link/Collateral/TND416-D.PDF.
[41] J. Mitrovics, H. Ulmer, U. Weimar, W. Göpel, Modular sensor systems for gas sensing and odor monitoring: the MOSES concept, Acc. Chem. Res. 31 (1998) 307–315.
[42] G. Müller, S. Beer, S. Paul, A. Helwig, Novel chemical sensor applications in commercial aircraft, Proc. Eng. 25 (2011) 16–22. Available from: https://doi.org/10.1016/j.proeng.2011.12.005.
[43] Federal Aviation Administration, Cabin Environmental Control Systems, Aviat. Maint. Tech. Handb. (2012) 1–62. Available from: https://www.faa.gov/regulations_policies/handbooks_manuals/aircraft/amt_airframe_handbook/media/ama_Ch16.pdf.
[44] National Research Council, The Airliner Cabin Environment and the Health of Passengers and Crew, (2002). doi.org/10.17226/10238.

[45] ILA Berlin Air Show 2014: Fraunhofer IBP presents Solutions for Improving Sustainability in the Avaition Industry, 2016, 1–4. https://www.ibp.fraunhofer.de/en/Press/Press_releases/20052014_ila.html.
[46] Boeing Commercial Airplanes-Cabin Air and Bleed Contaminants, (2016) 272. https://www.google.com/url?sa = t&rct = j&q = &esrc = s&source = web&cd = 1&ved = 2ahUKEwikpODQqsLdAhUGL1AKHTKXDFMQFjAAegQICBAC&url = https%3A%2F%2F787updates.newairplane.com%2Fgetmedia%2F0a6b6209-6b19-4dce-9afb-81ba391102de%2FCabin-Air-Contaminants-Boeing-backgrounder&usg = AOvVaw2JTBB5eDWBNKGIy1u38iy5.
[47] C.M. Ott, R.J. Bruce, D.L. Pierson, Microbial characterization of free floating condensate aboard the mir space station, Microb. Ecol. 47 (2004) 133–136. Available from: https://doi.org/10.1007/s00248-003-1038-3.
[48] C. Stenzel, Deployment of precise and robust sensors on board ISS—for scientific experiments and for operation of the station, Anal. Bioanal. Chem. 408 (2016) 6517–6536. Available from: https://doi.org/10.1007/s00216-016-9789-0.
[49] Airbus Defence and Space, Eine Nase für olfaktorische Fingerabdrücke, (2013). http://www.space-airbusds.com/de/nachrichten/eine-nase-fur-olfaktorischeFingerabdrücke'%22.
[50] A.T. Hodgson, S.J. Nabinger, A.K. Persily, Volatile organic compound concentrations and emission rates measured over one year in a new manufactured house, Indoor Environ (2004) 34.
[51] T. Thomas, (Hrsg.), Automobil-Sensorik: Ausgewählte Sensorprinzipien und deren automobile Anwendung, Spinger, Verlag, 2016.
[52] D.H. Kim, J.Y. Yoon, H.C. Park, K.H. Kim, CO_2-sensing characteristics of SnO_2 thick film by coating lanthanum oxide, Sens. Actuat. B Chem. 62 (2000) 61–66. Available from: https://doi.org/10.1016/S0925-4005(99)00305-6.
[53] M. Lee, A new process for fabricating CO_2 -sensing layers based on $BaTiO_3$ and additives, Sens. Actuat (2000) 293–299. Available from: https://doi.org/10.1016/S0925-4005(00)00447-0.
[54] European Commission, The INDEX project Summary on recommendations and management options, 2004.
[55] United States Environmental Protection Agency., Volatile Organic Compounds' Impact on Indoor Air Quality, 2017.
[56] S. Herberger, Beyond CO_2 : sensing VOCs in indoor air protects health, Saves Energy (2009) 1–7. Available from: https://www.google.com/url?sa=t&rct=j&q=&esrc=s&source=web&cd=1&ved=2ahUKEwi0ooSoq8LdAhWLLVAKHQqLArAQFjAAegQIBBAC&url=http%3A%2F%2Fwww.safe-detect.com%2Fimage%2Fusers%2F125583%2Fftp%2Fmy_files%2Farticles%2FIAQ-ArticleFrom%2520Apliedsensorf071010.pdf%3Fid%3D7135573&usg=AOvVaw2Ur5GQIS8jHW08PyZ28CtB.
[57] S.E. Boslaugh, Sick building syndrome, Online Encyplopaedia Britannica in: Britannica, 2016, p. 1301499. https://www.britannica.com/science/sick-building-syndrome
[58] U. S. Environmental Protection Agency, I. Environments Division, Indoor Air Facts No. 4 Sick Building Syndrome, EPA—Air Radiat. (6609J), Res. Dev. (1991) 1–4. Available from: https://doi.org/10.1136/oem.2003.008813.
[59] D. Mickaël, B. Bruno, C. Valérie, L. Murielle, P. Cécile, R. Jacques, et al., Indoor air quality and comfort in seven newly built, energy-efficient houses in France, Build. Environ. 72 (2014) 173–187. Available from: https://doi.org/10.1016/j.buildenv.2013.10.017.
[60] M. Großklos, Use of VOC sensors for air quality control of building ventilation systems, J. Sens. Sens. Syst. 4 (2015) 159–168. Available from: https://doi.org/10.5194/jsss-4-159-2015.

[61] R. Furchgott, Smart Doesn't Always Mean an Easy Home, New York Times. (2016).
[62] A. Teregulova, In the Future, Our Homes Will Read Our Minds, Fortune. (2016). http://fortune.com/2016/10/07/homes-of-the-future/.
[63] H.G. Moon, Y.S. Shim, D.H. Kim, H.Y. Jeong, M. Jeong, J.Y. Jung, et al., Self-activated ultrahigh chemosensitivity of oxide thin film nanostructures for transparent sensors, Sci. Rep. 2 (2012) 1–7. Available from: https://doi.org/10.1038/srep00588.
[64] Indoor Air Monitor, (2015). https://www.airqualitymonitor.ch/?gclid = EAIaIQobChMIsfn2wcem3QIVWouyCh0uuAtUEAAYAyAAEgJPNPD_BwE.
[65] hiddenwires, iControl Networks' 2014 State of the Smart Home Report Ranks Security as Top Feature, Icontrol Netw. (2014) 16. http://www.hiddenwires.co.uk/news/article/icontrol-networks-2014-state-of-the-smart-home-report-ranks-security-as-top-feature.
[66] J. Grey, The 10 most dangerous jobs in Australia Mybusiness (2013) 9–11. http://www.mybusiness.com.au/news/1045-the-10-most-dangerous-jobs-in-australia
[67] Ghent University, ProeTEX-PROtection E-TEXtiles:MicroNanoStructured Fibre Systems for Emergency-Disaster Wear (2006-2010), (2010). http://www.ugent.be/ea/textiles/en/research/projects/afgelopenprojecten/proetex.htm.
[68] P.M. Sivakumar, S. Balaji, V. Prabhawathi, R. Neelakandan, P.T. Manoharan, M. Doble, Effective antibacterial adhesive coating on cotton fabric using ZnO nanorods and chalcone, Carbohydr. Polym. 79 (2010) 717–723. Available from: https://doi.org/10.1016/j.carbpol.2009.09.027.
[69] B. Xu, Z. Cai, Fabrication of a superhydrophobic ZnO nanorod array film on cotton fabrics via a wet chemical route and hydrophobic modification, Appl. Surf. Sci. 254 (2008) 5899–5904. Available from: https://doi.org/10.1016/j.apsusc.2008.03.160.
[70] Z.H. Lim, Z.X. Chia, M. Kevin, A.S.W. Wong, G.W. Ho, A facile approach towards ZnO nanorods conductive textile for room temperature multifunctional sensors, Sens. Actuat. B Chem. 151 (2010) 121–126. Available from: https://doi.org/10.1016/j.snb.2010.09.037.
[71] E. Comini, Integration of metal oxide nanowires in flexible gas sensing devices, Sensors (Basel) 13 (2013) 10659–10673. Available from: https://doi.org/10.3390/s130810659.
[72] G. Mueller, A. Hackner, S. Beer, J. Goebel, Solid-state gas sensors: sensor system challenges in the civil security domain, Materials (Basel) 9 (2016). Available from: https://doi.org/10.3390/ma9010065.
[73] R.J. Mainiero, J.H. Rowland Iii, M.L. Harris, M.J. Sapko, Behavior of Nitrogen Oxides in the Product Gases from Explosive Detonations, 2006, https://www.cdc.gov/niosh/mining/works/coversheet713.html.
[74] B. Ruhland, T. Becker, G. Müller, Gas-kinetic interactions of nitrous oxides with SnO_2 surfaces, Sens. Actuat. B Chem. 50 (1998) 85–94. Available from: https://doi.org/10.1016/S0925-4005(98)00160-9.
[75] S. Maurer, R. Makarow, J. Warmer, G. Holl, P. Kaul, Heterogenes Sensorsystem zum Nachweis von Explosivstoff-typischen Merkmalen durch thermische Aktivierung, Dresdener Sens. Symp. (2015). https://doi.org/10.5162/12dss2015/2.3.
[76] R. Shenoi, Chemical warfare agents, Clin. Pediatr. Emerg. Med. 3 (2002) 239–247. Available from: https://doi.org/10.1016/S1522-8401(02)90036-4.
[77] A.A. Tomchenko, G.P. Harmer, B.T. Marquis, Detection of chemical warfare agents using nanostructured metal oxide sensors, Sens. Actuat. B Chem. 108 (2005) 41–55. Available from: https://doi.org/10.1016/j.snb.2004.11.059.
[78] D. Matatagui, M.J. Fernández, J. Fontecha, J.P. Santos, I. Gràcia, C. Cané, et al., Discrimination and classification of chemical warfare agent simulants using a Love-

wave sensor array, Proc. Eng. 25 (2011) 23–26. Available from: https://doi.org/10.1016/j.proeng.2011.12.006.
[79] S.C. Lee, H.Y. Choi, S.J. Lee, W.S. Lee, J.S. Huh, D.D. Lee, et al., The development of SnO_2-based recoverable gas sensors for the detection of DMMP, Sens. Actuat. B Chem. 137 (2009) 239–245. Available from: https://doi.org/10.1016/j.snb.2008.12.051.
[80] Ministry for Culture and Heritage, First Frozen Meat Shipment Leaves New Zealand, NZHistory, 2016, pp. 1–3. https://nzhistory.govt.nz/first-shipment-of-frozen-meat-leaves-nz
[81] R. Pirog, T. Van Pelt, K. Enshayan, E. Cook, Food, fuel, and freeways: an Iowa perspective on how far food travels, fuel usage, and greenhouse gas emissions, Leopold Cent. Sustain. Agric (2001) 1–37. Available from: https://www.leopold.iastate.edu/sites/default/files/pubs-and-papers/2011-06-food-fuel-and-freeways-iowa-perspective-how-far-food-travels-fuel-usage-and-greenhouse-gas-emissions.pdf.
[82] G. Borrud, Germany launches initiative to reduce food waste, 2016, pp. 16–18. http://dw.com/p/14K1s
[83] K. Persaud, G. Dodd, Analysis of discrimination mechanisms in the mammalian olfactory system using a model nose, Nature. 299 (1982) 352–355.
[84] E.A. Baldwin, J. Bai, A. Plotto, S. Dea, Electronic noses and tongues: applications for the food and pharmaceutical industries, Sensors. 11 (2011) 4744–4766. Available from: https://doi.org/10.3390/s110504744.
[85] W. Yongwei, J. Wang, B. Zhou, Q. Lu, Monitoring storage time and quality attribute of egg based on electronic nose, Anal. Chim. Acta. 650 (2009) 183–188. Available from: https://doi.org/10.1016/j.aca.2009.07.049.
[86] S. Labreche, S. Bazzo, S. Cade, E. Chanie, Shelf life determination by electronic nose: application to milk, Sens. Actuat. B Chem. 106 (2005) 199–206. Available from: https://doi.org/10.1016/j.snb.2004.06.027.
[87] S. Limbo, L. Torri, N. Sinelli, L. Franzetti, E. Casiraghi, Evaluation and predictive modeling of shelf life of minced beef stored in high-oxygen modified atmosphere packaging at different temperatures, Meat Sci. 84 (2010) 129–136. Available from: https://doi.org/10.1016/j.meatsci.2009.08.035.
[88] S. Balasubramanian, S. Panigrahi, C.M. Logue, C. Doetkott, M. Marchello, J.S. Sherwood, Independent component analysis-processed electronic nose data for predicting Salmonella typhimurium populations in contaminated beef, Food Control. 19 (2008) 236–246. Available from: https://doi.org/10.1016/j.foodcont.2007.03.007.
[89] S. Limbo, N. Sinelli, L. Torri, M. Riva, Freshness decay and shelf life predictive modelling of European sea bass (Dicentrarchus labrax) applying chemical methods and electronic nose, LWT—Food Sci. Technol. 42 (2009) 977–984. Available from: https://doi.org/10.1016/j.lwt.2008.12.011.
[90] M. Lebrun, A. Plotto, K. Goodner, M.N. Ducamp, E. Baldwin, Discrimination of mango fruit maturity by volatiles using the electronic nose and gas chromatography, Postharvest Biol. Technol. 48 (2008) 122–131. Available from: https://doi.org/10.1016/j.postharvbio.2007.09.010.
[91] E. Abad, S. Zampolli, S. Marco, A. Scorzoni, B. Mazzolai, A. Juarros, et al., Flexible tag microlab development: gas sensors integration in RFID flexible tags for food logistic, Sensors Actuators, B Chem. 127 (2007) 2–7. Available from: https://doi.org/10.1016/j.snb.2007.07.007.
[92] J. Kukkola, M. Mohl, A.-R. Leino, G. Tóth, M.-C. Wu, A. Shchukarev, et al., Inkjet-printed gas sensors: metal decorated WO3 nanoparticles and their gas sensing properties, J. Mater. Chem. 22 (2012) 17878. Available from: https://doi.org/10.1039/c2jm32499g.

[93] M. Rieu, M. Camara, G. Tournier, J.P. Viricelle, C. Pijolat, N.F. De Rooij, et al., Inkjet printed SnO_2 gas sensor on plastic substrate, Proc. Eng. 120 (2015) 75–78. Available from: https://doi.org/10.1016/j.proeng.2015.08.569.

[94] Markets and Markets, Press release: Smart Packaging Market worth $39.7 Billion by 2020, (2014). http://www.marketsandmarkets.com/PressReleases/smart-packaging.asp (accessed 04.12.16).

[95] K.H. Eom, K.H. Hyun, S. Lin, J.W. Kim, The meat freshness monitoring system using the smart RFID tag, Int. J. Distrib. Sens. Networks. 2014 (2014). Available from: https://doi.org/10.1155/2014/591812.

[96] G. Neri, First fifty years of chemoresistive gas sensors, Chemosensors. 3 (2015) 1–20. Available from: https://doi.org/10.3390/chemosensors3010001.

[97] S. Wutke, History of the Breathalyzer (2014) 1–6. https://guardianinterlock.com/blog/history-breathalyzer/. (accessed 04.12.17).

[98] A. Staerz, U. Weimar, N. Barsan, Understanding the potential of WO_3 based sensors for breath analysis, Sensors. 16 (2016) 1815. Available from: https://doi.org/10.3390/s16111815.

[99] M. Anderson, Technology Device Ownership, Pew Res. Cent. (2015). http://www.pewinternet.org/2015/10/29/technology-device-ownership-2015.

[100] J. Sherman, Spendy but Indispensable: breaking down the full $650 cost of the Iphone 5, Digit Trends 2016. (2016) 1–11. http://www.digitaltrends.com/mobile/iphone-cost-what-apple-is-paying/.

[101] Press Release: Cambridge CMOS Sensors to demonstrate air-quality monitoring and breath analysis on mobile devices, 2015. https://www.cambridgenetwork.co.uk/news/air-quality-monitoring-and-breath-analysis-on-mobile-devices9962/.

[102] L. Pauling, A.B. Robinson, R. Teranishi, P. Cary, Quantitative analysis of urine vapor and breath by gas-liquid partition chromatography, Proc. Natl. Acad. Sci. U. S. A. 68 (1971) 2374–2376. Available from: https://doi.org/10.1073/pnas.68.10.2374.

[103] M. Righettoni, A. Amann, S.E. Pratsinis, Breath analysis by nanostructured metal oxides as chemo-resistive gas sensors, Mater. Today. 18 (2015) 163–171. Available from: https://doi.org/10.1016/j.mattod.2014.08.017.

[104] K.-I. Choi, S.-J. Hwang, Z. Dai, Y. Chan Kang, J.-H. Lee, Rh-catalyzed WO3 with anomalous humidity dependence of gas sensing characteristics, RSC Adv. 4 (2014) 53130–53136. Available from: https://doi.org/10.1039/C4RA06654E.

[105] K.M. Paschke, A. Mashir, R.A. Dweik, Clinical applications of breath testing, Med. Rep. 2 (2010) 56. Available from: https://doi.org/10.3410/M2-56.

[106] P.I. Gouma, K. Kalyanasundaram, A selective nanosensing probe for nitric oxide, Appl. Phys. Lett. 93 (2008) 1–4. Available from: https://doi.org/10.1063/1.3050524.

[107] M. Righettoni, A. Tricoli, S. Gass, A. Schmid, A. Amann, S.E. Pratsinis, Breath acetone monitoring by portable Si:WO3 gas sensors, Anal. Chim. Acta 738 (2012) 69–75. Available from: https://doi.org/10.1016/j.aca.2012.06.002.

[108] M. Righettoni, A. Ragnoni, A.T. Güntner, C. Loccioni, S.E. Pratsinis, T.H. Risby, Monitoring breath markers under controlled conditions, J. Breath Res. 9 (2015) 47101. Available from: https://doi.org/10.1088/1752-7155/9/4/047101.

[109] A. Tangerman, E.G. Winkel, Intra- and extra-oral halitosis: finding of a new form of extra-oral blood-borne halitosis caused by dimethyl sulphide, J. Clin. Periodontol. 34 (2007) 748–755. Available from: https://doi.org/10.1111/j.1600-051X.2007.01116.x.

[110] I. Kim, S. Choi, S. Kim, J. Jang, Smart Sensors for Health and Environment Monitoring, 2015. https://doi.org/10.1007/978-94-017-9981-2.

[111] F.J. Hughes, R. McNab, Oral malodour—a review, Arch. Oral Biol. 53 (Suppl. 1) (2008) S1−7. Available from: https://doi.org/10.1016/S0003-9969(08)70002-5.
[112] A. Tangerman, E.G. Winkel, The portable gas chromatograph OralChroma™: a method of choice to detect oral and extra-oral halitosis, J. Breath Res. 2 (2008) 17010. Available from: https://doi.org/10.1088/1752-7155/2/1/017010.
[113] JIA Expert Committee on Technical Standards for Gas Alarms, Extension of Validity Period of Residential Town Gas Alarms, Minato-ku, Tokyo, 1985.
[114] Y. Matsuura, K. Takahata, Stabilization of SnO_2 sintered gas sensors, Sens. Actuat. B. Chem. 5 (1991) 205−209. Available from: https://doi.org/10.1016/0925-4005(91)80248-I.
[115] N. Yamazoe, K. Suematsu, K. Shimanoe, Two types of moisture effects on the receptor function of neat tin oxide gas sensor to oxygen, Sens. Actuat. B Chem. 176 (2013) 443−452. Available from: https://doi.org/10.1016/j.snb.2012.08.060.
[116] Technical Committee of Gas Alarm Industries Association of Japan, History of Technology of Gas alarms, Minato-ku, Tokyo, 2011.

Conclusion and outlook

6

Nicolae Barsan[1] *and Klaus Schierbaum*[2]
[1]Institute for Theoretical and Physical Chemistry, University of Tübingen, Tübingen, Germany, [2]Department of Materials Science, Institute of Experimental Condensed Matter Physics, University of Düsseldorf, Düsseldorf, Germany

The understanding of *charge carrier transport* in metal oxide gas sensors, that is, the transport of ions or/and electrons in an applied electrical field between the contacts, is essential for tailoring sensing properties and for making use of advanced materials and novel concepts. Let us take the "prototypical" oxygen ion conducting solid-electrolyte, yttria-stabilized zirconia, as an example. In Section 2 we have seen that YSZ shows a broad electrolytic domain which makes it applicable in a large range of oxygen partial pressures. The sintered ceramic bulk YSZ is well characterized with respect to its temperature- and dopant-concentration-dependent ion mobility although there are controversial issues concerning the nature of defect interaction at high doping concentrations [1] and the exact oxygen vacancy diffusion mechanism [2]. One interesting finding is that the mobility exhibits a maximum at a certain dopant concentration, which, however, sets an upper limit for ion conductivity at a given temperature apart from structural stability problems caused by higher doping levels. Since conductivity enhancement is technologically relevant for lower operational temperatures of solid electrolytes in many applications, research has been focused on the grain-boundary control and the lattice strain-control of ion conduction [3]. YSZ nanomaterials have been proposed for potentiometric oxygen sensors, but experimental evidence for a significant improvement in comparison with conventional YSZ has not been clearly established (see, e.g., references in [4]). The lower conductivity across grain boundaries may be problematic. Interfacial systems like thin YSZ films on insulators are also promising candidates for increased ion conductivity [5] and have been studied, for example, for solid electrolyte microsensors [6]. The thermal expansion mismatch between solid electrolyte films and substrates, however, seems to cause major stability problems, not only for YSZ but also for other ion conducting materials. Ceramic-type YSZ remains the most applied ion conducting metal oxide for gas sensing, which can be utilized for a variety of different detection modes, as discussed in Section 2.

Semiconducting oxides from the main and transition group metals are widely used for electronic conductivity sensors. For these oxides, structural disorder and related phenomena have been reviewed in Section 3. Various experimental techniques can be used to identify different types of intrinsic and extrinsic (by unintentional and intentional doping) defects, which cause localized states in the band structure of these materials and make them mostly n-type and in few instances *p*-type conducting. Three different groups of rules considered for doping concepts

Gas Sensors Based on Conducting Metal Oxides. DOI: https://doi.org/10.1016/B978-0-12-811224-3.00016-0

have been rationalized in terms of three groups of rules, that is, (1) preventing Fermi-energy-induced spontaneous compensation, (2) maintaining the approximate host material stoichiometry and avoiding partial decomposition of the host lattice into other solid-state phases, and (3) considering local lattice-dopant bond formation. So, similarly to YSZ, questions on charge carrier transport remain for semiconducting oxides that have not yet been answered conclusively and is valid, particularly, for grain-boundary determined materials of currently applied SMOX sensors.

Charge transfer coupled to fluxes (e.g., of molecules or electrons) is the key phenomenon that governs the properties of both metal oxide gas sensors based on ionic or electronic conductivity. To explain one of the consequences, we consider a sufficiently large system or an open nonequilibrium system so that the chemical reactions of the gas species at the sensor surface or bulk do not change the volume fractions or partial pressures of gas components, including those of the analyte gas (es). For simplicity, we assume a steady-state scenario with an ideal sensor in which the sensor signal (e.g., the measured resistance or the potential) is an unequivocal function of the partial pressures of the analyte gas(es) and the interfering components, for example oxygen or water. A scientific theory, describing the behavior of the sensor, must be testable so that it predicts the sensor signal for a given set of parameters including temperature and partial pressures. Obviously, the mathematical equations that describe the signal of solid-state and semiconducting gas sensors can be quite different. For a Nernstian-type potentiometric device (e.g., a lambda probe in an exhaust gas stream), the signal-determining chemical reaction is

$$O_2\left(p''_{O_2}\right) \rightleftarrows O_2\left(p''_{O_2}\right) \tag{6.1}$$

and the measured potential drop between both platinum electrodes is given by the Nernst equation

$$E = \frac{RT}{zF}\ln\frac{p''_{O_2}}{p'_{O_2}} \tag{6.2}$$

(for symbols, compare Eq. 9 in Section 2). For a semiconductor CO gas sensor, for example, the chemical reaction is

$$CO(p_{CO}) + \frac{1}{2}O_2(p_{O_2} \approx const) \longrightarrow CO_2(p_{CO_2}) \tag{6.3}$$

The following equation

$$S = \left(1 + \frac{\frac{k_{react}}{k_{des}} \cdot p_{CO}}{\kappa \cdot p_{H_2O}}\right)^{\frac{1}{(\alpha+\gamma)}} \tag{6.4}$$

has been derived to model the signal of tin oxide-based sensors in humid conditions (compare Eq. 36 in Section 3). The number of variables and constants are evidently quite different and it seems that both relationships operate in different leagues. The Nernst equation is well confirmed and applicable to various systems. In certain conditions, for example, in exhaust gases, deviations of the measured voltages from the ideal law, however, have to be taken into account. On the other hand, in studies on SMOX sensors, experimental sensor signals are rarely evaluated in terms of Eq. (6.4) with the consequence that consistent parameter sets are not available for a variety of other gases and sensor materials. Although the driving force for the chemical reactions (1) and (3), that finally controls the sensor signals, is related to the free energy of the system, ΔG, there is a noticeable difference. For the lambda probe, (1) is a *highly reversible reaction* whereas at the surface of the semiconductor sensor the *nearly irreversible reaction* (2) exists. The experimental determination of elementary processes of such mostly catalyzed reactions on SMOX sensors has been performed using different approaches as reviewed in Section 4. Whereas surface science or computational chemistry studies are suitable for idealized systems (such as surfaces of single crystals and well-order thin films), operando methods are a more applicable way to determine relevant surface species during gas sensor operation. Here, different complementary spectroscopic techniques are combined with electrical characterization. Such approaches have been used successfully to understand certain sensing mechanisms of SMOX sensors and are further supported by advanced computational atomic-scale calculations. The mixed-potential sensor, that operates at lower temperatures or with a so-called non-Nernstian electrode does not establish equilibrium of the different analytes, represents a novel concept for solid-electrolytes that has substantially extended the applicability of ion conducting metal oxides. The Butler–Volmer equation of electrochemistry may serve as the fundament for a theoretical description of sensor signals. The involved complicated reactions and how they are coupled to potential-determining steps at the electrode/electrolyte interface are known only for a few systems.

Although a detailed understanding of *charge carrier transport* and *charge transfer coupled to fluxes* is lacking for most metal oxide gas sensors, empirical work in numerous research laboratories, both in the industry and in academic research institutions, has led to manifold practical implications of ionically and electronically conductive metal oxides. The most relevant applications are treated in Sections 2 and 5 of this book. They reveal an exceptional variety and, without any doubt, some of them form an integral part of modern technologies.

Last, but not least, is the question of future prospects of research in the area of metal oxide gas sensors. From the materials point-of-view, semiconducting metal oxides have been studied extensively. Today there is a broad range of fabrication technologies available, including concepts from nanophysics and nanochemistry. Section 4 has demonstrated the versatility of different methods and approaches to deposit metal oxides on various substrates and structures. Obviously, for solid-state electrolytes the applicability of these methods is limited, mainly due to higher operational temperatures that cause a higher thermal stress between the deposited film and the substrate. Hence, clearly fewer studies have been conducted with ion conducting metal oxide film deposition.

In summary, this book provides a comprehensive guide on metal oxide-based gas sensors. As the world becomes more connected and digitalized, sensors will begin to play an indispensable role in numerous applications. Sensors based on metal oxides are an attractive option. There are experts from a wide spectrum of fields, engineering to material science, working in the field of metal oxide gas sensors. As their application becomes more common and the technology becomes more sophisticated, it is important that each member of the field maintain a basic understanding of all aspects of sensors. Due to the wide breadth of this book, it is the ideal guide for experts of one field to review aspects from other relevant areas.

References

[1] C. Ahamer, A.K. Opitz, G.M. Rupp, J. Fleig, J. Electrochem. Soc. 164 (7) (2017) F790–F803.
[2] R. Pornprasertsuk, P. Ramanarayanan, C.B. Musgrave, F.B. Prinz, J. Appl. Phys. 98 (2005) 103513.
[3] G. Korotcenkov, in: N. Pryds, V. Esposito) (Eds.), Metal Oxide-Based Thin Film Structures, Elsevier, Amsterdam, 2017, p. 420.
[4] C. López-Gándara, F.M. Ramos, and A. Cirera, Journal of Sensors, https://doi.org/10.1155/2009/258489.
[5] J. Jiang, X. Hu, W. Shen, C. Ni, J.L. Hertz, Appl. Phys. Lett. 102 (2013) 143901.
[6] A. Dubbe, Sens. Actuat. B 88 (2003) 138–148.

Index

Note: Page numbers followed by "*f*" and "*t*" refer to figures and tables, respectively.

FSC
www.fsc.org
MIX
Papier | Fördert gute Waldnutzung
FSC® C083411

Kolibri 360 GmbH
Ferdinand-Jühlke-Straße 7
99095 Erfurt, Deutschland
produktsicherheit@kolibri360.de